T0297700

Overview of Industrial Process Automation

Overview of Industrial Process Automation

Second edition

KLS Sharma

Automation Education and Training, Bengaluru, India

ELSEVIER

AMSTERDAM • BOSTON • HEIDELBERG • LONDON • NEW YORK
OXFORD • PARIS • SAN DIEGO • SAN FRANCISCO • SINGAPORE
SYDNEY • TOKYO

Elsevier
Radarweg 29, PO Box 211, 1000 AE Amsterdam, Netherlands
The Boulevard, Langford Lane, Kidlington, Oxford OX5 1GB, United Kingdom
50 Hampshire Street, 5th Floor, Cambridge, MA 02139, United States

**Cover page background image courtesy: Yokogawa, India (www.yokogawa.com/in) and Bosch
Rexroth, India (www.boshrexroth.co.in)**

Notices
Knowledge and best practice in this field are constantly changing. As new research and
experience broaden our understanding, changes in research methods, professional practices,
or medical treatment may become necessary.

Practitioners and researchers must always rely on their own experience and knowledge in
evaluating and using any information, methods, compounds, or experiments described herein. In using
such information or methods they should be mindful of their own safety and the
safety of others, including parties for whom they have a professional responsibility.

To the fullest extent of the law, neither the Publisher nor the authors, contributors, or
editors, assume any liability for any injury and/or damage to persons or property as a matter of products
liability, negligence or otherwise, or from any use or operation of any methods, products, instructions, or
ideas contained in the material herein.

Library of Congress Cataloging-in-Publication Data
A catalog record for this book is available from the Library of Congress

British Library Cataloguing-in-Publication Data
A catalogue record for this book is available from the British Library

ISBN: 978-0-12-805354-6

For information on all Elsevier publications visit
our website at https://www.elsevier.com/

Working together
to grow libraries in
developing countries

www.elsevier.com • www.bookaid.org

Publisher: Joe Hayton
Acquisition Editor: Sonnini R Yura
Editorial Project Manager: Ana Claudia Abad Garcia
Production Project Manager: Mohanambal Natarajan
Cover Designer: Greg Harris

Typeset by TNQ Books and Journals

Dedication

This book is dedicated to ABB India, where the author learned and practiced automation for over 24 years.

Contents

About the Author

KLS Sharma graduated from University of Mysuru, India, and received his masters and doctoral degrees from the Indian Institute of Technology, Delhi, India.

He has worked for the following organizations:

- Technical manager, Electronics Corporation of India, Hyderabad, India
- Assistant vice president, ABB, Bengaluru, India
- Consultant, Honeywell Technology Solutions Lab, Bengaluru, India
- Professor, International Institute of Information Technology, Bengaluru, India
- Professor emeritus, MS Ramaiah Institute of Technology, Bengaluru, India
- Member, Campus Connect Program, Automation Industry Association of India, Delhi, India
- Distinguished visiting professor, National Institute of Technology Karnataka, Surathkal, India

His current positions are:

- Advisor, Automation Education and Training, Bengaluru, India
- Principal consultant, Advanced Engineering Group, Infosys, Bengaluru, India
- Member, Editorial Advisory Board, A&D Magazine on Industrial Automation, Pune, India

Association

- Senior member, International Society of Automation (ISA)

Awards and Recognitions

- ISA Celebrating Excellence Member's Choice Award for Student Mentor of the Year 2015.

Foreword

The industrial revolution replaced the need for human muscles; the computer revolution substituted the routine functions of our brains. Today we operate robots on Mars that can vaporize rocks by laser to determine their composition, and soon we will be ordering pizza while sitting in our hydrogen-fueled driverless cars and the drone-delivered pizza will be waiting for us in the driveway by the time our smart car parks itself.

Similar advances took place in the field of industrial automation. 50 years ago we were tuning single loop pneumatic controllers and were perfectly satisfied with using filled bulbs for temperature and orifices for flow measurement. At that time the main job of the instrument department in the plant was to clean plugged pressure taps and open stuck control valves, and our control panels were full of push buttons, blinking lights, and manual loading stations. Over the past decades our tools changed into self-checking digital components, wireless transmission, and redundant safety backup systems that provide ease of configuring complex algorithms and generating dynamic displays. Today automation is minimizing operating and energy costs while maximizing both safety and profitability of our industries.

Yet the totality of the automation field is poorly understood because that knowledge is fragmented, since our colleagues are working only in particular segments of the field. Few have an overall view of the totality of the automation profession. This book of Dr. Sharma serves to fill that gap by describing the totality of this field.

We live at a time when cultural attitudes concerning automation are changing as we debate the proper role of machines in our lives. Whereas in everyday life we accept the spread of automation, that our mobile phones can do just about anything except sharpen our pencils (but who needs pencils or handwriting anymore, for that matter?), industrial attitudes change much more slowly. Our industries are still mostly run by human operators, and although they are assisted by machines, it is the operator who usually has the "last word." In other words, automation is seldom used to prevent human errors.

Safety statistics tell us that the number one cause of all industrial accidents is human error. One could refer to Three-Mile Island, where operators poured water into the instrument air supply; the BP accident, where there was no automation to keep the drill pipe straight; the ferry accident in Korea, where safety overrides were not provided to prevent the captain from turning sharply into a fast ocean current; or airplane accidents, where pilots are allowed to fly into mountains or attempt to land at wrong speeds. Yet we know that in addition to its other contributions, automation can overrule the actions of panicked or badly trained operators who often make the wrong decisions in emergencies.

To achieve this higher level of industrial safety, an override safety control (OSC) layer of safety automation is being added to our control systems, one that cannot be turned off or overruled by anything or anybody. With this design, if the plant conditions enter a highly accident-prone life- or safety-threatening region of operation, the uninterruptible safe shutdown of the plant is automatically triggered. The functioning of this layer of automation is free from possible cyber attacks because it is not connected to the Internet at all. In short, once the OSC layer of protection is activated, the plant is shutting down and nothing and nobody can prevent that.

In addition to safety improvements, advances in standardization are also taking place. Just as it occurred in the "analog age," a global standard is now evolving for digital communication that could link all digital "black boxes" and could also act as a "translator" for those automation devices that were not designed to "speak the same language." Naturally, this standardization should apply to both wired and wireless systems, thereby eliminating "captive markets" and allowing the easy mixing of different manufacturers' products in the same control loop. This trend is most welcome because once completed, it will allow the automation and process control engineers to once again concentrate on designing safe and optimized control systems and not worry about the possibility that the "black boxes" of the different suppliers might not be able to talk to each other. Therefore it is hoped that the "Babel of communication protocols" will shortly be over.

The automation profession can simultaneously increase gross domestic product and industrial profitability without building a single new plant, just by optimizing existing ones. We can achieve that goal while also reducing both pollution and energy consumption, solely through applying the state-of-the art of automation. We can increase productivity without using a single pound of additional raw material and without spending a single additional BTU of energy. We can also protect our industries not only from human errors but also from sabotage or cyber terrorism by replacing manual control with OSC-type automatic safety controls.

Our profession can do all this and much more, but to do so it is necessary for the people entering or working in this field to have an overall view of this profession, and this book of Dr. Sharma serves that goal.

May 1, 2016   **Béla Lipták**
Stamford CT, USA   Liptakbela@aol.com

Béla Lipták is president of the consulting firm Lipták Associates, PC.

Future of Industrial Automation

Extract from the article by Jim Pinto

"As the second decade of the 21st century moves forward, technology continues to accelerate and is generating rapid changes in industrial automation & control systems.

Industrial automation can and will generate explosive growth in the next decade, with several new inflection points:

- Nanotechnology will bring tiny, low-power, low-cost sensors and nano-scale electromechanical systems. The Internet-of-Things (IoT)—Europeans call it Industry 4.0—will be pervasive and change the face of industrial automation, generating vast productivity gains.
- Wireless sensors and distributed peer-to-peer networks will give rise to major new software applications—tiny operating systems in wireless nodes will allow nodes to communicate with each other in a larger complex adaptive systems.
- Today's pervasive smart-phone has more power than a super-computer of just a couple of decades ago. The use of WiFi-connected tablets, smartphones and mobile devices with new real-time monitoring software (Apps) will be used widely in industrial plants and process control environments.
- Conventional Real-time control systems will give way to multi-processors and complex adaptive systems

These are the waves of the future. The leaders will not be the traditional automation majors, but new start-ups with innovative ideas that will generate industrial productivity gains to drive significant growth."

June 28, 2016   **Jim Pinto**
Carlsbad, CA, USA   jim@jimpinto.com

Jim Pinto is an international speaker, technology futurist, automation consultant, and writer (http://JimPinto.com).

Preface to the First Edition

During my 33-year career in the computer and automation industry and subsequently my 7 years in academic institutions, I have observed a gap between academia and industry regarding the automation domain. These observations are based on my time spent training new recruits in Indian industry and later, on my teaching experience in Indian academic institutions. One of the ways this gap can be bridged is by introducing the basics of modern automation technology to those who are beginning careers in automation. This includes students and persons in industry who are switching to the automation domain. Prior knowledge of automation provides these beginners with a better and quicker start.

In many academic institutions, curriculum is being upgraded in instrumentation/control engineering courses to prepare students for careers in the automation industry. Currently the automation industry spends considerable time and money training and preparing new recruits for the job. The situation is more or less the same for persons switching to the automation domain in industry. This motivated me to write this book introducing the principles of automation in a simple and structured manner.

This book teaches beginners the basics of automation, and it is also intended as a guide to teachers and trainers who are introducing the subject. It addresses the current philosophy, technology, terminology, and practices within the automation industry using simple examples and illustrations.

The modern automation system is built out of a combination of technologies, which include the following:

- sensor and control
- electronics
- electrical drives
- information (computer science and engineering)
- communication and networking
- embedded
- digital signal processing
- control engineering, and many more

Current automation technology is one of the few engineering domains that use many modern technologies. Among these, information, communication, and networking technologies have become integral parts of today's automation. Basic subsystems of a modern automation system are **instrumentation**, **control**, and the **human interface**. In all of the subsystems, the influence of various technologies is visible. By and large, the major providers of automation use similar philosophies in forming their products, systems, and solutions.

As of now, most of the information on modern automation exists in the form of technical documents prepared by automation companies. These documents are usually specific to their products, systems, solutions, and training. This knowledge has not yet been widely disseminated to the general public, and the books that are available deal with specific products and systems. Most of this industry material is difficult for beginners to understand. It is good for next-level reading after some exposure to the basics.

Automation of the industrial process calls for **industrial process automation systems.** These involve designing, developing, manufacturing, installing, commissioning, and maintaining automation systems, which calls for the services of qualified and trained automation engineers. In addressing the basic concepts of automation, this book provides a starting point for the necessary education and training process.

Over the years, considerable advances have taken place in hardware technologies (mainly in electronics and communication). However, in computer-based automation systems, hardware interfaces have remained virtually the same, except that control has become more powerful owing to the availability of large memory and increased processor speeds. In other words, the memory and speed constraints present in earlier systems are not an issue today. In view of this, many aspects of hardware interface and function have been taken over by software, which does not need special interface electronics. Today, complete operation and control of industrial processes are by software-driven automation systems. Furthermore, this software provides maximum support to the user. The user is now only required to configure and customize the automation system for a particular process. Therefore the emphasis in this book is placed on hardware, engineering, and application programming.

This book is intended for:

* students beginning automation careers
* teachers of automation and related subjects
* engineers switching to automation careers
* trainers of automation

How to read the book:

* Follow the book from beginning to end, because its sequence is structured as a guided tour of the subject.
* Skip appendixes if you already have the background knowledge. They are provided for the sake of completeness and to create a base for easy understanding of the book.

Benefits the book provides for readers:

* does not call for any prior knowledge of automation
* presents a guided tour of automation
* explains the concepts through simple illustrations and examples
* makes further study easy
* prepares the reader to understand technical documents in the automation industry

Based on my experience in training, teaching, and interacting with trainees and students, I have specially formulated and simplified the illustrations and discussions in this book to facilitate easy understanding.

The book is organized as follows:

Chapter 1: Why Automation?—Industrial process, Undesired behavior of process, Types and classifications of process, Unattended, manually attended, and fully automated processes, Needs and benefits of automation, Process signals

Chapter 2: Automation System Structure—Functions of automation subsystems, Instrumentation, control, and human interface, Individual roles

Chapter 3: Instrumentation Subsystem—Structure and functions, Types of instrumentation devices, Interface to control subsystem, Interfacing standards, Isolation and protection

Chapter 4: Control Subsystem—Functions and structure, Interfaces to instrumentation and human interface subsystems

Chapter 5: Human Interface Subsystem—Construction, Active display and control elements, Types of panels, Interface to control subsystem

Chapter 6: Automation Strategies—Basic strategies, Open and closed loop, Discrete, continuous, and hybrid

Chapter 7: Programmable Control Subsystem—Processor-based subsystem, Controller, Input–output structure, Special features—communicability and self-supervisability

Chapter 8: Hardware Structure of Controller—Construction of controller, Major functional modules, Data transfer on the bus, Structure and working of functional modules, Integration

Chapter 9: Software Structure of Controller—Difference between general-purpose computing and real-time computing, Real-time operating system, Scheduling and execution of tasks, Program interrupt

Chapter 10: Programming of Controller—Programming of automation strategies, higher-level languages, IEC 61,131-3 standard, Ladder logic diagram, Function block diagram

Chapter 11: Advanced Human Interface—Migration of hardwired operator panel to software-based operator station, Layout and features, Enhanced configurations, Logging station, Control desk

Chapter 12: Types of Automation Systems—Structure for localized and distributed process, Centralized system, Decentralized/distributed system, Remote/networked system, Multiple operator stations, Supervisory control and data acquisition

Chapter 13: Special-Purpose Controllers—Customization of controller, Programmable logic controller, Loop controller, Controller, Remote terminal unit, PC-based controller, Programmable automation controller

Chapter 14: System Availability–Availability issues, Improvement of system availability, Cold and hot standby, Standby/redundancy for critical components

Chapter 15: Common Configurations–Configurations with operator stations, Supervisor stations, Application stations

Chapter 16: Advanced Input–Output System—Centralized input–output, Remote input–output, and Fieldbus input–output, Data communication and networking, Communication protocol

Chapter 17: Concluding Remarks—Summary, Application-wise classification of automation systems, Data handling, Future trends

Appendixes: Hardwired Control Subsystems, Processor, Hardware–Software Interfacing, Basics of Programming, Advanced Control Strategies, Power Supply System, Further Reading.

I wish to emphasize that the content in this book is mainly a result of my learning, practicing, teaching, and training experience in automation areas in ABB India, where I worked for over 24 years. I would also like to mention the following organizations, where I gained valuable automation teaching and training experience:

International Institute of Information Technology, Bangalore, India (http://www.iiitb.ac.in)
National Institute of Technology Karnataka, Surathkal, India (http://www.nitk.ac.in)
Axcend Automation and Software Solutions, Bangalore, India (http://www.axcend.com)
Honeywell Technology Solutions Lab, Bangalore, India (http://www.honeywell.com)
Emerson Process Management, Mumbai, India (http://www.emerson.com)
M S Ramaiah Institute of Technology, Bangalore, India (http://www.msrit.edu)

For their help with this book, I respectfully and gratefully acknowledge the kind guidance and support of my senior colleagues:

Prof. S.S. Prabhu, senior professor at IIIT/Bangalore, former professor at IIT/Kanpur, and a veteran on control systems and power systems.
Prof. H.N. Mahabala, former professor at IIT/Kanpur, IIT/Chennai, and IIIT/Bangalore, a veteran on information technology, and a founder of computer education in India.

I also acknowledge the help of my student, Mrs. Celina Madhavan, who developed the automation program examples for the book and reviewed the manuscript.

In addition, the following professionals supported me at every stage of preparation of the manuscript with their valuable suggestions and input:

Prof. R. Chandrashekar, IIIT/Bangalore, India.
Mr. Hemal Desai, Emerson Process Management, Mumbai, India.
Mr. Shreesha Chandra, Yokogawa, Bangalore, India.

I also gratefully acknowledge the support and encouragement of Prof. S. Sadagopan, director of IIIT/Bangalore, and Mr. Anup Wadhwa, director of Automation Industry Association of India (AIA).

Finally, I would like to thank my wife, Mrs. Sumitra Sharma, for her kind encouragement and support through it all.

March 1, 2011   **KLS Sharma**
Bangalore, India

Preface to the Second Edition

Greetings,

This is a continuation of the preface given in the first edition, also reproduced in this edition. Most of my observations enumerated in the preface of the first edition continue to be applicable for this edition as well, although a lot of progress has taken place in this area over the past 5 years, and hence not repeated here. Since the publication of the first edition of this book in August 2011, more advances have taken place in industrial automation technology. The purpose of this edition is to cover these new developments to the extent possible. As stated earlier in the first edition, industrial automation technology is one of the few engineering domains that employ many modern technologies such as:

Engineering technologies:

- sensor and control elements
- control engineering
- electronics, embedded, and digital signal processing
- micro-electromechanical and nano-electromechanical systems
- electrical, mechanical, etc.

Information technologies:

- computer science and engineering
- data communication and networking, etc.

Among these, information technology has been the enabling and driving force.

In line with the first edition, this edition is also primarily intended for:

- students about to embark on industrial automation careers
- teachers of industrial automation and related subjects
- engineers switching to industrial automation careers
- trainers of industrial automation

I suggest readers complete all chapters in the book from beginning to end. The sequence of chapters is structured as a guided tour of the subject. Readers may skip the appendixes if they are already familiar with the legacy technologies and basics. The appendixes are included for the sake of completeness and to create a base for easy understanding of the book.

Benefits of this edition for readers are that they:

- provide working knowledge for engineers without prior knowledge of industrial automation and present a guided tour of industrial automation
- explains the concepts of industrial automation through simple illustrations and examples
- make further study of industrial automation easy
- prepare understanding of technical documents in the industrial automation industry

I believe in imparting the basic knowledge and concepts of any subject to give readers a feel for the subject to prepare them for further learning on their own. The first edition followed this philosophy and this edition continues it.

Based on my experience in training, teaching, and interacting with trainees, students, and professionals, I have continued the same style followed in the first edition in this edition with specially formulated and simplified illustrations and discussions to facilitate easy understanding. This edition is organized in three parts.

Part 1: Basic Functionalities

Chapters 1–12 are repeated from the first edition with updated and reformatted content. These chapters introduce the reader to the basic concepts of automation:

1. **Why automation?** explains industrial processes and its management to deliver products and services of required quality, consistency, and cost-effectiveness. Also, the needs met by industrial automation and the resulting benefits are discussed.
2. **Automation system structure:** explains the basic subsystems in industrial automation systems. Their individual functions and interfacing among themselves and with the process for the management of industrial processes are also discussed.
3. **Instrumentation subsystem:** explains the interface between the process and control subsystem for the conversion of physical process signals into electronic signals and vice versa.
4. **Human interface subsystem:** explains the facility for manual interaction with the process to observe the status of process parameters and effect their control, if required.
5. **Control subsystem:** explains the heart of the industrial automation system, which not only coordinates the functions of all the subsystems but also manages the industrial process.
6. **Automation strategies:** explains the control philosophy specific to the individual processes that is required for its management to achieve the desired results.
7. **Programmable control subsystem:** explains the realization of the control subsystem with soft-wired or microprocessor technology.
8. **Data Acquisition and Control Unit (DACU) - Hardware:** explains the hardware modules (basic and functional) for realizing the DACU.
9. **Data Acquisition and Control Unit (DACU) - Software**: explains the software structure and application programming for implementing the automation strategy.
10. **Advanced human interface:** explains the application of computer-based platforms for realizing the software-based human interface that provides many more functionalities than are available in hardware-based operator panels.
11. **Types of automation systems:** explains different types of industrial processes based on their physical arrangements and specific automation systems for their management.
12. **Special-purpose Data Acquisition and Control Units (DACUs):** explains different types of DACUs that are optimized for different application requirements.

13. **System availability enhancements:** explains typical standby/redundancy schemes for all types of automation systems to increase their availability.
14. **Common configurations:** explains configurations that are commonly employed in distributed control systems and network control systems that are need based.
15. **Customization:** explains the realization of a tailor-made industrial automation system from a common platform.

Part 2: Extended Functionalities

Chapters 16–21 are added to this edition to take the traditional industrial automation system into the connected world integrated with enterprise and business processes.

16. **Data communication and networking:** explains the deployment of this technology to manage inter and intra system data exchange in industrial automation systems in a secure way.
17. **Fieldbus technology:** explains moving from conventional centralized input–output (I/O) to fieldbus I/O to eliminate control and signal cabling and field asset management.
18. **Safety systems:** explains the management of safety in process plants and machines to avoid or reduce damage to the people, property, and their environment in case of a hazard.
19. **Management of industrial processes:** explains the different kinds of industrial processes such as manufacturing (process plants and factories) and utility (civic and backbone) and their management.
20. **Information technology–operation technology (IT-OT) convergence:** explains the integration of OT (plant and automation system) and IT (business systems) to derive operational and business excellence.
21. **Concluding remarks:** Summarizes basic, extended, and some additional functionalities, and a peek into some emerging trends in industrial automation technology

Part 3: Appendixes

As mentioned, these are provided for the sake of completeness and to create a base for easy understanding of the relation between hardware and software in DACU.

1. **Hardwired control subsystem:** explains legacy technologies in the realization of control subsystems to implement all types of automation strategies.
2. **Processor:** explains the architecture of a hypothetical processor-based DACU with all of its basic and functional modules
3. **Hardware-software interfacing:** explains how software in a DACU is interfaced to hardware in a DACU to execute the automation strategy.
4. **Basics of programming:** explains the programming of DACU, starting from machine level to higher level and its conversion into executable code.

Acknowledgment

I wish to acknowledge that the content in this book is mainly the result of my learning, practicing, teaching, and training experience in industrial automation in ABB India, where I worked for over 24 years followed by teaching and training experience in

industry and academic institutions. I would also like to mention the following organizations, where I gained valuable teaching and training experience in the industrial
automation area:

Training and teaching programs in:

- ABB, Bengaluru, India (www.abb.com)
- Automation Industry Association of India, Campus Connect, Delhi, India (www.aia-india.org)
- Axcend Automation and Software Solutions, Bengaluru, India (www.axcend.com)
- Emerson Process Management, Mumbai, India (www.emerson.com)
- Honeywell Technology Solutions Lab, Bengaluru, India (www.honeywell.com)
- Infosys, Bengaluru, India (www.infosys.com)
- International Society of Automation, Bengaluru, India (www.isabangalore.org)
- International Institute of Information Technology, Bengaluru, India (www.iiitb.ac.in)
- M S Ramaiah Institute of Technology, Bengaluru, India (www.msrit.edu)
- National Institute of Technology Karnataka, Surathkal, India (www.nitk.ac.in)
- Power Engineering and Technology Consultants, Bengaluru, India (www.poetconsultants.in)

Faculty development and short student programs in:

- Adi Shankara Institute of Engineering Technology, Kochi, India (www.adishankar.ac.in)
- BMS College of Engineering, Bengaluru, India (www.bmsce.acin)
- BVB College of Engineering and Technology, Hubballi, India (www.bvb.edu)
- Dayanand College of Engineering, Bengaluru, India (www.dyanandasagar.edu)
- Dr. Ambedkar Institute of Technology, Bengaluru, India (www.dr-ait.org)
- Malnad College of Engineering, Hassan, India (www.mcehassan.ac.in)
- Manipal Institute of Technology, Manipal, India (www.manipal.edu/mit)
- MS Ramaiah Institute of Technology, Bengaluru, India (www.msrit.edu)
- PSG College of Technology, Coimbatore, India (www.psgtech.edu)
- RNS Institute of Technology, Bengaluru, India (www.rnsit.ac.in)
- RV College of Engineering, Bengaluru, India (www.rvce.edu.in)
- SDM College of Engineering and Technology, Dharwad, India (www.sdmcet.ac.in)
- Shri Siddhartha Institute of Technology, Tumukuru, India (www.ssit.edu.in)
- Siddaganga Institute of Technology, Tumukuru, India (www.sit.acin)
- Vellore Institute of Technology, Vellore, India (www.vit.ac.in)
- Vidya Vikas Institute of Technology, Mysuru, India (www.vidyavikas.edu.in)

*This edition is also primarily a collection of information from various sources
(books, websites, scholarly articles, white papers, technical documents, technical
discussion with experts, technical presentations, industry literature, etc. in the public
domain) and a presentation as a structured and guided tour on industrial automation
for easy and quick understanding by readers. The sequence of presentation is based
on the way I have understood the subject and interpreted it. I express my indebtedness
and gratitude to all professionals who have made their rich knowledge and experience
on industrial automation technology available in the public domain.*

Notwithstanding this, I would like to specifically and gratefully acknowledge:

Béla Lipták, well-known as the *father of modern industrial process automation*, for his
wonderful forward to this book.

Jim Pinto, well-known *critic, commentator, and futurist* in industrial automation, for his thoughts on the *future of industrial automation* for this book.

Ana Claudia Garcia, editorial project manager, Elsevier, for her excellent and comforting support all through the preparation of the manuscript.

Furthermore, I would like to acknowledge kind support and encouragement from:

Fellow professionals from the International Society of Automation, Bengaluru, India.
Colleagues from Infosys, Bengaluru, India.
Friends and colleagues (past and present) from industrial automation companies and academic institutions.

Finally, I would like to thank my wife, Sumitra Sharma, for her kind patience, encouragement, and support through it all.

July 1, 2016   **KLS Sharma**
Bengaluru, India   kls.sharma@hotmail.com

Acknowledgments

Institutional Support

The following organizations have extended their full support for this book with technical information and images of their products, systems, and solutions.

Companies	Location	Country	Address	Products
3S-Smart Software Solutions	Kempten	Germany	www.codesys.com	IEC 61131-1 Programming software
Adept Fluidyne	Pune	India	www.adeptfluidyne.com	Strip-chart recorder
Adept Technologies	San Ramon, CA	United States	www.adept.com	Industrial robots
Adobe Metal Products	Bengaluru	India	www.adobemetals.com	Mimic panels
Aplab	Mumbai	India	www.aplab.com	Float-boost chargers, uninterruptible power supply system
Avcon Controls	Mumbai	India	www.avconindia.com	Solenoid valve
B&R Automation	Pune	India	www.br-automation.com	Safety programmable logic controller (PLC)
Behr GmbH & Co. KG	Charleston, SC	United States	www.mahle.com	Coolant module assembly
Control Dynamics	Vadodara	India	www.mimicpanels.net	Annunciator windows, Mosaic panels
DiFacto	Bengaluru	India	www.difacto.com	Robotic end-effectors
ELCOM	Mumbai	India	www.elcom-international.com	Push button switch
Emerson Process Management	Mumbai	India	www.emerson.com	Control valve, flow transmitters, visual display units, control desk, control center, controller, remote input–output, fieldbus devices, wireless field devices, safety PLC
Exide Industries	Kolkata	India	www.exide.co.in	Battery banks
Fabionics	Bengaluru	India	www.fabionics.co.in	Distribution panel
Fuji Electric	Tokyo	Japan	www.fujielectric.com	Variable-speed drive, PLC, single loop controller
General Industrial Controls	Pune	India	www.gicindia.com	Electronic counter, PLC

Continued

—Cont'd

Companies	Location	Country	Address	Products
Ingenious Technologies	Bengaluru	India	www.aquamon.in	Water level controller and indicator
Integral Systems	Bengaluru	India	www.integralsys.in	Thumb wheel switch
Jai Balaji	Chennai	India	www.jaibalaji.firm.in	Limit switch, indication lamps
JSW Steels	Vidyanagar	India	www.jsw.in	Hot strip mill
Kevin Technologies	Ahmadabad	India	www.kevintech.com	Temperature transmitter
Levcon Controls	Kolkota	India	www.levcongroup.com	Flow switch
Manikant Brothers	Mumbai	India	www.watermeter.co.in	Water meter
MECO Instruments	Navi Mumbai	India	www.mecoinst.com	Analog panel meter, digital panel meter
Megacraft Enterprises	Mumbai	India	www.megacraft.net	Potentiometer
OEN India	Kochi	India	www.oenindia.com	Supervision relay, control relay, voltage switch
Omron	Kyoto	Japan	www.omron.com	PLC, push button switches, indication lamps, limit switches
OPTO 22	Temecula, CA	United States	www.opto22.com	PC-based controller, programmable automation controller
Pankaj Potentiometers	Bengaluru	India	www.pankaj.com	Potentiometer
Pepperl-Fuchs India	Bengaluru	India	www.pepperl-fuchs.com	Proximity switch, safety barriers
Renu Electronics	Pune	India	www.renuelectronics.com	Operator terminals
Rishabh Instruments	Nashik	India	www.rishabh.co.in	Energy meter
Rockwell Automation	Milwaukee, WI	United States	www.rockwellautomation.com	Safety PLC
Sahyadri Electro Controls	Bengaluru	India	www.secoindia.com	Electronic buzzer
Schneider-Electric	Bengaluru	India	www.schneider-electric.com	Safety PLC
Secure	Gurgaon	India	www.securetogether.com	Intelligent energy meter
Shridhan Automation	Bengaluru	India	www.shridhan.com	Level transmitter, level switch
SICK India	Bengaluru	India	www.sick.com	Technical protective devices
Siemens	Mumbai	India	www.siemens.com	Remote terminal unit (RTU)
Strategi Automation	Bengaluru	India	www.strategiautomation.com	Industrial robots
Sunlux Technologies	Bengaluru	India	www.sunluxtech.com	Field-programmable gate array controller, embedded controller, Internet of things—enabled device
VA Tech WABG	Chennai	India	www.wabag.in	Sewage treatment plant
V-Guard Industries	Kochi	India	www.vguard.in	Voltage stabilizer
Yokogawa India	Bengaluru	India	www.yokogawa.com	Pressure transmitter, Single loop controller, controller, RTU, fieldbus devices, wireless field devices, safety PLC

Technical Support

The following experts have extended their full support for this book with technical input and have reviewed the topics for their correctness and completeness.

Name	Organization	Location	Country	Address
General				
Amit Tiwari	Rockwell Automation	Delhi	India	www.ra.rockwellautomation.com
Jitendra Vasista	JSW Steels	Vidyanagar	India	www.jsw.in
PV Sivaram	B&R Automation	Pune	India	www.br-automation.com
Rajesh Nair	Siemens	Mumbai	India	www.siemens.com
Rajesh Rathi	Control Infotech	Bengaluru	India	www.control-infotech.com
Sachin Kulkarni	Bosch India	Ahmadabad	India	www.boshrexroth.co.in
Sameer Gandhi	Omron	Mumbai	India	www.omron.com
Siva Prakash	Schneider Electric	Bengaluru	India	www.schneider-electric.com
Suhas Bhide	Emerson	Delhi	India	www.emersom.com
Sunil Nambiar	Yokogawa	Bengaluru	India	www.yokogawa.com/in
Sunil Shah	Communications, Diagnostics, and Controls	Bangalore	India	www.cdcontrolsit.com
TV Gopal	VA Tech WABG	Chennai	India	www.wabag.in
Chapter 15: Customization				
BK Suresh	Consultant	Bengaluru	India	bettesuresh@gmail.com
Girish Ayya	Avadhoot Automation	Bengaluru	India	www.avadhoot.in
Sunil Nambiar	Yokogawa	Bengaluru	India	www.yokogawa.com/in
Chapter 16: Data Communication and Networking				
Mallikarjun Kande	ABB	Bengaluru	India	www.abb.com/in
Prasanna Inamdar	Senseops Tech Solutions	Bengaluru	India	www.sense-ops.com
Ram Kerur	Sunlux Technologies	Bengaluru	India	www.sunluxtech.com
Vijay Murugan	Schneider Electric	Bengaluru	India	www.schneider-electric.com

Continued

—**Cont'd**

Name	Organization	Location	Country	Address
Chapter 17: Fieldbus Technology				
Prasanna Inamdar	Senseops Tech Solutions	Bengaluru	India	www.sense-ops.com
Chapter 18: Safety systems				
Jan Baldauf	SICK	Düsseldorf	Germany	www.sick.de
Rajendran Menon	Rockwell	Delhi	India	www.ra.rockwell.com
S Vinod	Yokogawa	Bengaluru	India	www.yokogawa.com
Shabad Khan	Honeywell	Pune	India	www.honeywell.com
Chapter 19: Management of Industrial Processes				
Ajay Gopalaswamy	Difacto	Bengaluru	India	www.difacto.com
Arun Menon	Strategi Automation	Bengaluru	India	www.strategiautomation.com
BS Sugunanada	ABB	Bengaluru	India	www.abb.com
Deven Patel	Wipro	Bengaluru	India	www.wipro.com
HR Shivaprakash	JSW Steels	Vidyanagar	India	www.jsw.in
K Srinivasan	VA Tech WABG	Chennai	India	www.wabag.in
KA Sampath Kumar	Newfield Engineers	Bengaluru	India	www.newfieldengineers.com
KV Rehani	Automation Consultant	Mumbai	India	kvrehani@gmail.com
Mark Sauls	Control Infotech	Morrisville, NC	United States	www.control-infotech.com
Srinivas Patnaik	ABB	Bengaluru	India	www.abb.com
Chapter 20: Information Technology–Operations Technology Convergence				
Bhaskar Rao	Technospehere	Bengaluru	India	www.technospehere.com
Krishnanada Shenoy	Infosys	Bengaluru	India	www.infosys.com
MK Kumar	Infosys	Mysore	India	www.infosys.com
Nampuraja Enose	Infosys	Bengaluru	India	www.insosys.com
Ram Kerur	Sunlux Technologies	Bengaluru	India	www.sunluxtech.com
Sunil Shah	Communications, Diagnostics, and Controls	Bangalore	India	www.cdcontrolsit.com
Vijay Murugan	Schneider Electric	Bengaluru	India	www.schneider-electric.com

Name	Organization	Location	Country	Address
Chapter 21: Concluding Remarks				
BR Mehta	Reliance Industries	Mumbai	India	www.ril.com
GK Raviprakash	Infosys	Bangalore	India	www.infosys.com
Sunil Shah	Communications, Diagnostics, and Controls	Bangalore	India	www.cdcontrolsit.com
Vikas Kukshya	Infosys	Bangalore	India	www.infosys.com

Media Support

The following publishing houses have extended their full support and have provided technical information and established contacts with automation companies and experts.

Editor	Publisher	Location	Country	Address
Publisher of A&D India Automation & Drives				
Shekhar Jitkar	Publish Industry India	Pune	India	www.industr.com/in/en/
Publisher of Automation & Controls Today				
Jigar Patel	Himani Infomedia 8	Ahmadabad	India	www.aandctoday.com

K.L.S. Sharma
kls.sharma@hotmail.com

July 1, 2016
Bengaluru, India

Why Automation?

1

Chapter Outline

1.1   Introduction

Over the past few decades, the emphasis in industry worldwide has been to produce goods or deliver services which are of quality, consistency, and cost-effective, to stay in the market. Quality, consistency, and competitiveness cannot be achieved without automating the process of manufacturing goods and of delivering services. In line with this trend, the application of automation today is omnipresent in almost all

Overview of Industrial Process Automation. http://dx.doi.org/10.1016/B978-0-12-805354-6.00001-3

applications, from deep water sea to space, and has gained the confidence of the world for achieving desired results. Over the years, automation technology has advanced along with various other technologies. The main driving and enabling technologies are information, communication, networking, and electronics.

This list below shows the evolution of automation technology over the past decades:

- 1940–1960: Pneumatic
- 1960–2000: Analog: Electrical, mechanical, and hydraulic
- 1980–1990: Digital: Proprietary
- 2000 onward: Digital: Open.

The technology has already moved to an **open network of embedded systems.**[1] This chapter gives a brief introduction to automation and explains why it is necessary to automate the production of goods and delivery of services to achieve the required quality, consistency, and cost-competitiveness for today's marketplace. Apart from these tangible requirements, there are many more complex requirements that cannot be achieved without automation.

1.2 Physical Processes

Physical process[2] is a series of actions, operations, changes, or functions that takes place within, bringing about changes or producing an output or a result. It is also a sequence of interdependent operations or actions which, at every stage, consumes one or more inputs or resources to convert them into outputs or results to reach a known goal or the desired end result.

Whatever we see and work within reality are all physical processes. Physical processes can be broadly divided into three categories:

- Natural processes,
- Self-regulated processes, and
- Industrial processes.

1.2.1 Natural

Natural processes[3] are presented by or produced by nature. The best example is a human body that generally does not need external assistance to regulate its body parameters (e.g., body temperature) irrespective of the effects of surrounding environmental conditions. The human body maintains or regulates all of its parameters. Typically, in natural processes no abnormal behavior is present in most of the conditions.

---

[1] Pinto, Jim. Automation Unplugged, ISA, 2004.
[2] http://www.wikipedia.org.
[3] http://www.wikipedia.org.

1.2.2   Self-regulated

Self-regulated processes[4] are not natural but do not need external assistance for regulation. The best example is a domestic geyser that is designed such that the water level in it is always maintained to be full irrespective of any condition. Another typical example is an irrigation dam in which the water automatically overflows whenever the inflow exceeds the storage threshold. All natural processes are self-regulated but the reverse is not true.

1.2.3   Industrial

Industrial processes[5] are systematic series of physical, mechanical, chemical, or similar kinds of operations that produce a result. They manufacture goods or deliver services. These processes are not always self-regulating and need external regulation on a continuous basis to produce desired results. Typical examples of industrial process are factories manufacturing goods or equipment providing services such as passenger lift, traffic signals, etc. They are also known as **Man-made Processes.**

Technically, all industrial processes can be broadly classified into three levels (application, operational, and physical).

1.2.3.1   Application

Regarding application, any industrial process can be either manufacturing that creates physical values to deliver goods/products as per specification or utilities that deliver services as per requirement.

Manufacturing processes are further divided into process plants and factories. Typical examples are:

- Process plants: Chemical, cement, power generation, fertilizer, cement, etc.; and
- Factories: Engineered products, automobiles, assembly lines, etc.

Utility processes are further divided into civic utilities and backbone utilities. Typical examples are:

- Civic utilities: Water distribution, power distribution, gas distribution, water/sewage treatment, etc.; and
- Backbone utilities: Power transmission, water transmission, oil/gas pipelines, etc.

1.2.3.2   Operational

Regarding operation, any application process can be continuous, discrete, or batch. Typical examples are:

- Continuous: Process plants, distribution of power, water, gas; treatment plants, etc.;
- Discrete: Factories, assembly lines, etc.; and
- Batch: Pharma, food, beverage, etc.

[4] http://www.wikipedia.org.
[5] http://www.wikipedia.org.

1.2.3.3 Physical

All processes are physically classified as either localized or distributed (even geographical). Typical examples are:

- Localized: Process plants, factories, electrical substations, pump stations, etc.; and
- Distributed: Distribution of water, power, and gas; transmission of water, power, and gas, etc.

   Chapter 19 will discuss more about the industrial process and its management. This chapter discusses only the physical classification of the industrial processes.

1.3 Localized and Distributed Processes

Among the industrial processes, the localized process is present in a relatively small physical area with all of its subprocesses or components closely interconnected. Conversely, the distributed process is present in a relatively larger physical area with its subprocesses or components distributed and loosely interconnected. A distributed process can also be termed a network of many localized processes distributed over a larger physical area and interconnected. Some simple comparative examples are:

- *Several water taps (localized processes) connected to a common water supply line (water supply system) in a building*

   Here, the coupling or networking of the water taps is through the water pipeline supplying the water to all the taps in a building. Each water tap is a localized process, whereas the group of water taps, networked through the common water supply pipe, is a distributed process.

- *Several electric bulbs (localized processes) connected to a common electric supply line (electric supply system) in a building*

   Here, the coupling or networking of the electrical bulbs is through the electricity supply cable supplying the electricity to all bulbs in a building. Each electrical bulb is a localized process, whereas the group of electrical bulbs, networked through the common electricity supply cable, is a distributed process.

   Fig. 1.1 illustrates the above concept.

   It is generally difficult to draw a fine line between localized and distributed processes because processes that fall in the border area can be treated either way and managed. **The distributed process network can even be spread over a large geographical area.**

   Figs. 1.2–1.4 illustrate some typical comparative industry examples.

   The distributed process treats the entire network of localized processes as a **single entity** (not as individual/independent localized subprocesses). Here, each localized process has some effect on the operation and performance on the other individual localized processeses in the network.

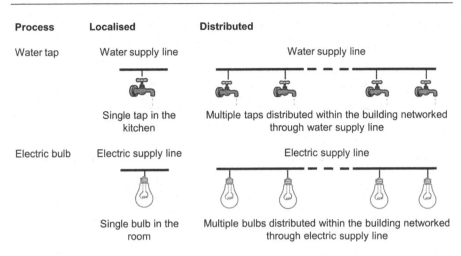

Process	Localised	Distributed
Water tap	Water supply line	Water supply line
	Single tap in the kitchen	Multiple taps distributed within the building networked through water supply line
Electric bulb	Electric supply line	Electric supply line
	Single bulb in the room	Multiple bulbs distributed within the building networked through electric supply line

Figure 1.1 Localized and distributed processes.

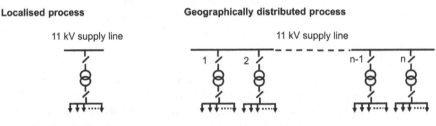

Localised process

11 kV supply line

400 V Outgoing lines
Domestic power distribution lines to
houses in a single locality

Single 11 kV/400 V pole mounted
substation in a
locality

Geographically distributed process

11 kV supply line

400 V Outgoing lines
Domestic power distribution lines to houses in many
different localities

Multiple 11 kV/400 V pole mounted substations networked
through 11 kV supply line within the distribution area
(multiple localities)

Figure 1.2 Power distribution system.

Localised process

Main water supply line

Domestic water distribution lines to
houses in a single locality

Single water distribution station
in a locality

Geographically distributed Process

Main water supply line

Domestic water distribution lines to houses in many
different localities

Multiple distribution stations networked through water
supply line within the distribution area (multiple localities)

Figure 1.3 Water distribution system.

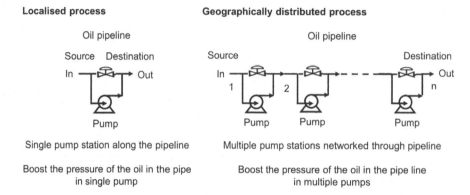

Figure 1.4 Oil transportation system.

1.4   Process Behavior

Following are some simple examples of the processes that are discussed here to demonstrate the behavior of industrial processes before we proceed further.

1.4.1   Water Tap

The function of the simple water tap in the house is to provide water when the tap is opened. The actions involved are to open the tap and wait for the water to flow. Here, the process is the water tap, the action is turning on the tap, and the output is the water. The intention is to get the water with the desired flow. This may not always happen for several reasons, such as low or no water pressure (external factors) or a clog in the pipe or in the tap (internal factors), resulting in either inadequate or no flow of water, leading to an undesired result.

1.4.2   Electric Bulb

Similarly, the function of a simple electric bulb in the house is to provide light when it is switched on. The actions involved are to switch the bulb on and wait for the bulb to glow. Here, the process is the electric bulb, the action is turning on the switch, and the output is the light. Once again, the intention is to get the light with the desired illumination. This may not always happen for several reasons, such as low or no voltage (external factors) or a faulty/broken filament (internal factors), resulting in either improper or no illumination, leading to an undesired result.

As seen, an undesired behavior is expected under unusual conditions in any industrial process owing to either external and/or internal factors that cannot be eliminated. Special efforts are required to overcome or minimize the effects of these factors to ensure that the process behaves the way we want or produces the desired results.

Because our discussion will focus on industrial processes and their management to get the desired results, the term "process" hereafter will refer only to industrial processes. These processes may also be called plants or process plants.

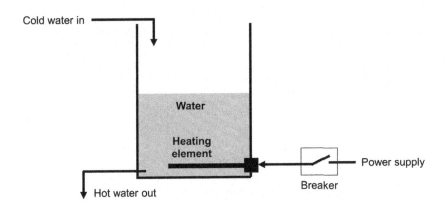

Figure 1.5 Water-heating process.

1.5 Process Management

The next obvious question is on managing the process to make it deliver the desired results. To understand and analyze this aspect, let us take a simple water heater as an example Fig. 1.5. The intention of the water heater is to have a heating process to maintain the temperature of the water at a desired value. There are three possibilities to achieve this.

In this example, internal disturbances are the quality of heating element, insulation, etc. whereas external disturbances are over-/under-voltage, outside temperature, etc., which are uncontrollable.

1.5.1 Unattended

In an unattended process, the actions involved are to:

1. Turn on the power supply to the water heater.
2. Wait for the water to heat up.

After switching on the power, water starts to heat up and continues even after crossing the desired level if not checked. Also, the heating time depends on the effects of internal and/or external factors, which are not in our control. In this setup, the heating process is totally uncontrolled and there is no guarantee that at any point of time we have the water with the desired temperature. Leaving the power switched on to the heater can also be a safety hazard if left unchecked.

Fig. 1.6 illustrates the schematic and operation of the unattended water heater system.

This approach is the simplest and most cost-effective approach. However, there is no mechanism to judge whether the temperature of the water is higher than, lower than, or equal to the desired level. Hence, the unattended processes not only do not produce the desired result but also can also lead to serious safety consequences such as overheating of the water, burning of the heating element, etc. Hence, **unattended management of a process always produces poor results.**

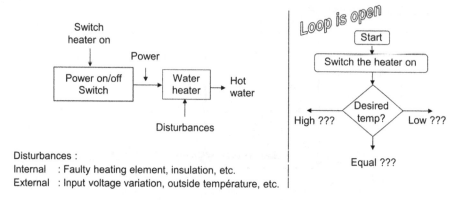

Disturbances :
Internal   : Faulty heating element, insulation, etc.
External   : Input voltage variation, outside température, etc.

Figure 1.6 Unattended process.

1.5.2 Attended

In an attended process, the actions involved are to:

a. Turn on the power supply to the water heater.
b. Check the temperature of the water.
c. Turn off the power supply to the water heater when the temperature of the water reaches/ crosses the desired level.
d. Repeat the steps [Go to (b)] periodically

With these steps, almost all drawbacks seen in the unattended process management are rectified, although with a lot of burden on the operator. The more frequently the operator checks, the better the result is, but the increased work for the operator is unproductive.

Fig. 1.7 explains the schematic and the operation of the attended water heater system.

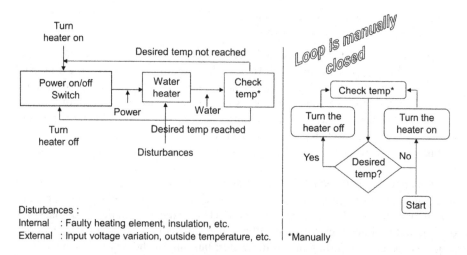

Disturbances :
Internal   : Faulty heating element, insulation, etc.
External   : Input voltage variation, outside température, etc.   *Manually

Figure 1.7 Attended process.

Furthermore, the effects of both the internal and external factors are taken care of without additional effort. Hence, **attended management of a process produces better, but average, results.**

1.5.3 Automated

In an automated process (with a temperature controller installed), the steps involved are:

a. Set the desired or reference temperature at which the temperature of the water is to be maintained.
b. Start the process by turning on the power to the heater.
c. The temperature controller continuously measures and checks the actual temperature and keeps the heater on if the actual temperature of the water is less than the reference temperature. The controller turns off the power to the heater if the actual temperature the water is either equal to or more than the reference temperature.
d. Repeat the cycle [Go to step (c)] as long as the process is on.

Fig. 1.8 illustrates the schematic and the operation of the automated water heater system.

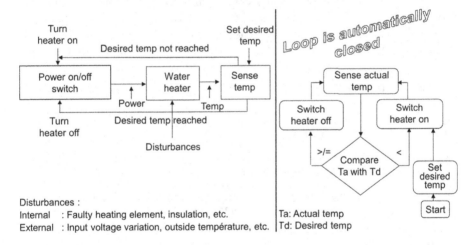

Figure 1.8 Automated process.

The cycle continues and the process management goes on with no manual intervention until the operator decides to stop the process. The operator's job is performed automatically by the temperature controller. Hence, **automated management of process always produces the best results.**

Fig. 1.9 illustrates the water heater system with automated temperature control.

1.6   Process Signals

A process signal[6] is a fluctuating or varying physical quantity (such as electrical, mechanical, chemical, thermal, etc.) whose variations represent the coded or embedded information of the process parameters. In practice, the process signals exist in one

---

[6] http://www.wikipedia.org.

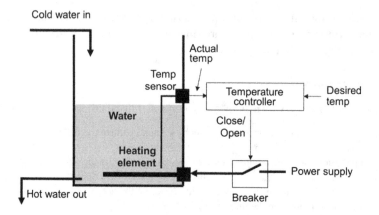

Figure 1.9 Automated water-heating process.

of the physical forms: discrete or digital, continuous or analog, and pulsating or pulse. Normally, all discussions on discrete or digital signals are equally valid for pulsating or pulse signals, because they are variations of discrete or digital signals.

Fig. 1.10 explains the nature and differences among the three types of physical signals.

Signal	Definition		Amplitude v/s Time
Discrete (Digital)	State changing discretely with time. The data here is the **state**.		
Continuous (Analog)	Amplitude changing continuously with time. The data here is the **amplitude**.		
Pulsating (Pulse)	State changing discretely and frequently with time. The data here is the **number of transitions of state**.		

Pulse signals are a variation of discrete/digital signal and the change of state may be periodic or aperiodic.

Figure 1.10 Physical signals.

The process signals are further divided into input and output signals. Signals that are sent by the process and received by the automation system (to understand the process behavior) are called input signals. Signals that are sent by the automation system (to change the process behavior) and received by the process are called output signals. **The input and output signals are always defined with reference to the automation system**.

Fig. 1.11 illustrates the relationship among the process, input–output signals, and automation system.

The input and output signals can be in any form, such as discrete, continuous, or pulsating. The form depends on which parameters of the process are being monitored and which parameters of the process are being controlled.

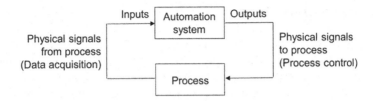

Figure 1.11 Input and output signals.

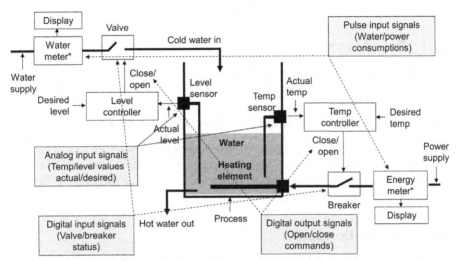

*Water and energy meters generate and output pulses proportional to the consumption of water and power

Figure 1.12 Process signals in water-heating process.

Fig. 1.12 illustrates various types of input and output signals in the water-heating process equipped with automated temperature/level control and water/power consumption reading.

1.6.1 Input Signals

Typical examples of input signals (measurements from process) are:

- *Discrete/digital*: Breaker status (on or off), control valve status (open or closed), etc.;
- *Continuous/analog*: Temperature value, pressure value, level value, flow value, etc.; and
- *Pulsating/pulse*: Energy consumption readings, water consumption readings, etc.

1.6.2 Output Signals

Typical examples of output signals (commands to process) are as:

- *Discrete/digital*: Open/close breaker, open/close control valve, start/stop motor, etc.;
- *Continuous/analog*: Vary control valve opening/closing, vary voltage output of the drive, etc.; and
- *Pulsating/pulse*: Stepper motor control, etc.

1.7   Automation Steps

A process automation system is an arrangement for automatic monitoring and control of the industrial process to obtain the desired results with no manual interventions. As discussed earlier, a typical automation system executes the automation steps sequentially, cyclically, and continuously, effecting corrections to the process operation and producing the desired results consistently.

Fig. 1.13 illustrates a typical automation cycle.

The basic automation steps are listed below:

1.7.1   Data Acquisition

The data acquisition step observes the behavior of the process by sensing or measuring the process parameters of interest. These parameters are called process inputs.

1.7.2   Data Analysis, Monitoring, and Decision Making

The data analysis, monitoring, and decision-making step analyzes the behavior of the process by comparing the acquired data from the process with the desired result. Then a decision is made about the new directives or commands that would be required to effect any corrections in the behavior of the process.

1.7.3   Control Execution

The control execution step executes the directives, which are commands decided by the data analysis and decision-making step to effect the corrections, if any.

Fig. 1.14 illustrates the automation steps in the water heater system.

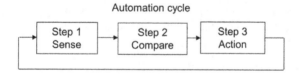

Figure 1.13 Typical automation steps and cycle.

1.8   Needs met by Automation

The basic need of any process is to produce goods or to deliver services while conforming to:

- Consistency,
- Quality, and
- Cost-effectiveness.

We have seen that only an automated process (with no human intervention) produces excellent results whereas unattended or attended processes generally do not always produce desired results. Although the foremost intention of any process is to

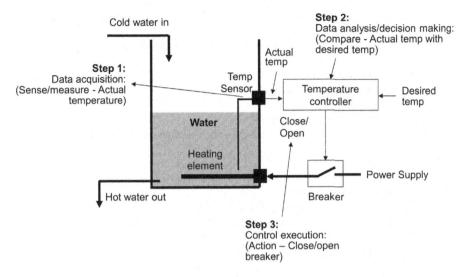

Figure 1.14 Automation steps in water-heating process. *temp*, temperature.

meet the desired result, there are other important intangible considerations to be met. Automation also manages the following:

- Hazardous processes (nuclear, high voltage, toxic, etc.) where human intervention is dangerous and is not desirable;
- Repetitive processes (traffic management, etc.) where continuous, repetitive manual operations can lead to failure owing to human fatigue;
- Sequential startup and shutdown of plants (power plants, chemical plants, etc.) while satisfying certain conditions at each step where sequential manual operation, with safety and/or other conditions to be satisfied, is highly time-consuming and prone to errors; and
- Complex processes (decisions based on heavy computing such as aircraft tracking and guidance) where manual computing and decision making within a short time is next to impossible.

There are many more complex situations with needs that can be met most efficiently through automation.

1.9 Benefits of Automation

In addition to attaining the desired quality, consistency, and cost-effectiveness of products and services, there are other benefits of automation.

1.9.1 Manufacturing Processes

The following are important benefits of automation in manufacturing processes:

- Reduction in production loss through a reduction in unproductive time through automated decisions for:
  - Restoring/restarting the plant or bringing the plant online faster during startup, after reconfiguration to new requirements, or after breakdown

- Keeping the plant online for maximum production (revenue generation) by increasing the overall availability of the plant
- Increasing the productivity of the process by making performance and repetitive tasks faster
- Optimization of resources through a reduction in:
  - Staff including dependence on highly skilled personnel
  - Inputs to process including energy
- Higher security, safety, and reliability for:
  - Plant personnel and equipment, because they do not come in contact with the working of the plant
  - Plant operation
- Faster response and result, because there is no human intervention required to:
  - Handle emergency situations
  - Reduce chances of human error and therefore defects
  - Ensure running of each process effortlessly and consistently every time it is run
- Compliance with:
  - Internal and external regulatory requirements (statutory, safety, and environmental), making auditing easier
  - Incredible flexibility and ease with which to make changes and adapt to new environment and requirements

1.9.2   Utility Processes

Almost all of these benefits for manufacturing processes are equally applicable to utility processes as well.

The above benefits lead to lesser overall operational costs or higher overall operational efficiency, increased productivity, and fewer chances of human error. This way, the production of goods or delivery of services is increased, contributing to higher revenue and profit.

1.10   Summary

In this introductory chapter, we have covered the basics of process and its automation. We have examined different types of processes, process classifications based on their structure and application, steps in automation, process needs to be met by automation, benefits of automation, and process signals and their classifications. These introductory concepts form the basis for further detailed discussions in the following chapters.

Automation System Structure

2

Chapter Outline

2.1   Introduction

In Chapter 1, we briefly discussed the automation functionalities, cycle, and steps. In this chapter, those concepts are further elaborated to define the physical structure of the overall automation system, the functions of its independent subsystems, and their interconnections.

2.2   Subsystems

The automation system is broadly divided into three subsystems: instrumentation, control, and human interface. They are interconnected, as illustrated in Fig. 2.1.

2.2.1   Instrumentation

The instrumentation[1] is the branch of engineering that deals with the measurement and control of process parameters. An instrumentation device measures the physical variable

[1] www.wikipedia.org.

Overview of Industrial Process Automation. http://dx.doi.org/10.1016/B978-0-12-805354-6.00002-5

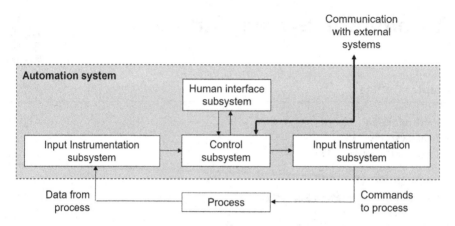

Figure 2.1 Basic structure of automation system.

and/or manipulates it to produce an output in an **acceptable form** for further processing in the next stage device. Instrumentation devices are not normally intelligent and are not capable of making decisions (exceptions are discussed later in the book).

The instrumentation subsystem primarily works in two directions: one on the input side of the control subsystem and the other on the output side. Hence the instrumentation subsystem can be further divided into input and output subsystems to explain the flow of data in the order they are received from a process, processed in the control subsystem, and sent to a process for actions:

1. **The input instrumentation subsystem** interfaces with the process to acquire data on the behavior of the process (measurement of process parameters) and sends them to the control subsystem **in an acceptable form**.
2. **The output instrumentation subsystem** interfaces with the process to send the command received from control subsystem to the process **in an acceptable form**, to change the behavior of the process (control of process parameters).

2.2.2 Human Interface

The human interface subsystem[2] presents information to the operator or user on the state of the process and facilitates implementing the operator's control instructions to the process. The human interface subsystem is also called the human–machine interface, man–machine interface, human–system interface, and so forth. The human interface subsystem is a facility for the user or operator to interact directly with the process, via **the control subsystem**, for:

1. Direct monitoring of process parameters to know what is happening inside the process (monitoring of process behavior), and
2. Direct control of process parameters by forcing a change, if required, by issuing manual commands (controlling of process behavior).

[2] www.wikipedia.org.

2.2.3 Control

The control subsystem[3] is a mechanism or device for automatically manipulating the output of a process and for managing, commanding, directing, or regulating the behavior of the process to achieve the desired result. The control subsystem is the **heart** of the automation system. It is an intelligent device capable of making decisions.

The control subsystem manages data flow to/from the instrumentation subsystems for process monitoring and control, and to/from the human interface subsystem for direct interaction with the process; it can exchange data bidirectionally with other external compatible systems if required.

Apart from data analysis and decision making, the control subsystem performs the following functions with respect to other subsystems:

1. Acquires data on process parameters via the **input instrumentation subsystem** to monitor the behavior of the process continuously,
2. Issues commands to process via the **output instrumentation subsystem** to correct or change the behavior of the process,
3. Routes process data to the **human interface subsystem** displays for direct monitoring, and
4. Acquires direct commands from the **human interface subsystem** and routes them to process for the control of process parameters.

The following sections further explain the implementation of functions associated with the subsystems with specific reference to automation of the water-heating process.

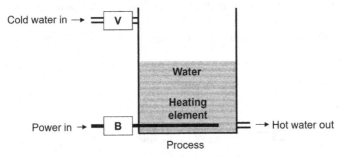

Figure 2.2 Basic water-heating process.

Fig. 2.2 illustrates the basic water-heating process with a manual control facility for letting cold water into the tank and switching power on to the heater. Here, the breaker controls the flow of power to the heating element while the valve controls the flow of water to the tank.

To implement automatic control of the water level, water temperature, and measurement of water and power consumption, the following functions need to be implemented:

1. Check the actual water temperature to know whether it is higher than, equal to, or less than the desired value.
2. Keep the breaker closed if the temperature is less than the desired value. Otherwise open the breaker.
3. Check the actual water level to know whether it is higher than, equal to, or less than the desired level.

[3] www.wikipedia.org.

4. Keep the valve open if the level is lower than the desired value. Otherwise close the valve.
5. Measure the water and power consumptions.

The following sections illustrate the application of automation subsystems to implement these steps.

2.3 Input Instrumentation Subsystem

2.3.1 Measurement of Data

Because they are electronics-based, modern automation systems understand and process signals only in electronic form. Hence in measuring data from the process, instrumentation devices need to convert physical process signals (electrical, mechanical, chemical, thermal, etc.) into their electronic equivalents so that the control subsystem can accept and process them. Their conversion into electronic form is carried out **with no loss of information**.

Required input instrumentation devices are:

1. Temperature and level sensors to measure the water level and temperature,
2. Flow switch and supervision relay to measure the valve and breaker status, and
3. Water and energy meters to measure water and power consumption.

Because they are physical signals, level and temperature are measured, converted into their electronic equivalents, and sent to the control subsystem as analog inputs by temperature and level sensors.

Fig. 2.3 illustrates the measurement of water temperature and level.

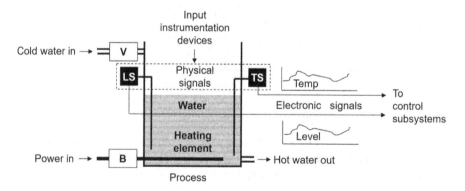

Figure 2.3 Measurement of water level and temperature.

Valve and breaker status are measured **indirectly**. The presence or absence of water flow in the pipe indicates valve status, open or closed, whereas the presence or absence of power flow in the line indicates breaker status, on or off. These status measurements are converted into their electronic equivalents and sent to the control subsystem as digital inputs by a flow switch and supervision relay. The flow switch detects the presence or absence of water flow in the pipe, whereas the supervision relay detects the presence or absence of power in the heater.

Fig. 2.4 illustrates the measurement of valve and breaker status.

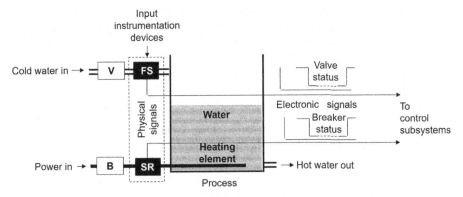

Figure 2.4 Measurement of status of valve and breaker.

In their physical form, water and power consumption are measured as a series of pulse inputs converted into their electronic equivalents and sent to the control subsystem by water and energy meters. The number of pulses over a period indicates consumption. The water meter measures the consumption of water, whereas the energy meter measures the consumption of power.

Fig. 2.5 illustrates the measurement of water and power consumption.

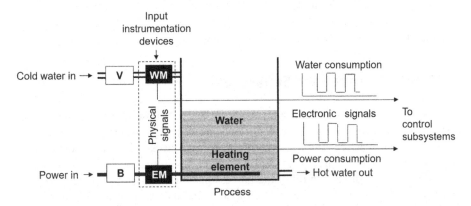

Figure 2.5 Measurement of water and power consumption.

2.4 Output Instrumentation Subsystem

2.4.1 Transfer of Control Command

Because the process understands only physical signals, electronic control signals generated by the control subsystem are converted into their physical equivalents (electrical, mechanical, etc.) in an acceptable form for the process to receive control commands from the control subsystem. Here also, the conversion or transformation is carried out **with no loss of information**.

Control commands (to open or close the valve and open or close the breaker) generated by the control subsystem in electronic form are converted into their physical equivalents (electronic to mechanical) and sent to the process by a solenoid control and control relay (digital outputs). The solenoid opens or closes the valve to allow or disallow water to the tank, while the control relay opens or closes the breaker to allow or disallow power to the heater.

Required output instrumentation devices are:

1. An on–off solenoid control to open and close the valve, and
2. An on–off control relay to open and close the breaker.

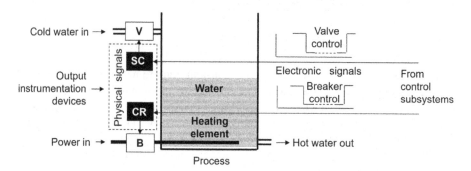

Figure 2.6 Control of valve and breaker.

Fig. 2.6 illustrates control of the valve and breaker.

2.5 Human Interface Subsystem

The human interface subsystem is the means by which the users or operators directly interact with the process for direct monitoring and control of the process.

The human interface subsystem is the interface between physical processes and the operator, via the control subsystem, for direct monitoring of the process parameters and to effect direct control of the process parameters. There is no need for any interface devices between the control subsystem and the human interface subsystem (unlike the need for an instrumentation subsystem between the process and the control subsystem). Signals between the control and human interface subsystems are electronic in both directions and therefore are compatible.

2.5.1 Direct Monitoring

Direct monitoring provision is made for observing process variables of interest on the displays whenever there is a need. In other words, operators can observe exactly what is happening inside the process through visual display of parameters of interest.

2.5.2 Direct Control

Direct control provision is made for manipulating or controlling process parameters of interest whenever there is a need. This action **overrides** the functions performed automatically by the control subsystem. The provision for setting reference values for limit checking is also part of the human interface subsystem.

The operator panel allows operators to perform the following manually:

1. Observe actual values of temperature and level,
2. Observe the actual status of the valve and breaker,
3. Observe the actual consumption of water and power,
4. Set reference values for temperature and level, and
5. Control (close or open) the valve and breaker.

Fig. 2.7 illustrates a typical hardware-based human interface or operator panel.

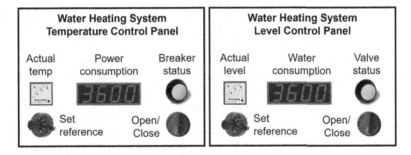

Figure 2.7 Operator panels for water-heating system.

2.6 Control Subsystem

The control subsystem performs the following steps continuously.

2.6.1 Data Acquisition

In this step, the control subsystem acquires process data in an electronic form from the input instrumentation subsystem, which in turn receives them from the process in a physical form.

2.6.2 Data Analysis and Decision Making

In this step, the control subsystem analyzes the acquired data by comparing them with preset reference values for any deviations and decides whether to effect changes in the behavior of the process or not.

2.6.3 Control Execution

In this step, the control subsystem issues commands to process in electronic form to output instrumentation devices, which in turn transfer them in a physical form to the process.

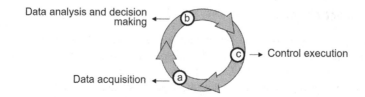

Figure 2.8 Typical automation cycle.

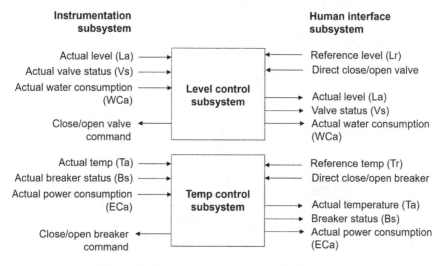

Figure 2.9 Inputs and outputs in control subsystems.

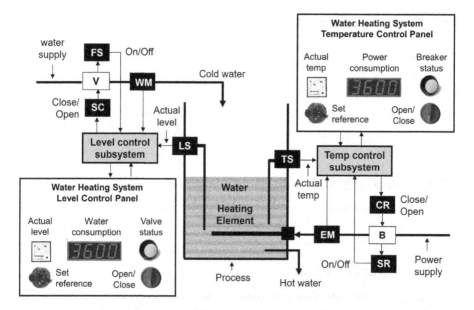

Figure 2.10 Automation system for water-heating process.

These steps are illustrated in Fig. 2.8 as a cyclic operation or automation cycle.

Fig. 2.9 illustrates the input–output to–from control subsystems to the instrumentation and human interface subsystems for level and temperature control in the water-heating process.

Fig. 2.10 illustrates the fully integrated automation system with all of its subsystems interconnected to monitor and control the water-heating process.

2.6.4 Communication

Because they are intelligent and communicable, modern control subsystems can communicate externally with other compatible systems. This will be discussed in detail in Chapter 7.

2.7 Summary

In this chapter, we discussed the overall structure of automation systems as well as their functional subsystems and interconnections. These functional subsystems are discussed further individually in greater detail in subsequent chapters.

Instrumentation Subsystems

3

Chapter Outline

3.1   Introduction

In the previous chapter, we discussed basic subsystems of the automation system. Of these subsystems, the instrumentation subsystem is the most important. Without the instrumentation subsystem, automation would not function. In this chapter, the instrumentation devices that comprise this subsystem are discussed in detail. As defined in Chapter 2, instrumentation is the branch of engineering that deals with the measurement and control of physical process parameters and instrumentation devices that provide an interface between the process and the control subsystem to facilitate the following:

- *Data acquisition*: The conversion of the physical signal generated by the process into an equivalent and compatible electronic signal (with no loss of data) acceptable to the control subsystem;
- *Control execution*: The conversion of the electronic control signal generated by the control subsystem into an equivalent and compatible physical signal (with no loss of data) acceptable to the process.

Overview of Industrial Process Automation. http://dx.doi.org/10.1016/B978-0-12-805354-6.00003-7

3.2 Structure

Instrumentation devices have different structures for different input and output signals; they arc different for continuous/analog, discrete/digital, and pulsating/pulse signals.

3.2.1 Continuous/Analog Devices

Fig. 3.1 illustrates the functional components and general structure of continuous/analog instrumentation devices for both input and output.

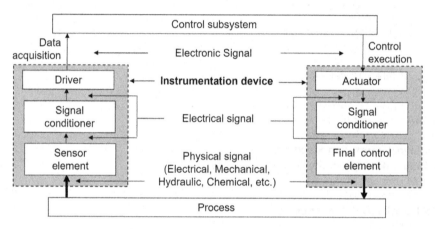

Figure 3.1 General structure of analog instrumentation device.

3.2.1.1 Measurement

Fig. 3.2 illustrates the general structure and input–output relationship of an analog input instrumentation device.

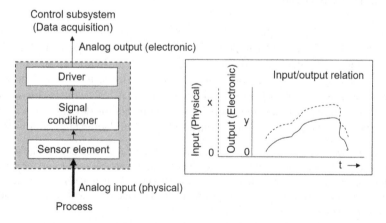

Figure 3.2 Analog input instrumentation device.

The analog input instrumentation device processes the received physical signal generated by the process with the following functional components:

- *Sensor*: It converts an analog or continuous physical signal into its electrical equivalent signal through the transduction process.
- *Signal conditioner*: Used in the intermediate stage, the signal conditioner prepares or manipulates (isolating, compensating, linearizing, amplifying, filtering, offset correcting, etc.) the weak and noisy sensor output to meet the requirements of the next stage for further processing.
- *Driver*: It strengthens the signal received from the signal conditioner to an appropriate form and drives it for transmission to the control subsystem.

Generally, the analog input instrumentation device has several names, such as sensor, transducer, and transmitter. **Sensor** outputs are in the order of millivolts or even less and are not suitable for transmission. The device is called a **transducer** if its output is suitable for transmission (in voltage or current form) over a relatively short distance. The device is called a **transmitter** if the output is suitable for transmission (in current form) over a relatively long distance.

Fig. 3.3 illustrates some industry examples of commonly used analog input instrumentation devices in the process industry (temperature transmitter, flow transmitter, level transmitter, and pressure transmitter). These devices receive physical signals (temperature, flow, level, and pressure) and generate corresponding electronic outputs in current form suitable for control subsystems. Furthermore, some devices support local display of the process parameter in engineering units to help field maintenance personnel.

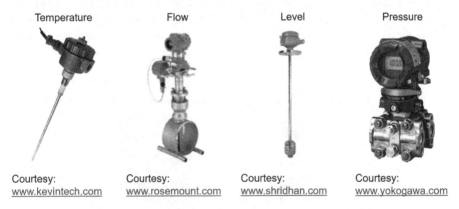

| Temperature | Flow | Level | Pressure |

Courtesy: www.kevintech.com   Courtesy: www.rosemount.com   Courtesy: www.shridhan.com   Courtesy: www.yokogawa.com

Figure 3.3 Analog input instrumentation devices: examples.

3.2.1.2 Control

Fig. 3.4 illustrates the structure and input–output relationship of an analog output instrumentation device for the transmission of continuous or analog signals (commands) to the process for control execution.

The analog output instrumentation device processes the electronic control signal generated by the control subsystem to produce the final value acceptable to the process. The associated components are the actuator, signal conditioner, and final control element.

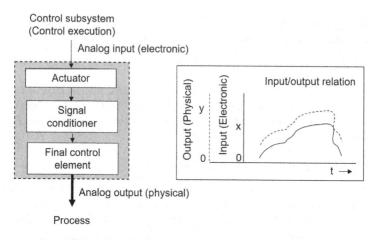

Figure 3.4 Analog output instrumentation device.

Although the signal conditioner has the same function as that of the analog input instrumentation device, the actuator and final control element have two functions:

- *Actuator*: Transforms the command output electronic signal from the control subsystem into a signal suitable for the signal conditioner;
- *Final control element*: Performs the final control action in the process.

Fig. 3.5 illustrates some industry examples of commonly used analog output instrumentation devices (variable control valve and variable speed drive). These devices receive the electronic signals in current or voltage form and generate a corresponding physical output (valve opening and motor speed) suitable for the process.

Variable control valve   Variable speed drive

Courtesy: Courtesy:
www.emerson.com   www.fein.fujielectric.com

Figure 3.5 Analog output instrumentation devices: examples.

3.2.2 Discrete/Digital Devices

Fig. 3.6 illustrates the functional components and general structure of a discrete/digital instrumentation device (both input and output).

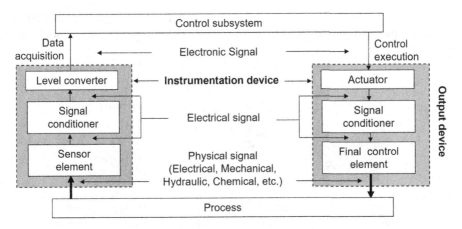

Figure 3.6 Digital input and output instrumentation devices: structure.

3.2.2.1 Measurement

Fig. 3.7 illustrates the structure and input–output relationship of a digital input instrumentation device used for data acquisition of discrete or digital signals from the process.

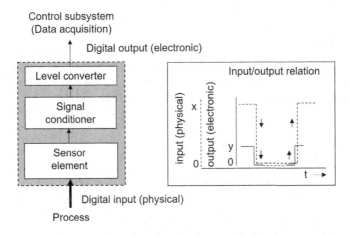

Figure 3.7 Digital input instrumentation device.

The digital input instrumentation device processes the physical signal generated by the process to make it suitable for the control subsystem. The associated components are a sensor, signal conditioner, and level converter. Although the roles of sensor and signal conditioner remain the same, the role of the level converter is to change the level of the electrical input (presence or absence) to produce an output (presence or absence) that meets the control subsystem requirements for processing.

Fig. 3.8 illustrates some examples in the industry of commonly used digital input instrumentation devices (limit switch, proximity switch, and supervision relay).

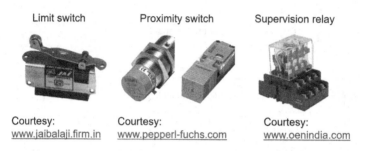

Figure 3.8 Digital input instrumentation devices: examples.

The limit switch receives the mechanical input signal (on–off) and converts it into an electronic output signal (on–off). Through its electromagnetic/optical field, the inductive/optical proximity switch senses the target and converts it into an electronic output signal (on–off). The supervision relay picks up voltage and gives a yes–no signal.

3.2.2.2 Control

Fig. 3.9 illustrates the structure and the input–output relationship of a discrete/digital output instrumentation device (digital control device) for the transmission of discrete or digital signals (commands) for the process or control execution.

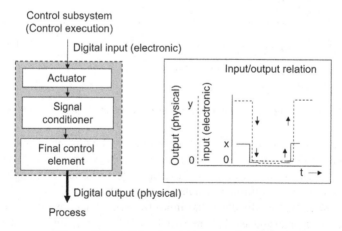

Figure 3.9 Digital output instrumentation device.

The digital output instrumentation device processes the electronic signal generated by the control subsystem, making it suitable for the process. The associated components are the actuator, signal conditioner, and final control element.

Fig. 3.10 illustrates some examples of commonly used digital output instrumentation devices in the industry (on–off solenoid valve and control relay).

On/Off solenoid valve   Control relay

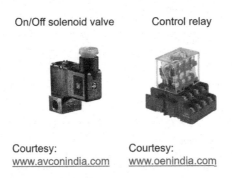

Courtesy: Courtesy:
www.avconindia.com   www.oenindia.com

Figure 3.10 Digital output instrumentation devices: examples.

These devices receive electronic signals in current or voltage form from the control subsystem and generate corresponding physical output (they open or close flow and open or close the breaker) suitable for the process.

3.2.3 Pulsating/Pulse Devices

Discussions of digital instrumentation devices apply to pulsating or pulse signals as well, because they are variations on digital signals. Pulses are received from the process in physical form by pulse input instrumentation devices and are sent to the control subsystem after conversion into a compatible electronic form. In the reverse direction, pulses in an electronic form are sent by the control subsystem and received by the pulse output instrumentation device, and are sent to the process after conversion into a compatible physical form. These operations are similar to those of digital input–output instrumentation devices. The only difference is that the input and output generally change state more frequently.

3.2.4 Switching Devices

Fig. 3.11 illustrates the structure and input–output relationship of a switching input instrumentation device used for data acquisition of a discrete or digital signal from the process on the input analog value violation of a set limit.

3.2.4.1 Measurement

A switching instrumentation device is an input device that receives continuous or analog input and produces discrete or digital output whenever the input exceeds the **locally**

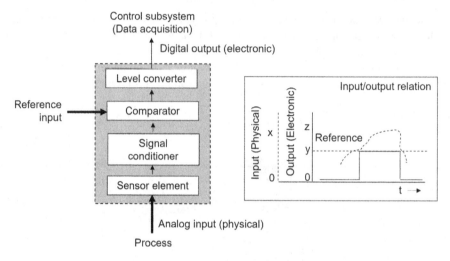

Figure 3.11 Switching instrumentation device.

preset reference value. This device is an example of a simple **stand-alone automation system** with its own instrumentation (analog measurement), control (comparing and decision making), and human interface (setting of reference value), all integrated into one. This device is also called a switch or stand-alone limit-checking device. This can be compared with the working of a **domestic pressure cooker** in which in steam is let off whenever the inside pressure is more than the reference value (as specified by the weight on the steam outlet).

Fig. 3.12 illustrates examples in the industry of three commonly used instrumentation switches (flow switch, level switch, and voltage switch).

Figure 3.12 Switching instrumentation devices: examples.

Whereas the functions of the level converter, signal conditioner, and sensor element are the same, the **comparator** produces a digital output (change of state) whenever the input analog signal violates the locally preset limit in the device and upon returning to normal.

These devices, with provisions for setting the limit values, receive the physical signals (flow, level, and voltage) and generate corresponding electronic output in current or voltage form suitable for control subsystems. These devices are less expensive compared with the transmitter equivalents.

3.2.5 Integrating Devices

The integrating instrumentation device is another type of special input device that receives continuous or analog input and produces pulsating or pulse output by integrating the input over time. It integrates the input value into proportional sequence pulses and sends it to the control subsystem in the same order in a compatible electronic form.

Fig. 3.13 illustrates the structure and the input–output relationship of the **integrator** (analog input–pulse output device), which is used for data acquisition.

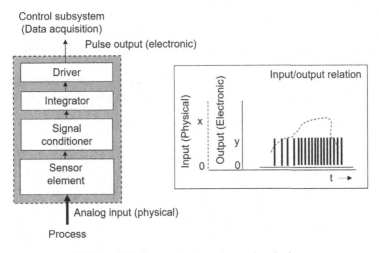

Figure 3.13 Integrating instrumentation device.

3.2.5.1 Measurement

Although the functions of the sensor, signal conditioner, and level converter are the same, the integrator produces a sequence of pulse output corresponding to the integrated value of the input.

Fig. 3.14 illustrates some examples in the industry of commonly used integrating instrumentation devices that compute consumption (water meter and energy meter). These devices, with provisions for local displays in engineering units, receive physical signals (water flow and voltage/current) and generate corresponding electronic pulse output in suitable current or voltage form for control subsystems.

Water meter   Energy meter

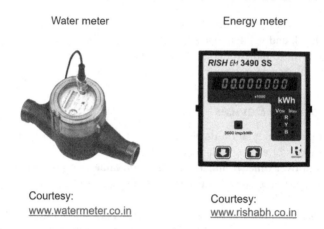

Courtesy: Courtesy:
www.watermeter.co.in   www.rishabh.co.in

Figure 3.14 Integrating instrumentation devices: examples.

3.3 Signal Interfacing Standards

To make instrumentation devices compatible and interoperable with control systems manufactured by different vendors, industry standards are defined for interfacing.

Fig. 3.15 illustrates interfacing standards applicable to all instrumentation devices from different manufactures for compatibility, replaceability, and interoperability amongst them.

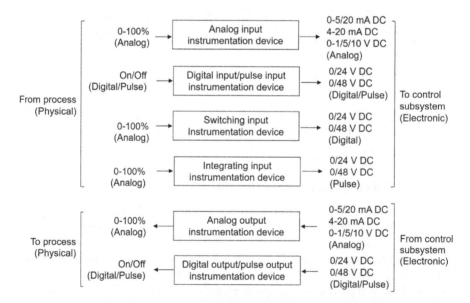

Figure 3.15 Signal interfacing standards.

The most commonly used standard for analog signals is **4–20 mA**, whereas it is **24 VDC** for digital signals.

All instrumentations devices (input and output) from different vendors follow this signal interface standard to comply with compatibility, replaceability, and interoperability with control subsystems that are also designed to follow these standards.

3.4 Input Data Reliability Enhancement

3.4.1 Analog

Analog input devices can be either self-powered or auxiliary powered. Self-powered devices use the power of the input signal itself to power the device. However, this has a serious shortcoming in that it is difficult to distinguish between zero input and device failure because the device output is zero in both cases. This is overcome by auxiliary-powered devices in which 4 mA is considered to be active zero. Self-powered devices are not possible in all cases. They are available only in electrical installations where input signals (current and voltage) can power the device. Auxiliary-powered instrumentation devices are expensive.

3.4.2 Digital

In the case of digital input, the instrumentation device generates **double-bit** or **complementary** signals (00 or 01 or 10 or 11) instead of **single-bit** signals (0 or 1) for inputting to the control subsystem. This calls for two digital input instrumentation devices connected to the same process input. The combinations 01 and 10 are accepted and used as valid data, whereas the combinations 00 and 11 are rejected as bad data (complementary errors). This doubles the input instrumentation devices and consumes additional resources in the control subsystem. This feature is normally used in switchgears (circuit breakers, isolators, etc.) in the electrical industry.

3.5 Isolation and Protection

Generally, the control center is not located close to the process, but instrumentation devices are installed close to the process equipment in the plant for technical reasons. Field cables (signal/control) are laid to send and receive electronic signals between the control center and the process.

Typical interconnections among the process, instrumentation, and control subsystem are illustrated in Fig. 3.16.

Fig. 3.17 shows a typical location of an instrumentation device in the process plant.

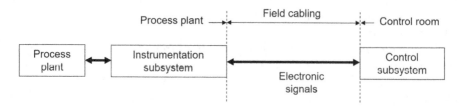

Figure 3.16 Automation system layout in plant.

Courtesy: www.emerson.com

Figure 3.17 Instrumentation device in process plant.

The presence of interconnecting field cables raises a few safety issues for process plants, control systems, and operating personnel. This section discusses safety issues (isolation and protection) and how to handle them. Typical examples are:

• An electrical fault in the input signal affecting the output signal, and vice versa; and
• An explosion caused by a spark in the plant in which potentially hazardous gas/vapor is present.

3.5.1 Isolation

The fault in the input signal can be caused by grounding, high voltage, a short circuit, faulty devices, etc. To avoid the effects of faults being transferred to the control center and/or instrumentation devices, there is a need for suitable isolating devices between the control center and the instrumentation devices. This arrangement protects the control subsystem, operating personnel, and instrumentation devices.

3.5.2 Protection

Ignitable sparks caused by high energy (beyond the acceptable limit) travel from the control center into the field area through the field signal/control cables. This situation is common in petrochemical complexes. The spark can lead to severe fire hazards and even explosions, because the environment may be charged with ignitable gaseous vapors. This calls for protection devices.

3.5.3 Solutions

Fig. 3.18 illustrates the preferred solution to overcome these safety problems by using appropriate isolation and protection units. These are called **intrinsic safety isolated barriers (ISB)**.

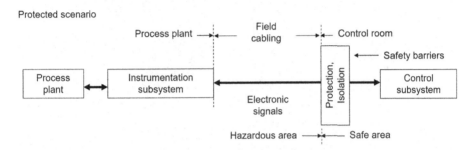

Figure 3.18 Application of barriers.

The ISB is an electronic device for:

* Isolating the signal in either direction; and
* Protecting the equipment by limiting current, voltage, and total energy delivered from a safe area to a hazardous area.

These circuits are not generally integral parts of the instrumentation subsystem, and they are installed in the control center. Fig. 3.19 illustrates a typical circuit schematic of an isolated safety barrier.

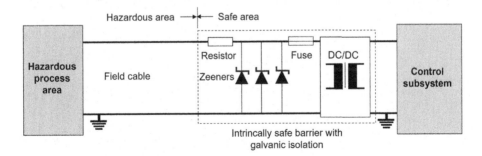

Figure 3.19 Safety barrier with isolation.

Fig. 3.19 also shows the roles of various electronic components:

* *Resistor*: limits current,
* *Fuse*: protects Zener diodes upon overcurrent,

- *Zener diode*: clips voltage upon overvoltage,
- *DC/DC transformer*: provides galvanic isolation.

In essence, safety barriers limit energy transferred from the safe area to the hazardous area. Fig. 3.20 illustrates an example from the industry of intrinsically safe barriers stacked and mounted on a DIN rail.

Single barriers   Barriers stacked on mother board
Courtesy: www.pepperl-fuchs.com

Figure 3.20 Intrinsically safe barriers.

3.6 Summary

In this chapter, we discussed in detail the construction of various types of instrumentation devices: namely, analog, digital, pulse, and special. Among these, we distinguished between input and output devices with industry examples. We also discussed interfacing standards to make instrumentation devices manufactured by different vendors compatible, interoperable, and replaceable with control subsystems. Finally, the chapter discussed the application of intrinsic safety barriers to protect equipment and people from possible high-voltage surges in control/signal cables.

Human Interface Subsystem

4

Chapter Outline

4.1   Introduction

In this chapter, the functions of the human interface subsystem are discussed in detail. Furthermore this chapter deals with the traditional hardwired human interface subsystem, which is still popular and cost-effective even in modern automation systems.

4.2   Operator Panel

The human interface[1], also called the human–machine interface, human–system interface, or man–machine interface, is the means by which operators or users interact manually with the process: they can force an action to control the process or observe the parameters of interest in the process. Another commonly used name for the human interface is the **operator panel**.

The basic function of the human interface system is to work directly with the process to manually control or assess its state over and above the functions performed automatically by the control subsystem. The structure of the human interface system facilitates operator control of the functions performed automatically by the control subsystem. The human interface subsystem is designed with active display and control components mounted suitably on a panel and wired to the control subsystem, as explained in the following sections.

---

[1] http://www.wikipedia.org.

Overview of Industrial Process Automation. http://dx.doi.org/10.1016/B978-0-12-805354-6.00004-9

4.2.1   Display

Commonly used active display elements on the operator panel are illustrated in Fig. 4.1.

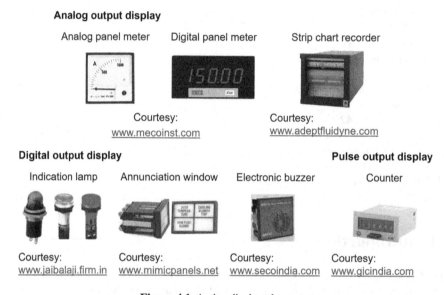

Figure 4.1 Active display elements.

These elements are mounted at appropriate locations on the panel for the continuous display of the value and status of process parameters. The functions of these active display elements are as follows:

* Because they are functionally the same, analog and digital panel meters display the current value of the analog parameter (value of temperature, level, flow, etc.) received from the control subsystem. A strip-chart recorder records, continuously against time, the value of analog parameters (values of temperature, level, flow, etc.) received from the control subsystem. The recorder can have more than one pen to record more parameters simultaneously against a common time reference.
* Because they are functionally the same, the indication lamp and annunciation window display the current status of the discrete parameters (status of the valve, breaker, etc.) received from the control subsystem. The annunciation window supports the display with a legend with back illumination.
* The counter/totalizer displays the accumulated value (consumption of water, power, etc.) received from the control subsystem.

In addition to this list, there can be many more active display elements, such as seven-segment displays, hooters, and buzzers etc.

4.2.2   Control

Commonly used active control elements on the operator are illustrated in Fig. 4.2.

Figure 4.2 Active control elements.

These elements are also mounted at appropriate locations on the panel for direct control of the process. The functions of these active control elements are as follows:

- The push button switch facilitates inputting the on–off command to the control subsystem, which can be either momentary or regulatory depending on the type of push button switch.
- The thumb wheel switch stack facilitates inputting the string of binary-coded decimal value into the control subsystem.
- The potentiometer facilitates inputting the analog value into the control subsystem.

All of the active components (display and control) from different vendors follow the signal interface standard indicated in Chapter 3, Section 3.3, to comply with compatibility, irreplaceability, and interoperability with control subsystems.

Chapter 5 on control subsystems discusses how display and control elements interface to the control subsystem.

4.2.3 Panel

The panel is a passive element and is just the base of the human interface system, which is made of painted sheet metal, painted fiberglass, or another similar material. The role of the panel is to hold the mounted active components. The active components are arranged on the panel in an order suitable to their operation; they are wired and connected to the control subsystem. The panel and its active components provide easy access for the operator to observe the process parameters of interest (both the states of discrete parameters and the values of continuous parameters) and to effect direct interaction with the process for monitoring and control.

4.3 Construction

There are two approaches to constructing the panel: basic and mosaic. The following sections explain the construction of different types of human interface subsystems. These approaches are illustrated in the example of the water-heating process.

4.3.1 Basic Approach

The basic and most cost-effective approach is to mount all of the active display and control components on the panel in an order that is convenient to their operation.

Fig. 4.3 illustrates this approach to the automation of the water-heating process.

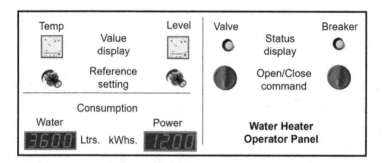

Figure 4.3 Traditional operator panel.

This approach allows room for the operator to make a mistake while identifying the active components (display or control), because they do not explicitly link to the process parameters. In other words, the object selection can be ambiguous.

4.3.2 Mimic Approach

The mimic approach overcomes the deficiency in the basic approach. It involves a panel with the process diagram depicted, and active control and display components positioned for easy and unambiguous identification.

Fig. 4.4 illustrates a mimic-based operator panel for the water-heating process.

In this approach, there is absolutely no room for the operator to make a mistake while identifying active components because they are placed in relation to the process parameters on the panel, and parameter selection is unambiguous.

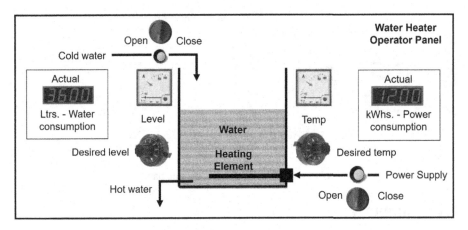

Figure 4.4 Mimic-based operator panel.

4.3.3 Types of Panels

Hardware-based operator panels can be of two types: sheet metal/fiberglass-based or mosaic tile–based. There is no difference between sheet metal–based and fiberglass-based except for the panel material.

Fig. 4.5 illustrates a typical sheet metal–based operator panel in an electrical substation.

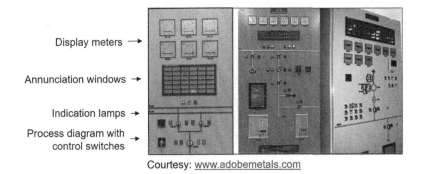

Courtesy: www.adobemetals.com

Figure 4.5 Sheet metal–based mimic panel.

The major problem with a sheet metal/fiberglass-based mimic panel is its inflexibility in terms of expanding or modifying the panel at a later stage. Compared with their sheet metal or fiberglass equivalents, mosaic tile–based mimic panels make it relatively easy to modify or extend the process diagram to add or remove active components. The greatest disadvantage of this panel is its cost.

Fig. 4.6 illustrates the structure of a typical mosaic-based panel in an oil installation.

Mosaic tiles   Mosaic mimic panel

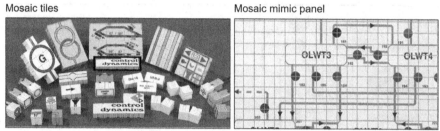

Courtesy: www.mimicpanels.net

Figure 4.6 Mosaic tile–based mimic panel.

4.4 Summary

In this chapter, we discussed the traditional human interface subsystem or operator panel and its structure with active display and control components. The major disadvantage of the operator panel (basic or mimic-based) is its inflexibility for expanding or modifying the panel to meet changing requirements. Although mosaic tile–based panels correct this issue to some extent, they are expensive.

Control Subsystem

Chapter Outline

5.1   Introduction

The control subsystem is the heart of automation systems; it coordinates and controls the functions of other subsystems, as described in Chapter 2. A control subsystem [1] is a device or a set of devices to manage, command, direct, or regulate the behavior of the process. The control subsystem is an essential mechanism for manipulating the output of a specific process to achieve the desired result. This subsystem performs the following functions:

- Data acquisition from the process via input instrumentation subsystem (process inputs) and from the human interface subsystem (direct inputs)
- Data analysis, monitoring, and decision making for corrections to the behavior of the process, if required
- Executions and transmission of control commands to the process (control execution) via output instrumentation subsystem and driving process parameters of interest for display on the human interface subsystem

5.2   Structure

In addition to its own functions (data analysis and decision making), or the **automation strategy,** the control subsystem has appropriate interfaces for interaction for data exchange with the instrumentation and human interface subsystems to perform its roles. In other words, the control subsystem is divided into automation strategy (responsible for managing automatic control) and interfaces for data exchange with instrumentation and human interface subsystems.

[1]www.wikipedia.com.

Overview of Industrial Process Automation. http://dx.doi.org/10.1016/B978-0-12-805354-6.00005-0

The general data structure of the control subsystem is illustrated in Fig. 5.1.

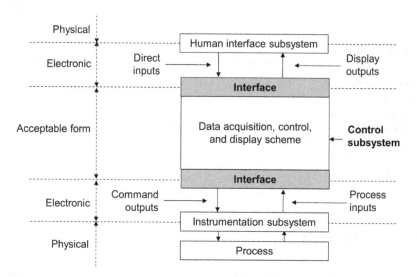

Figure 5.1 General data structure.

The interface shown in the figure is an arrangement to:

- Convert the data received from the process (via input instrumentation) and from human interface subsystems in electronic form to a form acceptable to the automation strategy in the control subsystem.
- Convert the data generated by the automation strategy in the control subsystem in electronic form and drive them to the process (via output instrumentation) and to the human interface subsystem in a form acceptable to them.

These conversions and transfers happen with no loss of data.

The automation strategy is responsible for managing the previously discussed functions through the interfaces and for performing automatic control of the process.

5.3   Interfacing

The following sections explain the interfacing of the instrumentation and human interface subsystems with the control subsystem.

5.3.1   General

To facilitate data transfer to and from the control subsystem, the interface is structured on an input–output (I/O) basis, as illustrated in Fig. 5.2.

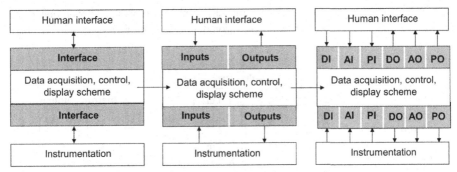

DI: Digital Input, AI: Analog Input, PI: Pulse Input
DO: Digital Output, AO: Analog Output, PO: Pulse Output

Figure 5.2 Interface mechanism.

Furthermore, these inputs and the outputs can be analog, digital, or pulse. This structure leads to two main categories, inputs and outputs; within each, there are three subcategories (digital, analog, and pulse). The interfaces consist of the following inputs and outputs:

- Digital inputs, analog inputs, and pulse inputs
- Digital outputs, analog outputs, and pulse outputs

Within this structure, there are six I/O-based interfaces for each human interface and instrumentation subsystem, as shown in Fig. 5.2. This explains the structure of interfacing at different levels. Requirements for the type and number of inputs and outputs are application-specific (flexible and modular).

The schematic of general-purpose control subsystem is illustrated in Fig. 5.3.

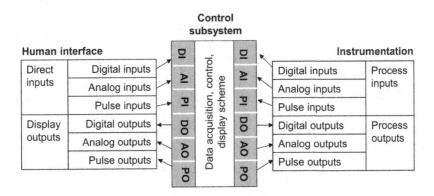

Figure 5.3 General purpose control susbsystem.

A variation of this general-purpose control subsystem is its modification with only data acquisition from process and display (no provision of process control), as shown in Fig. 5.4.

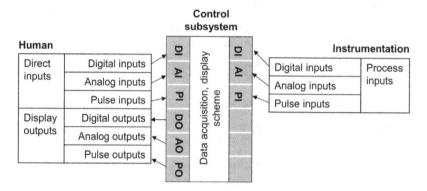

Figure 5.4 Control subsystem with only data acquisition and display.

The next question is how we interface instrumentation and human interface subsystems with the control subsystem. This is explained in the following sections.

Furthermore, the control subsystem complies with the signal interfacing standards discussed in Chapter 3, Section 3.3, for interfacing with instrumentation and the human interface subsystem.

5.3.2   Instrumentation

To illustrate the interfacing of the control subsystem with the instrumentation subsystem, let us return to the example of the water-heating process with two possibilities.

Fig. 5.5 illustrates Case 1, which employs analog input and output devices for temperature and level measurement and control:

- *Analog input instrumentation devices*: temperature and level measurements: for acquisition of temperature and level values, as analog inputs
- *Analog output instrumentation devices*: valve control (vary) and variable transformer control (vary) for control of water flow and power supply, as analog outputs
- *Integrating instrumentation devices*: water and energy meters: for acquisition of water and energy consumption as, pulse inputs

Fig. 5.5 also indicates the number of I/O signals between the instrumentation and control subsystems associated with this scheme.

Fig. 5.6 illustrates Case 2, which employs switching and digital output devices for measurement and control:

- *Switching instrumentation devices*: temperature and flow switches: for acquisition of the status of temperature and level, as digital inputs
- *Digital output instrumentation devices*: valve control (on–off) and breaker control (on–off): to control water flow and the power supply, as digital outputs
- *Integrating instrumentation devices*: water and energy meters: for the acquisition of water and energy consumption, as pulse inputs

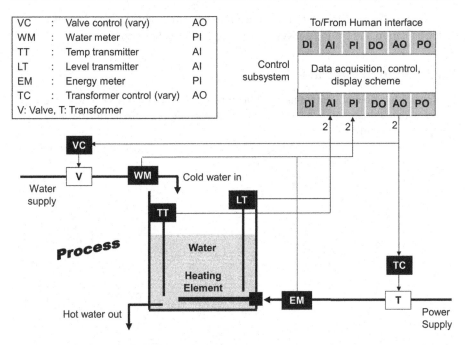

Figure 5.5 Interfacing of instrumentation devices with control subsystem: Case 1.

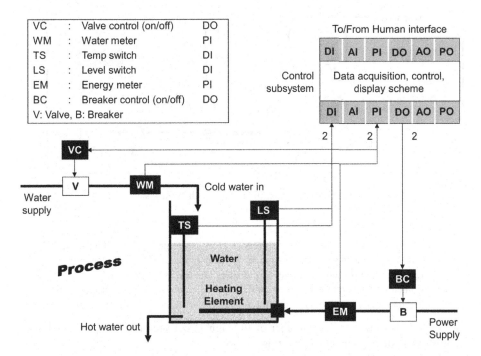

Figure 5.6 Interfacing of instrumentation devices with control subsystem: Case 2.

Both Fig. 5.5 and Fig. 5.6 also indicate the number of I/O signals between the instrumentation and control subsystems associated with the schemes.

5.3.3   Human Interface

No special interfacing equipment, such as for instrumentation subsystem, is required between the human interface subsystem and the control subsystem, because they exchange compatible electronic signals.

Fig. 5.7 illustrates interfacing that employs active display and control components mounted on the operator panel:

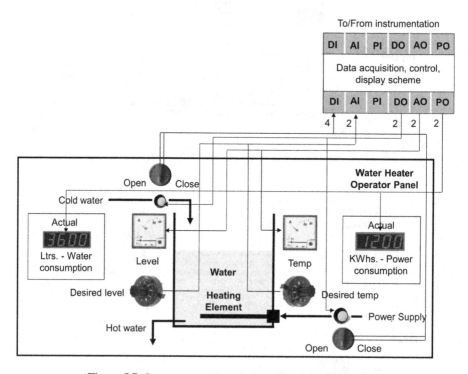

Figure 5.7 Operator panel interfacing with control subsystem.

- Control switches: valve and breaker - for control (on–off command), as digital inputs
- Indication lamps: valve and breaker - for display (on–off status), as digital outputs
- Analog meters: temperature and level - for display (value), as analog outputs
- Potentiometers: temperature and level - for setting (reference value), as analog inputs
- Counters/totalizers: water and power - for display (consumption), as pulse inputs

Fig. 5.7 also indicates the number of I/O signals between the operator panel and control subsystems associated with this scheme.

5.4 Summary

This chapter discussed in detail the mechanism for interfacing instrumentation and human interface subsystems with the control subsystem using inputs and inputs as the main criteria. This chapter also introduced the terminology **automation strategy** as the central part of the control subsystem that manipulates the inputs for data acquisition, analysis, monitoring, and decision making, and produces control outputs.

Automation Strategies

6

Chapter Outline

6.1   Introduction

In Chapters 3–5, we discussed the functions of individual subsystems and their associated interfacing. We also mentioned that together, these subsystems are made and interconnected to produce the desired result through functions called data acquisition, analysis, decision making, control, and display. The chapter on instrumentation subsystems discussed process data acquisition and control transfer schemes, whereas the chapter on human interface subsystems discussed process data display and the direct process control scheme. In this chapter, we discuss the internal workings of the control subsystem, namely, **data analysis, decision making, and control**, henceforth called

Overview of Industrial Process Automation. http://dx.doi.org/10.1016/B978-0-12-805354-6.00006-2

automation strategy, which produces the desired results with the support of other subsystems.

The automation strategy is a predefined and built-in scheme in the control subsystem to guide the automation system to achieve the desired results. As per the built-in strategy, it performs:

- Data acquisition through an input instrumentation subsystem
- Data processing, data analysis, and decision making
- Process control through an output instrumentation subsystem
- Data display and direct control through a human interface subsystem

This is also called an automation function or automation task. Physical processes, each of which is different from the other, need specific automation strategies, meeting different requirements to produce different desired results. In other words, each control subsystem is designed to meet a process-specific automation strategy.

An automation strategy operates on inputs from the process (via the input instrumentation) and from the human interface subsystem (for direct control). It analyzes the data and produces the required command output, as per the predefined criteria, and sends the output to the process (via output instrumentation) and to human interface subsystem to display the process parameters of interest.

Fig. 6.1 illustrates the transformation of functions of data acquisition, control, and display into an automation strategy:

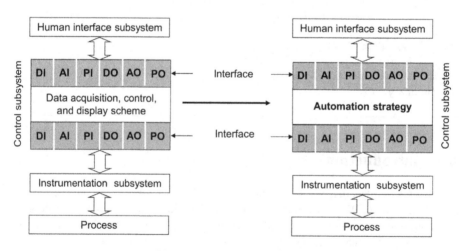

Figure 6.1 Automation strategy.

The implementation of automation strategies is explained in subsequent sections using the automation of a water heater as an example.

This chapter discusses the basic strategies followed by advanced strategies that are variations on the basic strategies.

6.2 Basic Strategies

Basic strategies for implementing automation schemes are:

- Open loop control
- Closed loop control or feedback control

An open loop control strategy supports **preknown** results (responses) to the control inputs. No assessment of the responses and corrections is possible for any internal and/or external disturbances.

This scheme is illustrated in Fig. 6.2.

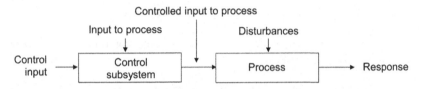

Figure 6.2 Open loop control.

Although it is used frequently, open loop control cannot always ensure the desired result. It is simple, economical, used in less demanding applications, and applied in both discrete and continuous process automation, as discussed in forthcoming sections.

Closed loop control, also known as feedback control, eliminates the shortcomings of open loop control. Here, the response or the actual result is continuously compared with the desired result, and the control output to the process is modified and adjusted to reduce deviations, thus forcing the response to follow the reference. The effects of disturbances (external and/or internal) are automatically compensated for. This scheme is superior, complex, expensive, and used for more demanding applications.

Fig. 6.3 illustrates this scheme.

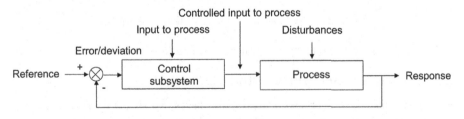

Figure 6.3 Closed loop control.

The following sections discuss the application of these two basic strategies in the automation of discrete, continuous, and hybrid processes.

6.2.1 Discrete Process Control

Discrete process control is employed for processes involving only discrete inputs and discrete outputs and their associated instrumentation devices. Discrete process control can be further classified into open loop control and sequential control with interlocks.

6.2.1.1   Open Loop Control

On–off commands are issued to produce the desired results in open loop discrete control. This scheme does not compensate for internal and/or external disturbances.

Fig. 6.4 illustrates the open loop discrete control scheme.

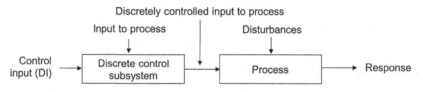

Figure 6.4 Discrete control: open loop.

Fig. 6.5 illustrates the application of open loop discrete control to control the temperature and level in a water heater.

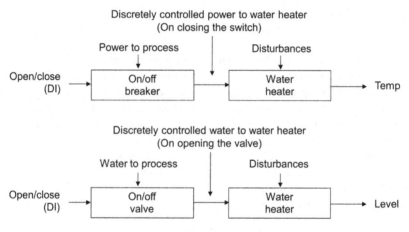

Figure 6.5 Open loop control in water heater (discrete).

As seen in Fig. 6.5, this is nothing more than on–off control of the valve and the breaker.

6.2.1.2   Sequential Control With Interlocks

Sequential control with interlocks addresses drawbacks in a simple open loop control. Like discrete open loop control, instrumentation devices for both data acquisition and control are discrete. In each step, sequential control with interlocks ensures that the desired intermediate conditions or **interlocks** are satisfied before executing the next step.

Fig. 6.6 illustrates sequential control with interlocks.

In a discrete open loop control strategy for water heater automation, there is one serious drawback: the control strategy assumes the availability of water in the tank before switching on power to the heating element. This may not always be the case for several reasons, such as no or insufficient water in the tank owing to a clogged water inlet pipe/valve. If this is not taken care of, it is possible that the heating element will be damaged. Therefore, it is necessary to allow sufficient time for the water to fill

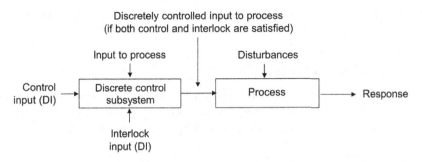

Figure 6.6 Sequential control with interlock.

before switching the power on. This strategy becomes safer if the power is switched on only after ascertaining the water level in the tank.

Fig. 6.7 illustrates the application of sequential control with interlocks for a water heater. It makes sure there is sufficient water in the tank before allowing the heating element to be turned on.

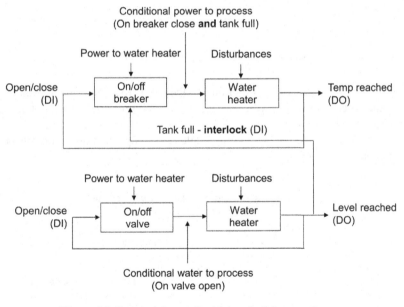

Figure 6.7 Sequential control with interlock in water heater.

As seen in Fig. 6.7, the breaker does not close, even if a command is given (either directly or automatically), unless the water level is full.

Here the control moves in steps, or sequentially, but only after satisfying certain conditions at every step until the desired result is reached. Here also, instrumentation devices for both data acquisition and control are discrete. Sequential control with interlocks is widely used to:

- Meet personnel and equipment safety in plants
- Start up and shut down complex plants

Some common examples of its use can be seen in the operation of passenger lifts and traffic signals.

6.2.2   Continuous Process Control

Continuous process control is employed in processes involving analog inputs and analog outputs with their associated instrumentation devices. Continuous process control can be further classified into open loop control and closed loop or feedback control.

6.2.2.1   Open Loop Control

Fig. 6.8 illustrates simple open loop continuous control.

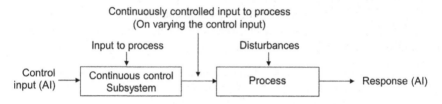

Figure 6.8 Continuous control: open loop.

In this case, the response is proportional to the input. To achieve the desired result, input to the process must simply be varied. Because it is open loop control, it does not compensate for disturbances.

Fig. 6.9 illustrates the application of simple open loop continuous control for the level and temperature of a water heater.

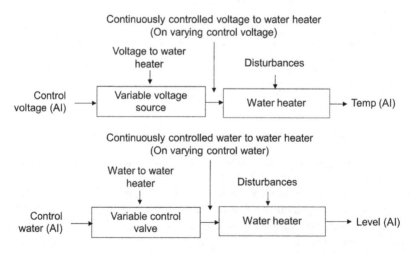

Figure 6.9 Open loop control in water heater (continuous).

Here, the variable voltage source increases or decreases power flow to the heating element proportional to the input. Similarly, the variable control valve increases or decreases water flow to the tank proportional to the input.

6.2.2.2   Closed Loop/Feedback Control

For closed loop control or feedback control, the need is to track the process output continuously compared with the reference or the desired output and to vary the control input proportionally to minimize deviation or error (output following the reference). Here also, both data acquisition and control and their associated instrumentation devices are continuous.

Fig. 6.10 illustrates closed loop control.

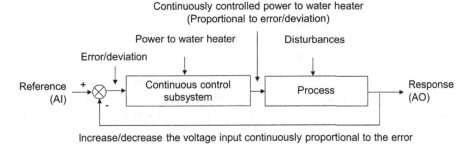

Figure 6.10 Continuous control: closed loop.

Fig. 6.11 illustrates the application of closed loop control to the temperature and level of a water heater.

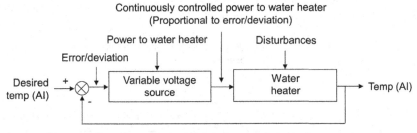

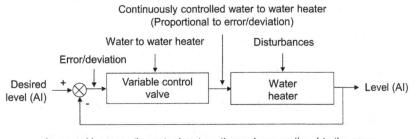

Figure 6.11 Closed loop control in water heater (continuous).

With the water heater as the process and the variable voltage source and variable control valve as control subsystems, the actual temperature and level are continuously measured and compared with their desired values to generate deviations. These deviations increase or decrease the power input proportional to the heating element and the water input to the tank, forcing the process to follow reference values.

6.2.3 Hybrid Process Control

Hybrid processes are a combination of both discrete and continuous processes involving both analog inputs and discrete outputs with their associated instrumentation devices. Control schemes, namely, two-step and two-step with dead-band, are discussed in the following sections.

6.2.3.1 Two-Step Control

Two-step control is a crude approach to continuous closed loop control; it employs continuous inputs for data acquisition and produces discrete outputs for control execution. This is similar to the function of a switching instrumentation device in the input instrumentation subsystem. The only difference is that in the switch, the reference value is set locally in the instrumentation device itself, whereas in two-step control the reference value is set in the human interface subsystem. Just as with the switch, the control is exerted discretely only if the process parameter deviates from the reference input. This is also called **on–off** or **bang–bang** control. Data acquisition and its associated instrumentation devices are continuous (analog), whereas the control and its associated instrumentation devices are discrete (digital).

Fig. 6.12 illustrates a two-step control scheme.

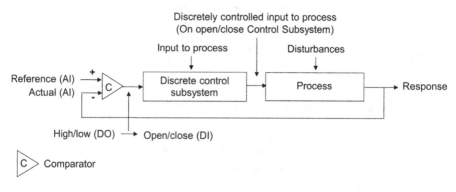

Figure 6.12 Two-step control.

Fig. 6.13 illustrates the application of two-step control to the temperature and level of a water heater.

Two-step control, which is a crude approach to continuous control, has a serious downside. The control command operates on the final control elements (valve and

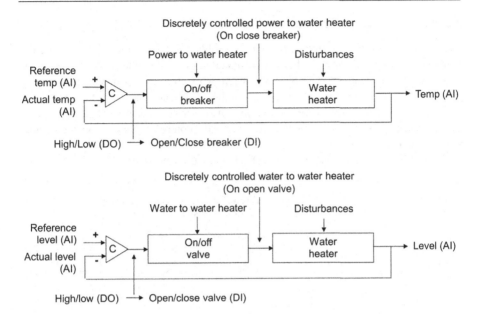

Figure 6.13 Two-step control in water heater.

breaker) even for minor deviations, forcing them to hunt or oscillate between their two positions around the reference value.

As illustrated in Fig. 6.14, this causes a lot of wear and tear on the final control elements, especially electromechanical ones.

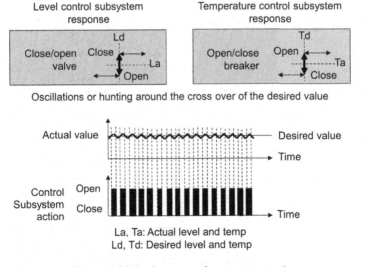

Figure 6.14 Performance of two-step control.

6.2.3.2 Two-Step Control With Dead-Band

To a major extent, the hunting or oscillation problem can be reduced by introducing a **dead-band** into the control scheme to take action only when the process value goes outside the preset dead-band. In other words, in two-step control with a dead-band, the process output is always forced to stay within the dead-band. The lower the dead-band is, the higher the oscillations are and the finer the control is. The higher the dead-band is, the smaller the oscillations are and the coarser the control is. **No correction is done within the dead-band**.

Fig. 6.15 illustrates a two-step control scheme with dead-band.

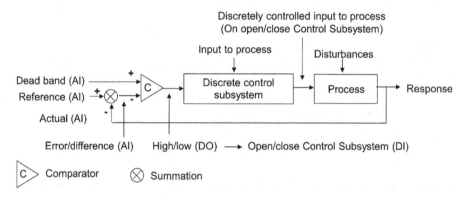

Figure 6.15 Two-step control with dead band.

Fig. 6.16 illustrates the application of two-step control with a dead-band for temperature and level control of a water heater.

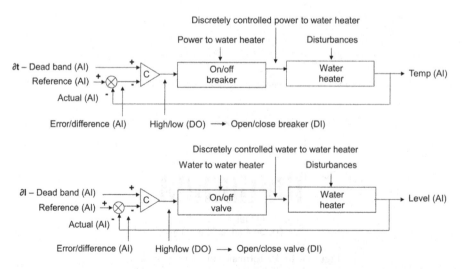

Figure 6.16 Two-step control with dead band in water heater.

Here, ∂l is the dead-band for the level; the level control subsystem closes the valve only when the actual level goes above **desired level + ∂l**. It opens the valve when the actual level goes below **desired level – ∂l**.

Similarly, with ∂t as the dead-band for temperature, the temperature control subsystem opens the switch only when the actual temperature goes above the desired **temperature + ∂t**. It closes the switch when the actual temperature goes below the desired **temperature – ∂t**.

Fig. 6.17 illustrates the performance of two-step control with a dead-band.

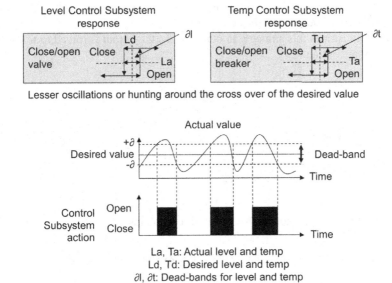

Figure 6.17 Performance of two-step control with dead band.

As seen, the lower the dead-band is, finer the control is, but with less stability. Conversely, the higher the dead-band is, the coarser the control is, but with higher stability.

Typical examples of two-step control with a dead-band are room air conditioners and refrigerators in which the preset band is factory set. Theoretically, two-step control with no dead-band or with a very small dead-band can produce almost continuous control, but this scheme is not feasible owing to oscillatory or unstable response, which leads to frequent switching operations in the final control elements. Therefore, two-step control with a dead-band does not perform true continuous control even with a decrease in dead-band. It produces undesired effects on the final control elements because of inherent hunting or oscillation.

6.3 Advanced Strategies

We discussed the basics of continuous closed loop control and hybrid control in previous sections. Closed loop control has some limitations in its applications.

This section discusses the extensions of basic closed loop control to overcome the limitations. Also, extensions of closed loop control to advanced control strategies such as feed-forward, cascade, and ratio control are discussed. The chapter concludes with a discussion on multiple-step control as an improvement over two-step control.

6.3.1   Continuous Control

As discussed in Section 6.2.2.2, in basic continuous closed loop control, the input to the process is continuously controlled to eliminate or minimize error or deviations between the desired output and the actual output.

Fig. 6.18 illustrates this strategy. In other words, the output is forced to follow the reference input continuously:

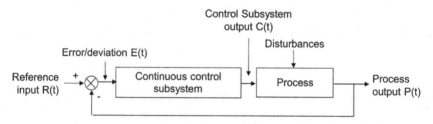

Figure 6.18 Continuous control strategy.

Following are the control system parameters, as shown in Fig. 6.18:

- $R(t)$: Reference or set-point input
- $P(t)$: Process output
- $E(t)$: Error/deviation equal to $R(t) - P(t)$
- $C(t)$: Control subsystem output

6.3.1.1   Response to Control Input

Closed loop continuous control discussed in Section 6.2.2.2 is basic or **proportional control**. This means that input to the control subsystem is proportional to the difference (deviation) between the actual output and the desired output at any point in time.

The typical response of the control subsystem, or the control subsystem output, to a unit step change is shown in Fig. 6.19 wherein the following terminologies employed:

- *Maximum overshoot*: $C(t)$ is the response of the control subsystem to the unit step input. C_{max} is the value of the control subsystem output at the maximum overshoot, and C_{ss} is its value at steady state. Maximum overshoot, which usually occurs at the first overshoot, is an indication of the relative stability of the control subsystem. A control subsystem with a large overshoot may indicate poor stability, and this is undesirable.
- *Delay time* (T_d): Delay time is defined as the time required for the response to reach 50% of its final value.

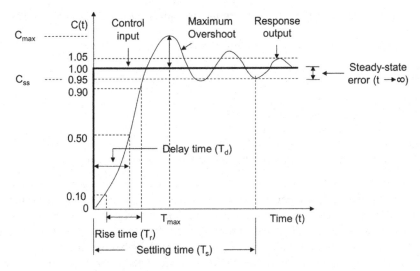

Figure 6.19 Control subsystem response to error input.

- *Rise time* (T_r): Rise time is defined as the time required for the response to rise from 10% to 90% of its final value.
- *Steady-state error*: Steady-state error is the desired value within which the control subsystem output finally stays. This is typically within a band of 5% in a good system.
- *Settling time* (T_s): Settling time is defined as the time required for the step response of the control subsystem to reach and stay within the desired value (steady-state value) of the specified band.

As seen in Fig. 6.19, after receipt of a control input (usually a step input), the control subsystem takes some time to settle down owing to its own dynamics. A good control subsystem response should provide the following:

- Fewer oscillations (more stable)
- Less settling time (faster response)
- Less delay time (faster reaction)
- Less steady-state error (more accurate)

In practice, traditional proportional control discussed earlier may not meet these listed requirements. The following sections discuss different means to achieve these requirements.

6.3.1.2 Proportional Control

Basic proportional control (P) is illustrated in Fig. 6.20.

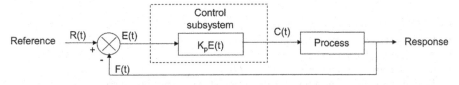

Figure 6.20 Proportional control.

The mathematical relationship governing the traditional P strategy is as follows:

$$C(t) = K_p E(t)$$

where

- $C(t)$ is the control subsystem output
- $R(t)$ is the reference (desired) input
- $F(t)$ is the feedback or process output
- $E(t)$ is the error/deviation equal to $\{R(t) - F(t)\}$
- K_p is the proportional gain between the error and the control subsystem output

Every incremental change in error $E(t)$ produces a corresponding incremental correction (amplification or attenuation) by proportional gain K_p. Pure P reacts to current error and does not settle at its target value. It retains a residual steady-state error (a function of the proportional gain and process gain). Despite the steady-state error, it is the P that contributes to the bulk of the output change. A high proportional gain results in a large change in the response for a given change in the error. If the proportional gain is too high, even though it reduces residual error, the system can become oscillatory. A small gain results in a small change in response, even to a large input error. If the proportional gain is too low, the control action may be too small to respond to disturbances.

6.3.1.3 Proportional and Integral Control

The problem of steady-state error is overcome in proportional and integral control (PI). The schematic of this control is illustrated in Fig. 6.21.

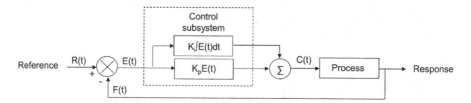

Figure 6.21 Proportional and integral control.

The mathematical relationship governing the PI strategy is as follows:

$$C(t) = K_p E(t) + K_i \int E(t)\, dt$$

where

- $C(t)$ is the control subsystem output
- $R(t)$ is the reference (desired) input
- $F(t)$ is the feedback or process output
- $E(t)$ is the error/deviation equal to $\{R(t) - F(t)\}$
- K_p is the proportional gain between the error and the control subsystem output
- K_i is the integral gain between the control subsystem input and control subsystem output

When added to P, integral control accelerates the movement of the process toward the reference and eliminates the residual/steady-state error that occurs with a proportional-only control subsystem. Integral control reacts to the sum of recent errors. However, because the control is responding to accumulated errors from the recent past, it can cause the current value to overshoot the reference value, creating a deviation in the other direction, leading to oscillation or instability. When the residual error is 0, the control subsystem output is fixed at the residual error (as seen in proportion-only control). If the error is not 0, the proportional term contributes a correction and the integral term begins to increase or decrease the accumulated value, depending on the sign of the error.

6.3.1.4 Proportional, Integral, and Derivative Control

A combination of proportional, integral, and derivative control (PID) takes care of problems associated with P and PI, and it produces the best results.

A schematic of PID control is shown in Fig. 6.22.

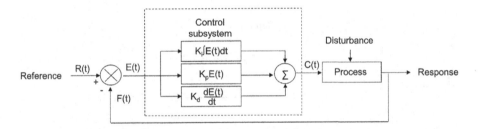

Figure 6.22 Proportional, integral, and derivative control.

The mathematical relationship governing the PID strategy is as follows:

$$C(t) = K_p E(t) + K_i \int E(t)\, dt + K_d \{ dE(t)/dt \}$$

where

- $C(t)$ is the controller output
- $R(t)$ is reference (desired) input
- $F(t)$ is feedback or process output
- $E(t)$ is error/deviation equal to $\{ R(t) - F(t) \}$
- K_p is proportional gain
- K_i is integral gain
- K_d is derivative gain

Derivative control is the reaction to the rate at which the error has been changing, and it slows the rate of change of the controller output. Hence the derivative control is used to reduce the magnitude of the overshoot produced by the integral component and to improve process stability. This is also referred to as rate or anticipatory control. This mode cannot be used alone because when the error is 0 or constant, it provides no output. For every incremental rate of change, this control provides a corresponding incremental change in output by K_d. In case of a process disturbance, P-only or PI actions cannot react fast enough in returning the process back to set-point without overshoot. Other characteristics are as follows:

- If the process measurement is noisy, the derivative term changes widely and amplifies the noise unless the measurement is filtered.
- The larger the derivative term is, the more rapid the controller response is to changes in the process value.
- The derivative term reduces both the overshoot and the settling time.

Fig. 6.23 illustrates the control subsystem's response to all three types of inputs (proportional, proportional with integral, and proportional with integral and derivative).

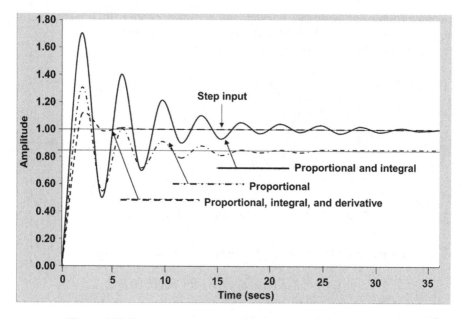

Figure 6.23 Response to proportional, integral, and derivative control.

Fig. 6.24 illustrates an application in the industry of PID control (speed controller in an automobile).

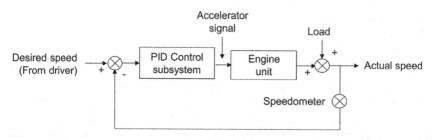

Figure 6.24 Example of proportional, integral, and derivative (PID) control.

6.3.1.5   Summary of Control Schemes

The following is a summary of the discussions of various types of closed loop control schemes for continuous process automation:

- P: Basic and essential but introduces steady-state error, speeds up process response, and reduces offset with higher gain. Larger proportional gain (K_p), typically means faster response and larger proportional compensation. An excessively large proportional gain leads to oscillations and instability in the process.
- PI: Eliminates steady-state error but introduces oscillations. Larger integral gain (K_i) typically implies quicker elimination of steady-state errors. The tradeoff is a large overshoot and oscillations.
- PID: Reduces oscillations. Larger derivative gain (K_d) typically decreases overshoot but slows down transient response. However, this may amplify the noise in its response because noise has high-frequency components. It provides an optimum response with proper selection of proportional, integral, and derivative gains.

The selection of what controller modes to use in a process is a function of the process characteristics. The process control loop only regulates the dynamic variables in the process. The controlling parameter is changed to minimize the deviation of the controlled variable. Tuning a control loop adjusts its control parameters (proportional gain/proportional band, integral gain/reset, derivative gain/rate) to optimal values for the desired controller response.

6.3.2   Feed-Forward Control

PID control subsystems act on the error (difference between the set-point and the response) to produce the control input to force the process to follow the set-point while compensating for the external disturbance.

Feed-forward control makes changes to inputs to the process to counter or preempt anticipated effects of the disturbance. These changes are based on prior knowledge of the effects of the disturbance before it is observed in the process output. Feedback control, when clubbed with feed-forward control, significantly improves performance of the PID control and ideally can entirely eliminate the effect of the measured disturbance on the process output. To implement feed-forward control, the time taken for the disturbance to affect the output should not be longer than the time taken for the feed-forward control subsystem to affect the output.

Fig. 6.25 illustrates feed-forward control.

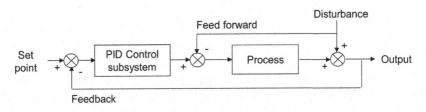

Figure 6.25 Feed-forward control.

Fig. 6.26 illustrates the application of feed-forward control to a two-element boiler.

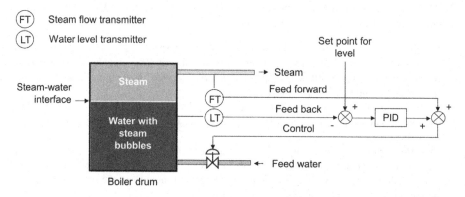

Figure 6.26 Example of feed-forward control.

In the boiler, the two variables (steam flow and drum water level) influence the feed water valve position. In the normal situation, the PID control subsystem regulates the drum water level based on its set-point without taking into consideration the effect of steam flow on the drum water level. In the feed-forward case, the PID control subsystem output is combined with the steam flow to compensate for the effect of steam flow on the drum water level. In other words, the feed-forward control better regulates the feed water to respond to any changes in the steam demand.

6.3.3 Cascade Control

In some cases, a controlled process may be considered to be two processes in series or a cascaded process in which the output of the first process is measured. The output of the master control subsystem is the set-point for the slave control subsystem. This control provides better dynamic performance.

Fig. 6.27 illustrates cascade control.

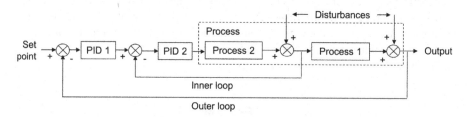

Figure 6.27 Cascade control.

The benefit of using the cascade configuration is that any disturbances within the inner loop can be quickly corrected by the outer loop without waiting for them to show up in the outer loop. This gives better (tighter and faster) overall control. The outer loop can be relatively tuned for a faster response. It responds to disturbances quickly and minimizes any fluctuations in its output caused by a disturbance. The outer loop still controls the final output.

Fig. 6.28 illustrates an application in industry of cascade control in heat exchangers.

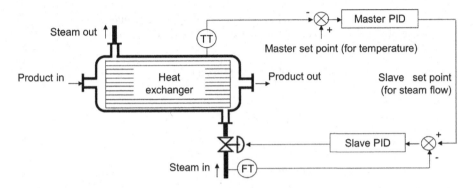

FT: Flow transmitter, TT: Temp transmitter

Figure 6.28 Example of cascade control.

Heat exchangers are used to heat or cool a process fluid to a desired temperature by steam. The output of the master control subsystem (temperature) manipulates the set-point of the slave control subsystem (steam flow). This eliminates the effects of feed or product disturbances to improve the performance of the control loop. Combinations of single- and multiple-loop control subsystems can be configured to work as cascade control subsystems.

6.3.4 Ratio Control

This is a technique to control a process variable at a set-point, which is calculated as a proportion (ratio) of an uncontrolled or lead input variable. The set-point ratio determines the proportion of the lead value to be used as the actual control set-point. The set-point ratio can be greater or lesser than the lead input.

Fig. 6.29 illustrates ratio control.

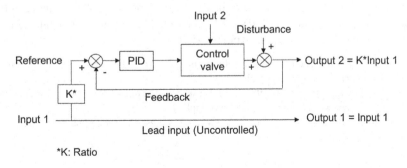

*K: Ratio

Figure 6.29 Ratio control.

Ratio control systems are used extensively to optimize the relationship of the mixture between two flows in processing plants. In other words, ratio control is used to ensure that two or more flows are kept at the same ratio even if the lead flow changes. Typical applications are:

- Blending two or more flows to produce a mixture with specified composition
- Blending two or more flows to produce a mixture with specified physical properties
- Controlling the air–fuel ratio

Fig. 6.30 illustrates ratio control (air–fuel mixer).

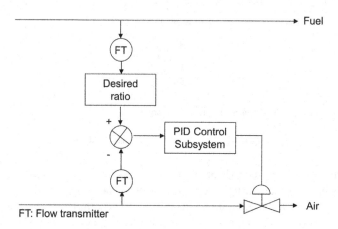

Figure 6.30 Example of ratio control.

6.3.5 Multiple-Step Control

Multiple-step control is similar to two-step control, as discussed in Section 6.2.3.1, but with at least one more intermediate step. This intermediate step further reduces the possibility of oscillations or hunting around the set-point. In other words, with more steps, finer or near continuous control is possible. Multiple-step control is best explained using the example of a domestic voltage stabilizer with a three-step control subsystem.

Fig. 6.31 illustrates the control schematic of a voltage stabilizer with three steps.

The function of a domestic voltage stabilizer is to provide regulated voltage output within the acceptable range of 220V ± 10%, or 198–242V, from an unregulated input voltage. The specifications of the stabilizer are as follows:

- The stabilizer is designed to work within the input range of 176–264V, and it gets cut off from the mains if the input is outside this range.
- If the input voltage falls below 198V, switch 1 is closed to output 110% of the input voltage (step 1). In other words, +10% correction is applied.
- If the input voltage is within 198–242V, switch 2 is closed to follow the input (step 2). In other words, no correction is applied.

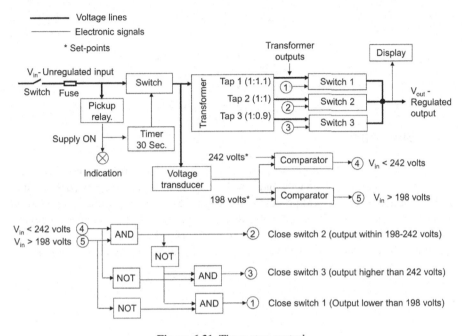

Figure 6.31 Three-step control.

- If the input voltage rises above 242 V, switch 3 is closed to output 90% of the input voltage (step 3). In other words, −10% correction is applied.
- The output is made to stay within the acceptable range (198–242 V) provided the input range is between 176 and 264 V.

Fig. 6.32 illustrates an example of a voltage stabilizer with five-step control, supply indication, and input–output voltage display.

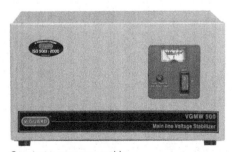

Courtesy: www.vguard.in

Figure 6.32 Example of voltage stabilizer with five-step control.

Other specifications are as follows:

- Input voltage range: 100–290 V AC
- Output voltage range: 200–250 V AC (from 100 to 280 V input)
- Low-voltage cutoff: 80 V input with 138 V output
- High-voltage cutoff: 300 V input with 269 V output

6.4 Summary

In this chapter, we described various control strategies commonly employed in automation systems. Advantages and disadvantages of basic strategies such as discrete, continuous, and hybrid process control were discussed using the example of a water-heating process. This chapter also discussed in detail advance control strategies (variations of basic strategies) such as PID, feed-forward, cascade, and ratio control schemes. The chapter ended with a discussion of the multiple-step control strategy.

Programmable Control Subsystems 7

Chapter Outline

7.1 Introduction

In their early days, automation systems used pneumatic, hydraulic, and mechanical components for both measurement and control. These were all slow, bulky, sluggish, unreliable, and prone to environmental conditions, and they required frequent maintenance, space, power, etc. Automation technology gradually moved toward electrical components and is currently based on electronics technology.

Modern control systems started out as hardwired systems, using technology that was available at that time. They began with relay technology, followed by solid-state technology (The reader is recommended to go through Appendix A before proceeding further.). Later, automation technology moved into **processor-based** (microprocessors or computers) or information technology. Advances in electronics, communication, and networking technologies had a vital role in making the entire control subsystem programmable, compact, power-efficient, reliable, flexible, communicable, and

Overview of Industrial Process Automation. http://dx.doi.org/10.1016/B978-0-12-805354-6.00007-4

self-supervising. All of the drawbacks, as discussed in Appendix A, associated with the hardwired control subsystem were overcome with this **soft-wired control subsystem** or **programmable control subsystem**, which also brought many more advantages.

7.2   Soft-wired or Microprocessor Technology

Microprocessor-based technology has all of the advantages of solid-state technology, because the former is also a solid-state technology. In addition, it has the advantages of modularity, extendibility, and flexibility for automation strategy modification and/ or extension (soft-wired strategy). Because the control subsystems are soft-wired, the connections within are established through software. However, this technology requires programming skills to realize the automation strategy.

7.2.1   Additional Features

By far, information technology is the greatest enabler and driver the world has ever invented. The central component in information technology is a microprocessor or computer. A microprocessor- or computer-based system, in contrast to a hard-wired system, stands out with the following additional features:

- Data processing: Ability to do arithmetic operations (number crunching such as addition, subtraction, multiplication, and division) and logical operations (such as AND, OR, and NOT) on data;
- Intelligence: Ability to perform decision-making operations (such as comparing and branching on equal to, greater than, less than, etc.) on data;
- Communicability: Ability to communicate with external compatible systems for data exchange; and
- Self-supervision: Ability to diagnose its own health and announce any abnormality.

These features of the microprocessor or computer built into the programmable control subsystem as well.

Fig. 7.1 illustrates the migration of a general-purpose **hardwired** control subsystem into a general-purpose processor-based, **soft-wired**, or **programmable** control subsystem.

The hardwired systems are structured around signals whereas soft-wired systems are based on **computer data** or simply **data**. The data structure in the programmable control subsystem is illustrated in Fig. 7.2.

During data acquisition, electronic inputs from the input instrumentation devices are converted into **data** for further processing within the programmable control subsystem. Similarly, during control execution, **data** are converted into electronic outputs for use by output instrumentation devices.

Fig. 7.3 illustrates the hardware structure of the programmable control subsystem with an in-built watchdog (self-supervision) facility supporting all types of process input and output.

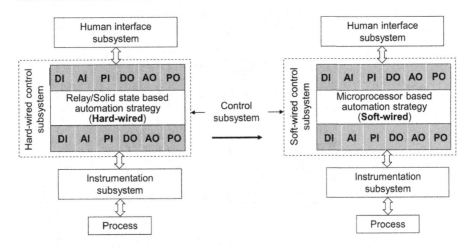

Figure 7.1 Migration to programmable control subsystem.

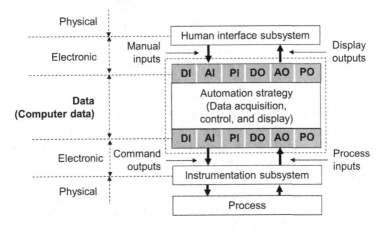

Figure 7.2 Data structure in programmable control subsystem.

The automation strategy presented here is in the form of a memory resident program operating on acquired data and other memory resident data, as per the stored strategy to produce the control outputs. The programmable control subsystem can be designed or configured for any type of automation strategy with the appropriate combination of input/output (I/O) modules and automation strategy program.

Like a typical computer system, the programmable control subsystem is designed with the following:

- A power supply subsystem consisting of a power supply module;
- A processor subsystem consisting of a processor, memory, and watchdog modules; and
- An I/O subsystem consisting of various types of I/O modules.

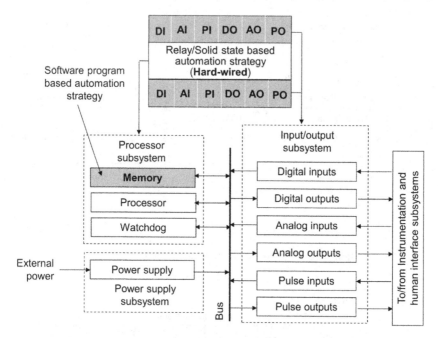

Figure 7.3 Hardware structure of programmable control subsystem.

All of these modules, known as **functional modules**, are physically assembled in a mechanical rack and integrated electronically over a **bus**. The rack holds the functional modules while the bus provides the path for communication and data exchange between the processor and other functional modules. The details are discussed in Chapter 8.

Because each I/O channel within an I/O module is logically independent, the instrumentation subsystem and human interface subsystem can share a single I/O module with multiple channels if required. The memory resident automation strategy takes care of the distinction, as illustrated in Fig. 7.3.

To be in line with information technology–based systems, the programmable control subsystem is addressed here by its more commonly used name, the **Data Acquisition and Control Unit (DACU)**.

7.2.2 Data Acquisition and Control Unit

DACU has three main variants, stand-alone, acquisition only, and communicable, which are discussed subsequently.

7.2.2.1 Stand-alone

Stand-alone DACU can send and receive **data only in an analog form (voltage and/or current)** to and from the instrumentation and human interface subsystems.

The basic DACU discussed earlier in Section 7.2.1 (illustrated in Fig. 7.3) is stand-alone. Here, DACU supports automation strategy.

7.2.2.2 Acquisition Only

A data acquisition-only unit (DAU) is a stripped-down version of DACU. This supports only data acquisition and display. These systems are typically used for laboratory automation applications. Here, the DAU supports only a display strategy.

Fig. 7.4 illustrates a typical configuration of DAU.

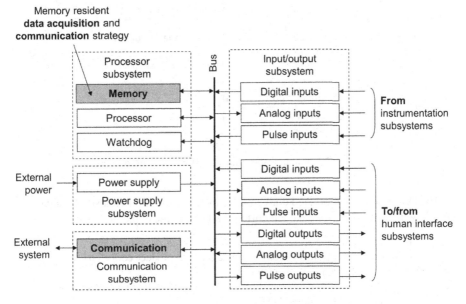

Figure 7.4 Data acquisition unit.

7.2.2.3 Communicable

Communicable DACU can exchange **data in digital form (computer data)** with external compatible systems. To facilitate this, the DACU is equipped with a communication module and the strategy for both automation and communication.

Fig. 7.5 illustrates the communicable DACU configuration.

A good example of the use of communicable DACU can be seen in managing a remotely located coffee maker built with stand-alone and communicable DACUs.

The coffee maker is located in a remote location and is supervised by the operator from his workplace, ensuring that:

- The bins are full of roasted coffee beans, milk, and water; and
- The machine is healthy and properly functioning (fault-free).

Whenever necessary, the user places a cup below the tap in the coffee maker, presses the start button, and waits for the cup to be filled with coffee.

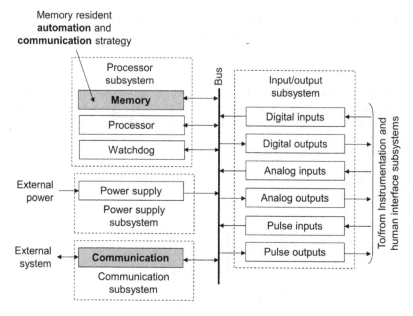

Figure 7.5 Communicable data acquisition and control unit.

The DACU-based automated coffee maker executes the following steps (sequential control with interlocks) to make the coffee available in the cup:

1. Start: Press start button.
2. : Grinds required quantity of roasted coffee beans to produce coffee powder.
3. : Heats water to required temperature.
4. : Passes hot water through freshly ground coffee powder to produce decoction.
5. : Collects required quantity of decoction.
6. : Mixes required quantity of milk into decoction.
7. : Heats mixture to required temperature.
8. Stop: Fills cup.

For this service to function smoothly and continuously, the coffee maker should always ensure:

- A sufficient quantity of water, milk, and roasted beans; and
- Proper working condition with no faults.

How are these conditions monitored, and what actions are taken if one or more of the requirements are not met?

Fig. 7.6 illustrates the solution with the **stand-alone DACU**-operated coffee maker.

In this approach, the operator needs to make periodic visits from his workplace to the remote location to make sure the coffee maker has sufficient quantities of milk, water, and beans in the bins, and to verify the healthy condition of the machine. The operator takes corrective (reactive) actions if necessary when any of these conditions is not met. With more visits by the operator the service becomes better but the efficiency of the operator becomes poorer.

Figure 7.6 Stand-alone data acquisition and control unit (DACU)-based coffee maker.

Fig. 7.7 illustrates the solution with a **communicable DACU**-operated coffee maker.

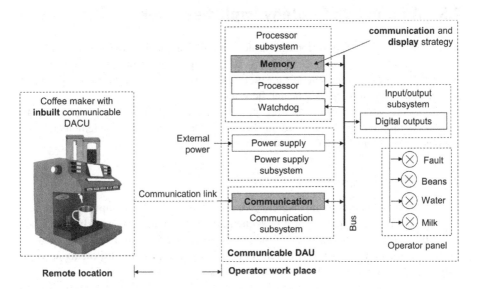

Figure 7.7 Remote management of communicable data acquisition and control unit (DACU)-based coffee maker.

The remotely located communicable DACU-based coffee maker continuously communicates with the communicable DAU located in the operator workplace and sends data about its current status. The DAU displays the received data on the operator panel, which is also located in the operator workplace. With this arrangement, the operator receives continuous online information about the status of the coffee maker at his workplace. The operator can take advance (proactive) actions.

The communicable DACU solution is expensive and technically more complicated but it provides better and more efficient service. Furthermore, this solution can be extended to support multiple coffee makers located in remote places but it is connected to a single operator panel in the operator's workplace over the communication line.

The DACU configuration shown in Fig. 7.4 is the general-purpose DACU configuration capable of executing multiple strategies (all types and variants) with multiple inputs, multiple outputs, multiple loops, and interaction among the loops, simultaneously and in parallel.

As seen in Fig. 7.5, DAU has no output module linked to the process; it can only acquire data from the process and cannot issue a command to the process. However, a human interface subsystem (operator panel) can have both input and output modules for interaction with the DAU.

The hardware structure of a DACU is described in Chapter 8 and its software structure is described in Chapter 9.

7.3   Automation Strategy Implementation

The following sections describe the implementation of various automation strategies (discrete, continuous, and hybrid) in a DACU using the example of a water-heating process. The implementation of an automation strategy using relay and solid-state systems (see Appendix A) is repeated here with DACU.

The example of the water-heating process, with its instrumentation and human interface subsystems, is employed in this section to discuss the implementation of various types of automation strategies in DACU.

7.3.1   Discrete Control

As discussed in Chapter 6, the discrete process needs only discrete inputs and outputs to support either open loop control or sequential control with interlocks. In both strategies, the hardware configuration of the DACU remains identical. However, software for the automation strategy is different in each case. Hence in the following section, only sequential control with interlocks is discussed.

7.3.1.1   Sequential Control With Interlocks

Fig. 7.8 illustrates the water-heating process with an instrumentation subsystem with discrete devices and a human interface subsystem with discrete panel components.

Table 7.1 provides details about the allocation of I/O channels and devices to the operator panel (control and display components) and process (input and output instrumentation devices).

Fig. 7.9 illustrates the configuration of a DACU to support discrete process automation with only digital inputs and outputs, with an automation program for sequential control with interlocks for the water-heating process.

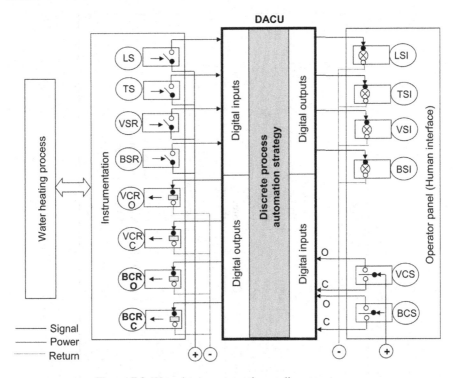

Figure 7.8 Water heater automation as discrete process.

Apart from the allocation of signals to different I/O channels, the implementation of software strategy also requires some memory locations allotted to remember momentary conditions temporarily (similar to hardware latching of momentary commands into level commands, as explained in Appendix A) and store intermediate results. These memory locations are called **flags**.

Table 7.2 describes memory locations allotted for flags and intermediate results.

Fig. 7.10 shows a flowchart for the execution of the automation strategy.

Appendix D explains further steps in detail to be followed to convert the flowchart into an executable code in DACU.

7.3.2 Continuous Control

As discussed earlier, the continuous process needs only continuous inputs and outputs (continuous control) to support both open loop and closed loop control. Hardware configuration of the DACU is the same for these two cases except for the number of I/O channels and the automation strategy. In view of this, only closed loop control is discussed here.

7.3.2.1 Closed Loop Control

Fig. 7.11 illustrates the water-heating process with an instrumentation subsystem with analog devices and a human interface subsystem with analog panel components.

Table 7.1 Allocation of input/output channel/device to operator panel and process (discrete control)

Source	Type	Channel	Device	Label	Signal
Digital input (DI) module (eight channels: DI0–7)					
Process	Instrumentation device	DI0	Switch	LS	Level status (full/not full)
		DI1	Switch	TS	Temperature status (reached/not reached)
		DI2	Supervision relay	VSR	Valve status (open/closed)
		DI3	Supervision relay	BSR	Breaker status (open/closed)
Operator panel	Control device	DI4, DI5	Control switch	VCS-O/C	Valve command (open/close)
		DI6, DI7	Control switch	BCS-O/C	Breaker command (open/close)
Digital output (DO) module (eight channels: DO0–7)					
Operator panel	Display device	DO0	Indication lamp	LSI	Level status
		DO1	Indication lamp	TSI	Temperature status
		DO2	Indication lamp	VSI	Valve status
		DO3	Indication lamp	BSI	Breaker status
Process	Instrumentation device	DO4	Control relay	VCR-O	Valve open command
		DO5	Control relay	VCR-C	Valve close command
		DO6	Control relay	BCR-O	Breaker open command
		DO7	Control relay	BCR-C	Breaker close command

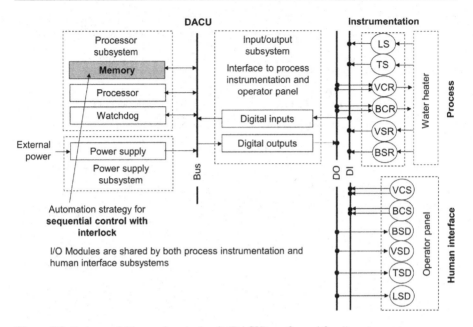

Figure 7.9 Data acquisition and control unit (DACU) configured for discrete process automation.

Table 7.2 Allocation of memory locations for special requirements (discrete control)

Flags	VCOC	To remember momentary valve pen command
	VCCC	To remember momentary valve close command
	BCOC	To remember momentary breaker open command
	BCCC	To remember momentary breaker close command
	LSS	To remember level status
	TSS	To remember temperature status
	VSS	To remember valve status
	BSS	To remember breaker status
	XXX	To store intermediate results

Table 7.3 provides the details of allocation of I/O channels and devices to the operator panel (control and display components) and process (instrumentation devices).

Fig. 7.12 shows the configuration of the DACU to support continuous process automation with only analog inputs and outputs with the automation program for closed loop control for the water-heating process.

Apart from the allocation of process signals to different I/O channels, the program requires some memory locations to store the intermediate results.

Table 7.4 describes the memory locations for flags and intermediate results.

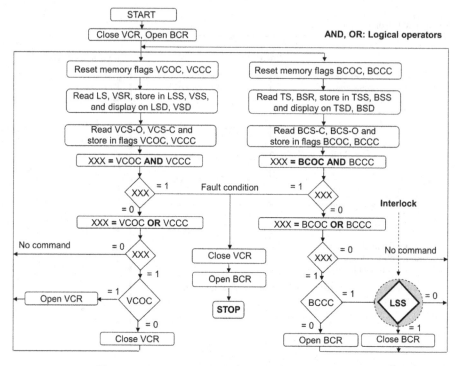

Figure 7.10 Sequential control with interlock: flowchart.

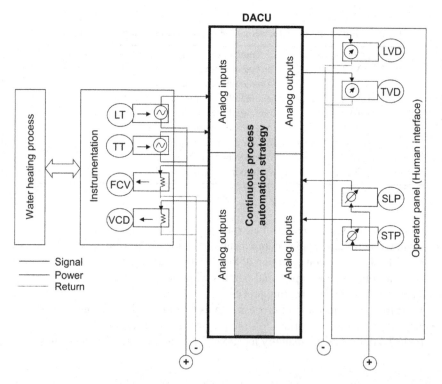

Figure 7.11 Water heater automation as a continuous process.

Table 7.3 Input/output channel/device allocation to operator panel and process (continuous control)

Source	Type	Channel	Device	Label	Signal
Analog input (AI) module (eight channels: AI0–7)					
Process	Instrumentation device	AI0	Transmitter	LT	Level value
		AI1	Transmitter	TT	Temperature value
Operator panel	Control device	AI2	Potentiometer	SLP	Level reference value
		AI3	Potentiometer	STP	Temperature reference value
		AI3–7	Not used		
Analog output (AO) module (eight channels: AO0–7)					
Operator panel	Display device	AO0	Display meter	LVD	Level value
		AO1	Display meter	TVD	Temperature value
Process	Instrumentation device	AO2	Flow control valve	FCV	Water input control
		AO3	Voltage control drive	VCD	Voltage input control
		AO4–7	Not used		

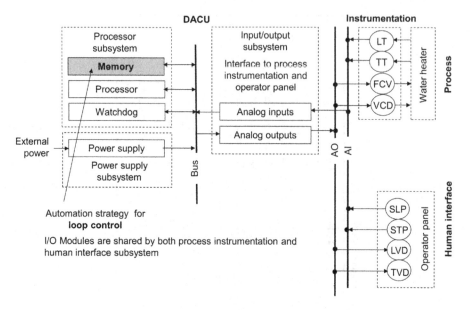

Figure 7.12 Data acquisition and control unit (DACU) configured for continuous process automation.

Table 7.4 Allocation of memory locations for special requirements (continuous control)

Name	Location	Function
Flags	LTV	To remember level value
	TTV	To remember temperature value
	SLPV	To remember level reference value
	STPV	To remember temperature reference value
	XXX	To remember intermediate results

Fig. 7.13 is a flowchart for the execution of the automation strategy.

Appendix D explains further steps in detail to be followed to convert this flowchart into an executable code in DACU.

7.3.3 Multiple Input/Multiple Output Control

Automation of the continuous water-heating process discussed so far has two closed loops: one for level control and the other for temperature control. In this example, each loop is independent and does not require interaction with the other loop. Each loop is an example of a single-input/single-output (SISO) process, and the DACU is SISO DACU. However, in practice, continuous processes have multiple loops operating parallel to each other and requiring interaction among them. This is a case of a multiple-input/multiple-output (MIMO) process, and the DACU is MIMO DACU.

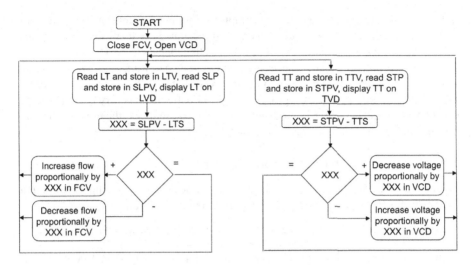

Figure 7.13 Closed loop control: flowchart.

Fig. 7.14 illustrates the schematics of both SISO and MIMO loops. The MIMO DACUs provide for control outputs as functions of one or more inputs.

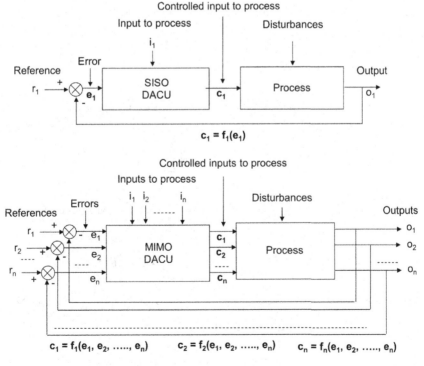

Figure 7.14 Single-input/single-output (SISO) and multiple-input/multiple-output (MIMO) process automation.

Unlike discrete process automation, implementation of a control subsystem for MIMO in a continuous process using hardwired technology is relatively complex and may not be possible if the number of I/O and interacting loops is large. However, because the DACU is processor-based, it can easily be programmed to work as an MIMO control subsystem because the data related to all loops are available in the common memory for use by any loop. Furthermore, the processor in the DACU, which is very fast, can handle many loops simultaneously while still meeting time requirements (the timing and duration in which each loop is allowed to complete its task).

7.3.3.1 Two-Input/Two-Output Control

To illustrate the MIMO concept, let us discuss the design of a two-input/two-output DACU for the water-heating process with the following requirements:

- Loop 1: Temperature control to maintain temperature at the reference value;
- Loop 2: Level control to maintain the level at the reference value; and
- Interaction between loops 1 and 2: Temperature control works only when the water level is above 25% of the reference level.

The setup of the continuous water-heating process with instrumentation and an interface or a DACU configuration (Figs. 7.7 and 7.8) remains the same. Only the strategy implementation is different. Here, the temperature control loop (loop 2) is supervised by the level control loop (loop 1).

Fig. 7.15 shows a flowchart of the execution of this automation strategy.

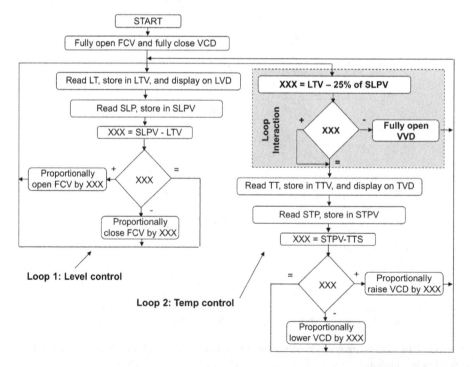

Figure 7.15 Two-input/two-output programming of data acquisition and control unit: flowchart.

Appendix D explains further steps in detail to be followed to convert this flowchart into an executable code in DACU.

7.3.4 Hybrid Control

As mentioned in Chapter 6, hybrid control has two versions:

- Two-step control and
- Two-step control with dead-band.

Two-step control is a combination of both discrete and continuous control philosophies, and it is a good and cost-effective approximation of true continuous control implemented by a closed loop control. Because the DACU hardware configuration is the same for both two-step and two-step with dead-band control, only the latter is discussed here.

7.3.4.1 Two-Step Control with Dead-Band

Fig. 7.16 illustrates the water-heating process with analog and digital instrumentation devices and analog human interface panel components.

Table 7.5 provides details about the allocation of I/O channels and devices to the operator panel (control and display components) and process (instrumentation devices).

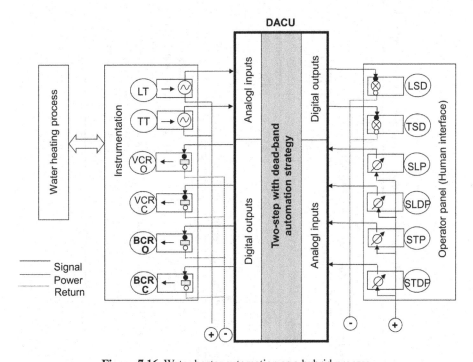

Figure 7.16 Water heater automation as a hybrid process.

Fig. 7.17 illustrates a DACU configuration for the implementation of two-step control with dead-band for the water-heating process.

Table 7.5 Input/output channel/device allocation to operator panel and process (two-step control with dead-band)

Source	Type	Channel	Device	Label	Signal
Analog input (AI) module (eight channels: AI0–7)					
Process	Instrumentation device	AI0	Switch	LT	Level status (full/not full)
		AI1	Switch	TT	Temperature status (reached/not reached)
Operator panel	Control device	AI2	Potentiometer	SLP	Level reference value
		AI3	Potentiometer	SLDP	Temperature reference value
		AI4	Potentiometer	STP	Level dead-band value
		AI5	Potentiometer	STDP	Temperature dead-band value
		AI6–7	Not used		
Digital output (DO) module (eight channels: DO0–7)					
Operator panel	Display device	DO0	Indication lamp	LSD	Level status
		DO1	Indication lamp	TSD	Temperature status
		DO2	Indication lamp	VSD	Valve status display
		DO3	Indication lamp	BSD	Breaker status
Process	Instrumentation device	DO4	Control relay	VCR-O	Valve command (open)
		DO5	Control relay	VCR-C	Valve command (close)
		DO6	Control relay	BCR-O	Beaker command (open)
		DO7	Control relay	BCR-C	Breaker command (close)

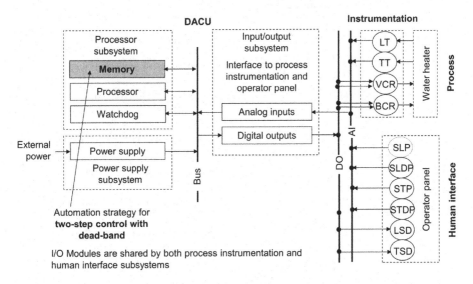

Figure 7.17 Data acquisition and control unit (DACU) configured for hybrid process automation.

Table 7.6 provides details about the allocation of I/O channels and devices to the operator panel (control and display components) and process (instrumentation devices).

Table 7.6 Allocation of memory locations for special requirements (two-step control with dead-band)

Name	Location	Function
Flags	LTV	To remember level value
	TTV	To remember temperature value
	SLPV	To remember valve reference
	SLDPV	To remember valve dead-band
	STPV	To remember temperature reference
	STDPV	To remember temperature dead-band
	LSDS	To remember level exceed
	TSDS	To remember temperature exceed
	XXX	To store intermediate results temporarily

Fig. 7.18 is a flowchart of the execution of the automation strategy.

Appendix D explains further steps in detail to be followed to convert this flowchart into an executable code in DACU.

7.3.5 Multiple-Step Control

As mentioned in Chapter 6, multiple-step control gives much better performance compared with two-step control. The higher the number of steps there are, better the regulation is. The next section illustrates a three-step controlled voltage stabilizer.

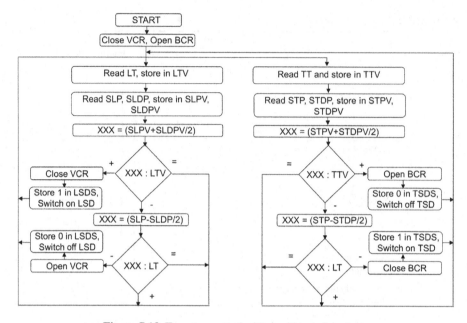

Figure 7.18 Two-step control with dead-band: flowchart.

7.3.5.1 Three-Step Control

Fig. 7.19 illustrates the voltage stabilizer process with analog and digital instrumentation devices and analog human interface panel components.

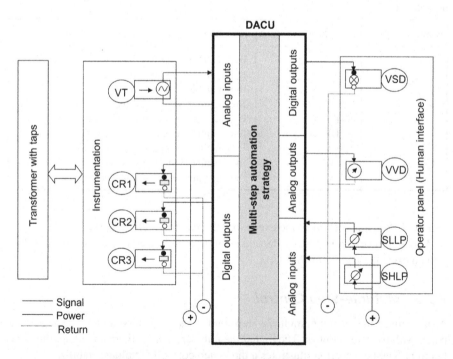

Figure 7.19 Voltage stabilizer automation as a multiple-step process.

Table 7.7 Input/output channel/device allocation to operator panel and process (multiple-step control)

Source	Type	Channel	Device	Label	Signal
Analog input (AI) module (eight channels: AI0–7)					
Process	Instrumentation device	AI0	Transducer	VT	Voltage value (input)
Operator panel	Control device	AI1	Potentiometer	SLLP	Voltage (low limit)
		AI2	Potentiometer	SHLP	Voltage (high limit)
		AI3–7	Not used		
Analog output (AO) module (eight channels: AO0–7)					
Operator panel	Display device	AO0	Display meter	VVD	Voltage (output)
		AO1–7	Not used		
Digital output (DO) module (eight channels: DO0–7)					
Operator panel	Display device	DO0	Indication lamp	VSD	Voltage status (input) display
Process	Instrumentation device	DO1	Control relay	CR1	Valve command (open)
		DO2	Control relay	CR2	Valve command (close)
		DO3	Control relay	CR3	Beaker command (open)
		DO4–7	Not used		

Table 7.7 provides details about the allocation of I/O channels and devices to the operator panel (control and display components) and process (instrumentation devices).

Fig. 7.20 illustrates the DACU configuration for the implementation of multiple-step control of the voltage stabilizer process.

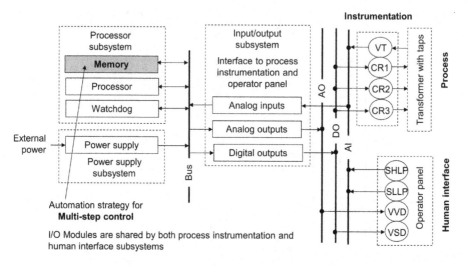

Figure 7.20 Data acquisition and control unit configured for multiple-step automation.

Table 7.8 provides details about the allocation of I/O channels and devices to the operator panel (control and display components) and process (instrumentation devices).

Table 7.8 Allocation of memory locations for special requirements (multiple-step control)

Name	Location	Function
Flags	VTS	To remember voltage value
	SLLPS	To remember low voltage limit
	SHLPS	To remember high voltage limit
	VSDS	To remember voltage status

Fig. 7.21 is a flowchart for the execution of the automation strategy.

The next section gives detailed steps for converting the flowchart into an executable code in DACU.

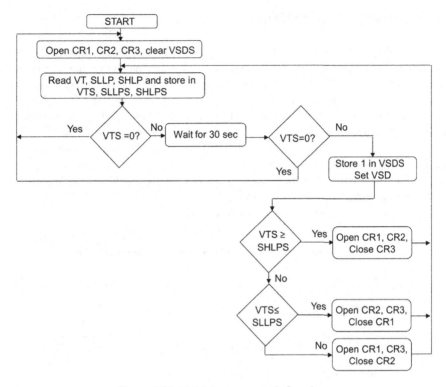

Figure 7.21 Multiple-step control: flowchart.

7.3.6 *Program Execution*

In the preceding sections, we developed flowcharts to execute the automation strategy. The following steps are applied to convert the flowcharts into a code executable by the DACU:

- Code the flowchart into a program using an appropriate automation application programming language.
- Simulate the program and test it in the host machine.
- Convert the automation application program into its machine-executable code in the host machine.
- Transfer or download the machine-executable code (download) from the host machine into the memory of the DACU.

When these steps are completed, the DACU is ready to execute the automation strategy, and it **behaves** exactly like a hardwired control subsystem until the strategy is reprogrammed and reloaded.

The formal application programming languages used to code industrial automation strategies and related steps to convert them into an executable code and download them onto DACU are covered in Chapter 9.

7.4   Upward Compatibility

Normally the process equipment, instrumentation subsystem, and human interface
subsystem do not quickly become obsolete; they last much longer than the control subsystems. There is often a need to change the control subsystem for a variety of reasons:
to increase its functionality, to extend its I/O capacity, because of the unavailability of
spares/service owing to obsolescence, and so forth. The old technology might become
obsolete or unable to meet the additional functionality, increase in I/O, performance,
etc. To provide for investment protection, new-generation control subsystems are
designed to replace the older-generation systems as long as the new-generation system
terminal configurations, functionality, physical parameters, and such are compatible
with ones being replaced (upward compatibility).

This upward transition is illustrated in Fig. 7.22.

Terminals for input/output connections to human interface subsystem

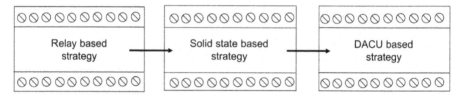

Terminals for input/output connections to instrumentation subsystem

⊘⊘⊘⊘ Screw terminals

Figure 7.22 Upward compatibility of data acquisition and control unit (DACU).

7.5   Summary

A general-purpose programmable control subsystem or DACU is modular, programmable
(soft-wired), and communicable with all types of I/O modules, and has customized
automation functions (software for data acquisition, analysis, decision making, control,
display, and communication). Functions supported include open loop control and all
variants of closed loop control with multiple I/Os with multiple loops, and with interactions among loops, display, and communication. This general-purpose DACU can be
customized to meet any specific automation applications. This chapter also discussed
special features of the DACU, namely, communicability and self-supervision.

Data Acquisition and Control Unit: Hardware

Chapter Outline

8.1   Introduction

In Chapter 7, we discussed the philosophy of the data acquisition and control unit (DACU) with specific reference to the implementation of various automation, display, and communication strategies. This chapter details the hardware construction of DACU in general and its functional modules in particular. A general-purpose DACU is illustrated in Fig. 8.1.

Overview of Industrial Process Automation. http://dx.doi.org/10.1016/B978-0-12-805354-6.00008-6

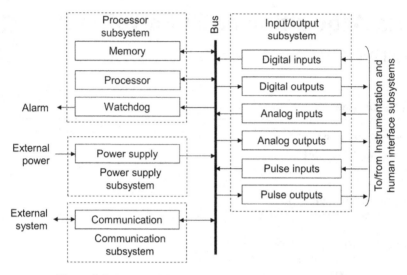

Figure 8.1 Data acquisition and control unit: logical structure.

The DACU is also computer-like equipment and functions like a computer with all of the necessary functional modules. A general-purpose computer employs devices such as keyboards and displays input–output (I/O) devices, whereas the DACU employs process I/O modules.

Fig. 8.2 illustrates the typical overall hardware structure of the DACU.

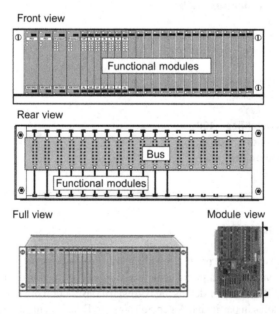

Figure 8.2 Data acquisition and control unit: physical structure.

The structure and functions of various modules are described in the following sections. In practice, many different arrangements and types of construction are available depending on the manufacturer. The one discussed here is simple and commonly implemented to best explain the concepts involved. Some variants in the construction are discussed in Section 8.7.

8.2 Basic Modules

The basic modules of a DACU are the rack and the bus, which hold the functional modules together mechanically, electrically, and logically. However, both are passive.

8.2.1 Rack

A rack is a mechanical structure that holds functional modules in their place to facilitate their physical and electrical connections to the bus. A rack can hold a limited number of functional modules. Hence, to accommodate more modules, additional racks are needed. Racks are of two types: main and extension or supplementary. Although they look similar, they are structurally different; the difference between them is covered in later sections.

Generally, the first few slots are reserved for specific functional modules such as the power supply, processor, memory, watchdog, and communication. In some designs, the placement of functional modules is flexible. For example, the power supply module is placed in the center of the rack to minimize voltage drops (better power distribution) on both sides.

The general structure of the rack is illustrated in Fig. 8.3.

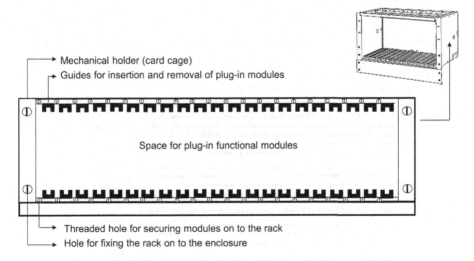

Mechanical holder (card cage)
Guides for insertion and removal of plug-in modules

Space for plug-in functional modules

Threaded hole for securing modules on to the rack
Hole for fixing the rack on to the enclosure

Figure 8.3 Rack.

8.2.2 Bus

The bus is also a passive electronic assembly used to supply power to the functional modules and to provide a communication path between the processor module and the other functional modules for data exchange.

Fig. 8.4 illustrates a typical bus.

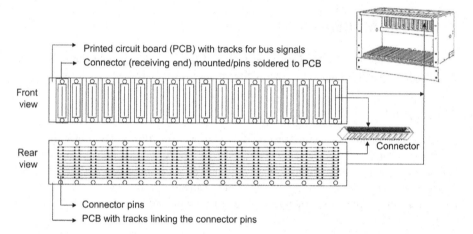

Figure 8.4 Bus: physical structure.

The bus considered here in its simplest form has lines or tracks to carry:

- Power to all functional modules (power lines)
- Addresses of memory and functional modules (address lines)
- Data in both directions (data lines)
- Read and write control (control lines)
- Interrupt and clock (special lines)

The number of lines or tracks on the bus in each category depends on the architecture of the processor (microprocessor/computer) employed in the DACU and the design of the processor module.

The logical structure of the bus is illustrated in Fig. 8.5.

Figure 8.5 Bus: logical structure.

8.3   Functional Modules

Functional modules perform specific functions and have bus connectivity for data exchange with the processor module. Typical functional modules are:

- Power supply
- Processor
- Memory
- I/O
- Communication
- Watchdog
- Bus driver/adapter

All functional modules are active modules. Fig. 8.6 illustrates a typical structure of the functional module.

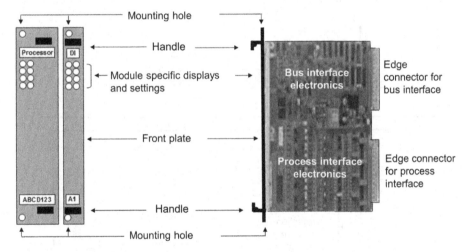

Figure 8.6 Functional module.

The top portion of the functional module has an electronic interface to the bus that enables the modules to:

- Communicate with the processor for data exchange
- Store module-related data temporarily, such as:
  - Data received from the process (via the instrumentation–human interface) before passing them on to the processor
  - Data received from the processor before passing them on to the process (via the instrumentation–human interface)
- Control and supervise the operation of the module

The bottom portion of the module has process interface electronics and facilitates linking of the module with the process signals (via instrumentation devices) and human interface signals either to convert incoming electronics signals to data or to convert outgoing data to electronic signals.

More discussion about the bus and process interfacing can be found in Appendix B.

The following sections describe the construction and function of various subsystems of the DACU and their associated functional modules.

8.3.1 Power Supply Subsystem

The DACU needs power from an external source to work. The power supply subsystem provides this through the power supply module.

8.3.1.1 Power Supply

Because they are electronic, functional modules of the DACU need power at specific voltages (+5 and +24 V DC) to work. To facilitate this, the power supply module takes external AC or DC power (230/110 or 24/48 V DC typically), converts it internally to regulated +5 and +24 V DC (required to operate DACU electronics), and supplies to all functional modules on the bus.

Fig. 8.7 is a functional schematic of a power supply module.

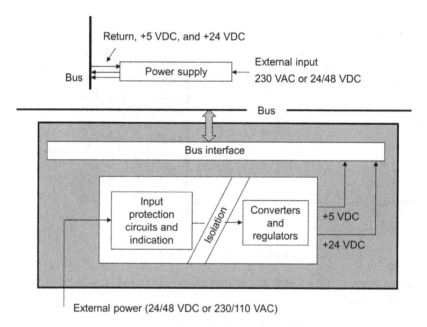

Figure 8.7 Power supply module.

8.3.2 Processor Subsystem

The processor subsystem has three modules: processor, memory, and watchdog.

8.3.2.1 Processor Module

The processor module is the heart of the DACU; it is built around a microprocessor and is responsible for executing memory resident control instructions, which are based on the automation strategy. This module is the master of bus communication and operates on other functional modules for data acquisition and control.

Fig. 8.8 is a functional schematic of the processor module.

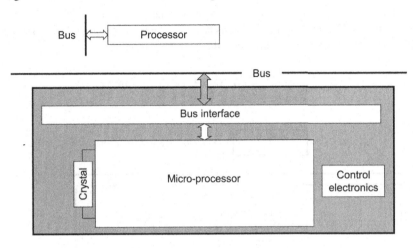

Figure 8.8 Processor module.

The processor module has a built-in clock through which the processor derives all necessary clocking and timing functions to synchronize its own activities, including keeping track of the time of the day. The processor also uses clock signals to synchronize the data exchange activities externally on the bus with the other functional modules placed on the bus. The bus interfacing electronics are located on the top portion of the processor module.

8.3.2.2 Memory Module

The memory module has two parts: a nonvolatile memory (read-only memory, or similar) for storing the system and automation programs and a volatile memory (random access memory) for processing and intermediate operational data storage.

Fig. 8.9 is a functional schematic of the memory module.

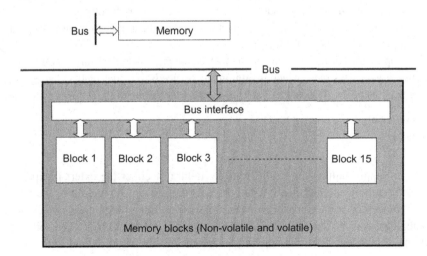

Figure 8.9 Memory module.

8.3.2.3   Watchdog Module

The watchdog or supervision module monitors the health of the DACU and makes an announcement if the DACU malfunctions (for both fatal and nonfatal failures).

Fig. 8.10 is a functional schematic of the watchdog module.

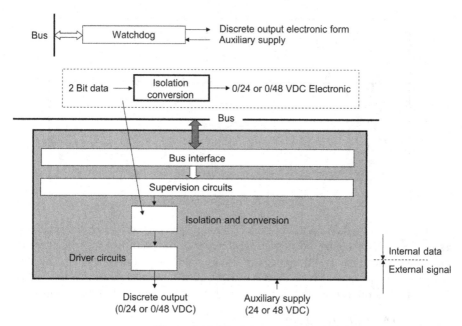

Figure 8.10 Watchdog module.

As long as the processor is healthy and functional, the DACU executes diagnostic programs in the background (whenever it is free from executing automation functions). It detects and announces any malfunctioning of functional modules. These faults, if any, are **nonfatal** in nature, because the DACU is healthy except for that faulty module. The processor, which is healthy, makes the announcement through the watchdog module's audiovisual facility.

Self-supervision deals with a situation in which the processor, the heart of DACU, fails for some reason. This fault, which is **fatal** in nature, needs to be detected and announced right when it happens. Typical fatal faults result from the failure of the following modules:

- Power supply
- Processor
- Memory

Power supply failure can result from the failure of either the external supply or the power supply module itself. The processor failure can be caused by hardware failure (processor or memory) or software failure (program going into an endless loop or hanging). Any of these fatal faults totally knocks off the DACU. The following sections discuss the detection and announcement of fatal and nonfatal faults in the

DACU. The watchdog **independently** announces the fatal fault when the processor is nonfunctional (owing to either its own failure or power failure).

8.3.2.3.1   Diagnostic Error Annunciation

The processor periodically executes diagnostic programs to check the health of all of its functional modules. Whenever a fault is noticed in any functional module, the processor announces it through audiovisual output. Diagnostic (nonfatal) error annunciation is illustrated in Fig. 8.11.

8.3.2.3.2   Power Failure Detection and Annunciation

Power to the DACU becomes unavailable owing to the failure of either the external power supply or the internal power supply module. In both cases, the effect is same: total failure of the DACU.

Fig. 8.11 illustrates the watchdog circuit or the arrangement for power failure detection and its annunciation over a common audiovisual output.

Here the central detecting element is a normally closed (NC) relay whose contact remains open as long as the relay coil receives +5V DC from the bus (meaning that the DACU is receiving the power). When the relay coil does not receive +5V DC (meaning either the external power supply has failed or the internal power supply module has failed), the relay contact drops down to close the annunciation circuit through an **auxiliary power supply**.

8.3.2.3.3   Processor Failure Detection and Annunciation

The processor can fail as a result of either hardware or software failure, and either case amounts to total DACU failure.

Here, as shown in Fig. 8.11, the central detecting element is once again an NC relay operated by a retriggerable multivibrator. The multivibrator generates an output pulse

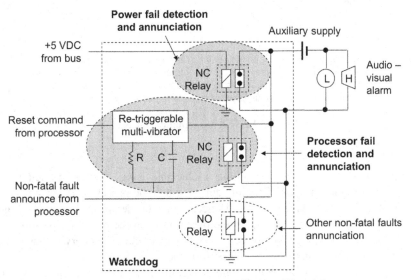

The elapse time is preset by suitable combination of R and C

Figure 8.11 Watchdog circuits.

of preset duration. This duration can be extended by retriggering the multivibrator before the preset time is elapsed, to remain high for the next duration. The duration starts when the trigger is applied. As long as the multivibrator keeps receiving the triggers periodically within the preset time durations, its output remains high and does not allow the relay contact to drop. If for any reason the multivibrator fails to receive the trigger before the end of the preset time, the relay contact drops down to drive the audiovisual output.

The processor, if healthy, periodically issues the trigger to the multivibrator. Keeping the multivibrator output high is the highest priority task the processor performs in real time. Whenever the processor fails owing to a hardware or software fault, it cannot perform this task, which leads to the output of the multivibrator becoming low, the relay contact closing, and the audiovisual alarm going high, which means the occurrence of a fatal fault.

Fig. 8.12 illustrates the timing sequence of triggering and retriggering of the multivibrator.

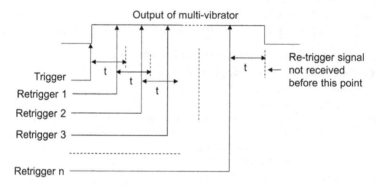

As long as the processor keeps sending re-trigger signals periodically before the elapse of preset time "t", the output of the multi-vibrator keeps its output high. If the processor fails to send re-trigger signal before the elapse of preset time "t", the multi-vibrator drops its output to low.

Figure 8.12 Watchdog timing sequence.

In case of a fatal error, the preset time is selectable using a resister–capacitor combination of the multivibrator. The time during which the processor is not supervised is limited to the preset time duration. Hence, the lower the preset time is, the higher the self-supervision is and the higher the overhead is on the processor performance. On the contrary, the higher the preset time is, the lower the self-supervision is and the lower the overhead is on the processor performance.

In this illustration, relay components are employed simply to explain the concept of the detection and annunciation of fatal and nonfatal conditions. However, the arrangement can also be with equivalent solid-state circuits for higher reliability and compactness.

8.3.3 Input–Output Subsystem

In the I/O subsystem, modules of different types interface with the DACU with the process via the instrumentation subsystem and human interface subsystem.

8.3.3.1 Digital Input and Output Modules

The digital input module acquires discrete electronic inputs from the instrumentation and human interface subsystems, converts them into a computer equivalent (data), and passes them on to the processor over the bus for further processing.

Fig. 8.13 is a functional schematic of this module.

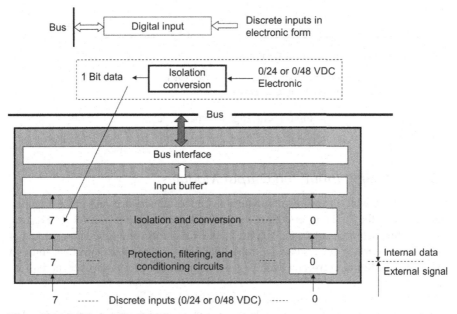

*Stores input data before transferring onto bus
Generally, a module has 8 or 16 or 32 input channels

Figure 8.13 Digital input module.

Similarly, the digital output module receives a computer equivalent (data) of the discrete outputs from the processor over the bus, converts it to its electronic equivalent of electronic output, and sends it to the instrumentation and human interface subsystems.

Fig. 8.14 is a functional schematic of this module.

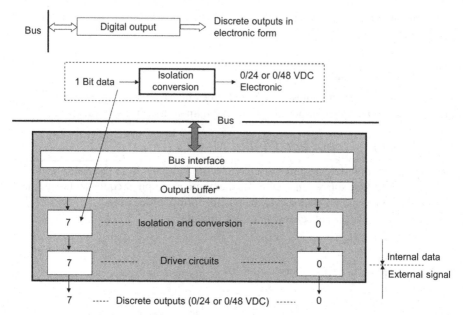

*Latches the output data before transferring into output channels
Generally, a module has 8 or 16 or 32 output channels

Figure 8.14 Digital output module.

8.3.3.2 Analog Input and Output Modules

The processor, which is a digital component, cannot understand and process analog signals unless they are converted into understandable digital equivalents. The analog input module acquires continuous inputs in electronic form from the instrumentation and human interface subsystems, converts them into a computer equivalent through an **analog to digital converter**, and passes this on to the processor over the bus for further processing.

Fig. 8.15 is a functional schematic of this module.

Similarly, the analog output module receives the computer equivalent of the analog signal in digital form from the processor over the bus, converts it into a continuous electronic equivalent through a **digital to analog converter**, and sends it to instrumentation and human interface subsystems.

Fig. 8.16 provides a functional schematic of this module.

8.3.3.3 Pulse Input and Output Modules

The pulse input module receives pulsating inputs from instrumentation and human interface subsystems in electronic form, counts them (serial to parallel conversion) into its computer equivalent (counter form), and passes the counter value to the processor over the bus for further processing.

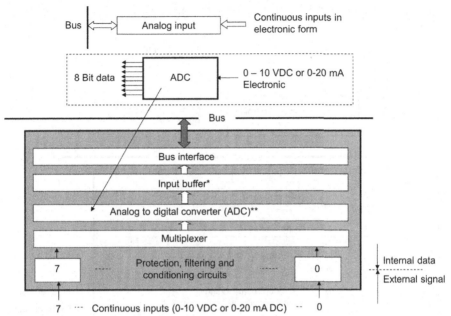

*Stores the input data before transferring onto bus
**ADC is shared by all the input channels sequentially
Generally, a module has 8 or 16 input channels

Figure 8.15 Analog input module.

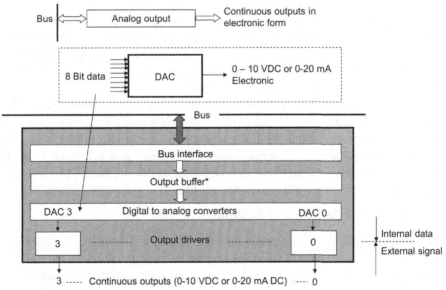

*Latches the output data before transferring to output channels
Generally, a module has 2 or 4 output channels

Figure 8.16 Analog output module.

Fig. 8.17 is a functional schematic of this module.

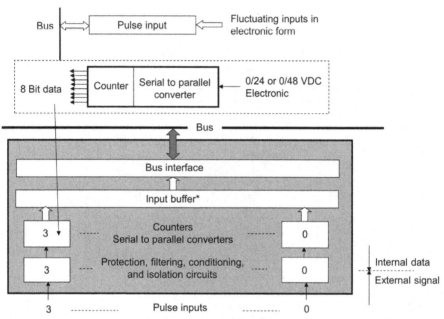

*Stores the counter value before transferring onto bus
Generally, a module has 4, 8, or 16 input channels

Figure 8.17 Pulse input module.

Similarly, the pulse output module receives the computer equivalent (counter value) of the pulses outputs (in counter form) from the processor over the bus, converts them into a pulse stream (parallel to serial conversion) in electronic form, and passes this on to the instrumentation and human interface subsystems.

Fig. 8.18 is a functional schematic of this module.

In practice, pulse inputs from the process (data acquisition) and pulse outputs to the process (control execution) are of very low frequency compared with electronic capabilities. Hence, the digital input and output modules, which are fast in response, can be employed for pulse input and output functions. Modern digital I/O modules, which were explained earlier, are designed to accept both digital and pulse I/O. For high-speed counting, special modules are commonly used in manufacturing automation.

8.3.3.4 Signal Capacity of Modules

Generally, I/O modules are designed to have the capacity of 2, 4, 8, 16, or 32 channels per module. The capacity in I/O modules depends on the physical size of the module, its component density, connector terminations, signal types (isolated or nonisolated), and so forth. A module of the same size with the same connector and without isolated (single-ended) signals can have twice the number of channels as a module with isolated (differential) signals.

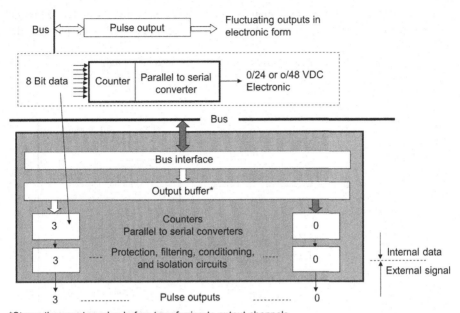

*Stores the counter value before transferring to output channels
Generally, a module has 4, 8, or 16 output channels

Figure 8.18 Pulse output module.

If instrumentation devices are powered by a common source, there is no need to isolate each signal. This also applies to signals from a human interface subsystem. This means more channels per module and less field cabling.

Fig. 8.19 shows a schematic of I/O modules with nonisolated signals.

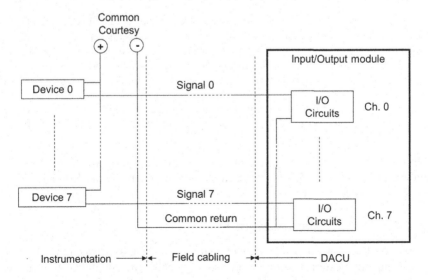

Figure 8.19 Input–output (I/O) modules with nonisolated signals.

If instrumentation devices or human interface components are powered by different sources, there is a need to isolate each input. This means fewer channels per module and more field cabling.

Fig. 8.20 shows a schematic of I/O modules with isolated signals.

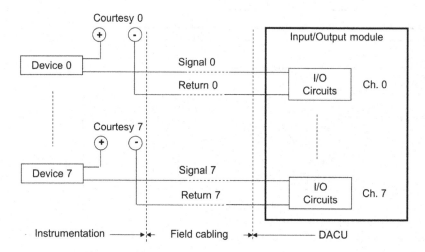

Figure 8.20 Input–output (I/O) modules with isolated signals.

8.3.4 Communication Subsystem

In an automation context, the communication subsystem has traditionally supported two types of interfaces (I/F) for data exchange with external compatible systems:

- *Serial I/F*: These are typically used for slow-speed links for data exchange in wide area networks. Generally, this is used for data transfer between a computer and peripherals such as printers. In automation, this is used to link measurement, monitoring, and communication equipment, etc. RS-232C (point-to-point) and RS 425 (point-to-multipoint) are typical examples.
- *Ethernet I/F*: These are typically used for high-speed links for data exchange in local area networks.
- Barring internal differences, these interfaces are essentially serial in the way in which data are transmitted and exchanged over communication media. However, they follow different standards and methodologies.

In the following discussion, without becoming too complex, we employ the commonly used terms **serial I/F** to denote a slow-speed interface and **Ethernet I/F** to denote a high-speed interface.

8.3.4.1 Communication Module

A communication module is for bidirectional communication between the processor and a compatible external device or system. Typically, the communication module receives the serial pulse stream from the communication media, converts it into parallel computer data (serial to parallel conversion), and passes these on to the processor over the bus for further processing. In the reverse direction, the communication module converts the parallel computer data received from the processor over the bus into a serial pulse stream (parallel to serial conversion) and passes this on to the communication media. This is similar to the combination of the pulse input and output module, except for the meaning of data.

Fig. 8.21 is a functional schematic of this module.

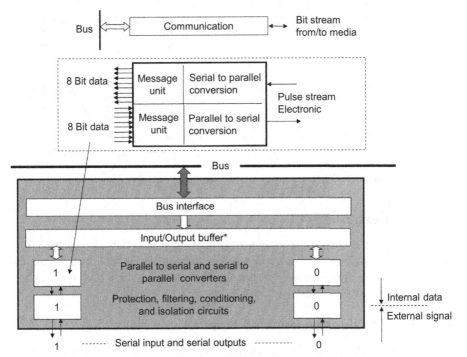

*Stores the outgoing message unit before transferring into converter and incoming message unit onto bus. Generally, a module has one or two channels

Figure 8.21 Communication module.

8.4 DACU Capacity Expansion

As stated, the basic rack has a limited number of slots to hold functional modules (mainly I/O modules). Whenever the application calls for more I/O modules, extension or supplementary racks are required to house the additional modules (capacity expansion). Supplementary racks only physically extend the bus controlled by the processor in the main rack. Bus amplifier/driver modules are used to extend the bus logically. The number of racks and functional modules depends on the ability of the processor module to handle them.

Fig. 8.22 shows a schematic of DACU capacity expansion through bus extension modules.

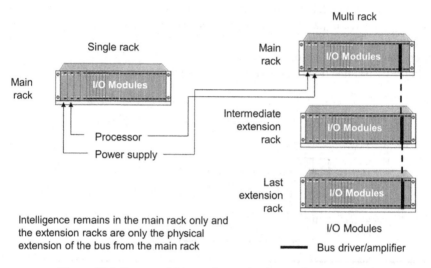

Figure 8.22 Data acquisition and control unit capacity expansion.

8.4.1 Bus Extension

Between racks, the bus can be extended either parallel or serial, as described subsequently. Extension modules are equipped with transreceivers that drive the digital signals in both the directions by strengthening and reshaping them. Two types of bus extension are possible, as discussed next.

8.4.1.1 Bus Extension (Parallel) Module

This module amplifies (repeats) all incoming signals in parallel and drives them in parallel in both directions. Apart from this, this module also drives the bus in the rack in which the module is located.

Fig. 8.23 is a schematic of the bus extension (parallel) module.

8.4.1.2 Bus Extension (Serial) Module

In the previous example, all bus signals are extended through parallel signal transmission (in both directions) from one rack to the other with a bus extension (amplifier/repeater) module. The trend now is to go for serial extension of the bus to reduce the hardware and facilitate increasing the distance between racks. Here the bus extension is through serial signal transmission in both directions (serial to parallel and parallel to serial conversions with signal amplifiers/repeaters), which is built into the bus itself. This arrangement facilitates remote placement of the extension racks, because serial signals can be driven over longer distances.

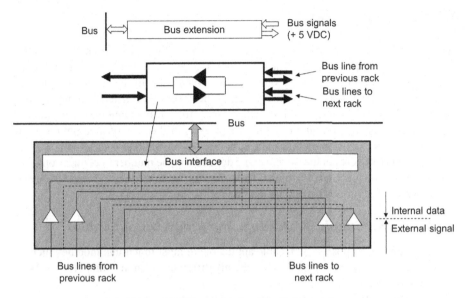

Figure 8.23 Bus extension (parallel) module.

Fig. 8.24 illustrates the bus extension (serial) module.

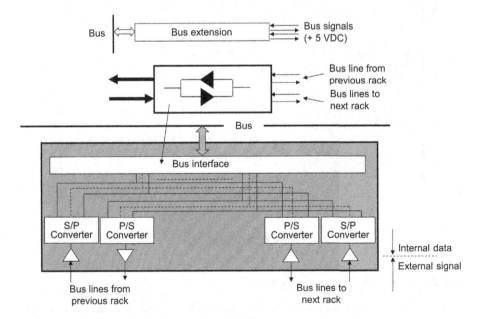

Figure 8.24 Bus extension module (serial).

The electronics associated with this module does not need to be in a functional module form and can be part of the rack itself.

8.5   System Cables

A system cable is a prefabricated cable that terminates the module signals within the enclosure for further extending. The most commonly used system cable is a multicore signal cable with a connector on one end (for connecting to the functional module mounted in a DACU) and a terminal block at the other end (for terminating the signal inside the DACU enclosure) for further wiring to instrumentation devices, human interface components, or similar devices. However, in the case of a communication module and bus extension modules, the structure or construction is different.

The number of cores in the system cable depends on the number of connections that are present on the functional modules. The following sections describe the structure of system cables for various functional modules.

8.5.1   Power Supply, Watchdog, and Input–Output Cables

This is a general-purpose system cable applicable to most functional modules such as the power supply, watchdog, and I/O. The only difference is in the number of cores in the multicore cable.

Fig. 8.25 illustrates the typical structure of the system cable for these modules.

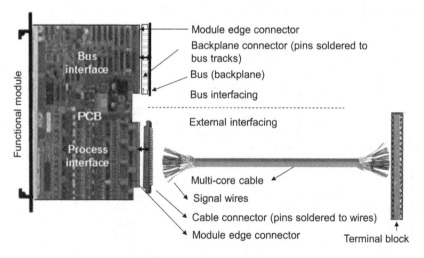

Figure 8.25 System cable.

8.5.2   Communication Cable

The system cables for a communication module employ two types of interfaces, which extend signals between the communication module and the communication equipment: RS-232C for a serial interface and Cat 5 for an Ethernet interface.

Fig. 8.26 illustrates the typical structure of system cables for two types of communication interfaces.

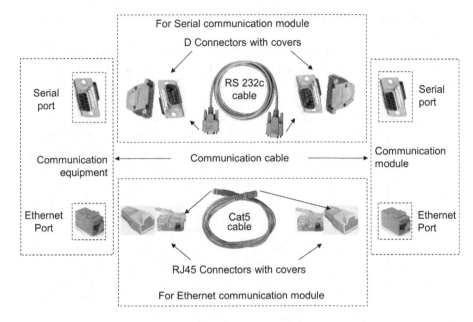

Figure 8.26 Communication cables.

8.5.3   Bus Extension Cable

Fig. 8.27 illustrates the typical structure of system cables for bus extension to extend bus signals from one rack to another.

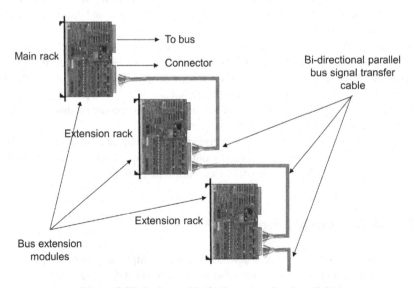

Figure 8.27 System cable for bus extension (parallel).

8.6   Integrated Assemblies

Fig. 8.28 illustrates a functional schematic of an integrated processor module (processor module physically and logically integrated with memory, watchdog, and communication modules over the internal bus).

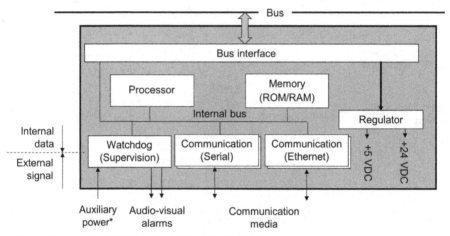

*Auxiliary power is required to drive the audio-visual alarms

Figure 8.28 Integrated processor module.

Memory, watchdog, and communication modules need fast communication with the processor, so a separate **internal bus** is provided for these modules. This arrangement saves the external bus for the exclusive use of functional modules for data exchange with the processor.

An integrated module can have more than one serial and Ethernet interface, to provide redundancy or connect DACU to more than one external system. In addition, it is common to have an exclusive serial interface to connect the DACU to a computer-based terminal to program and diagnose the DACU.

Fig. 8.29 illustrates the front view of the integrated processor module with single and dual ports for each serial and Ethernet port.

Fig. 8.30 illustrates a completely integrated DACU with instrumentation devices. This arrangement can have either an individual power supply module in each rack or a common power supply powering all racks within the cabinet.

8.7   DACU Construction

The hardware structure of a DACU discussed so far is simple and commonly followed. There are different types of hardware construction for DACUs. However, the logical structure of a DACU remains the same in all types of construction.

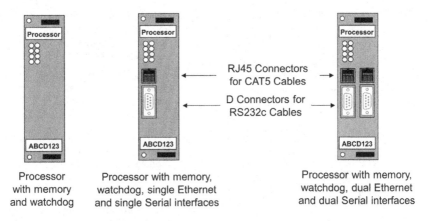

Processor with memory and watchdog

Processor with memory, watchdog, single Ethernet and single Serial interfaces

Processor with memory, watchdog, dual Ethernet and dual Serial interfaces

Figure 8.29 Integrated processor module: front view.

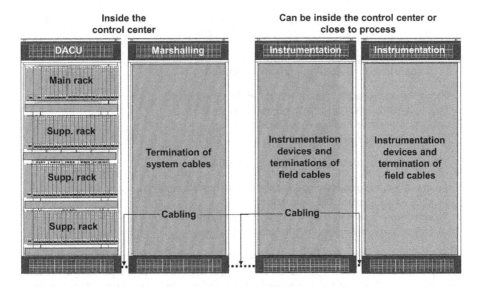

Figure 8.30 Data acquisition and control unit (DACU) integrated with instrumentation devices.

Fig. 8.31 illustrates two different types of DACU construction.

8.8   Data Exchange on Bus

To explain these basic concepts, the processor module can be seen as the master of the bus, whereas all other functional modules, such as I/O, communication, and watchdog,

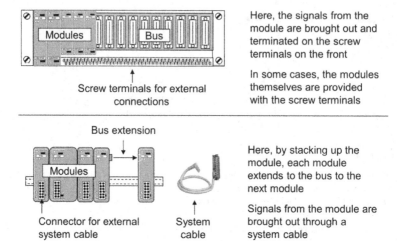

Figure 8.31 Physical variants of data acquisition and control unit.

are slaves on the bus. Data exchange on the bus is totally controlled by the processor module. There are two types of data exchange sequences.

From the slave module to the processor module:

- Processor module places the address of the slave module on the bus.
- Processor module places the read control signal on the bus.
- Addressed slave module places the data on the bus.
- Processor module accepts the data from the bus.

From the processor module to the slave module:

- Processor module places the address of the slave module on the bus.
- Processor module places the data on the bus.
- Processor module places the write control signal on the bus.
- Addressed slave module accepts the data from the bus.

Slave functional modules cannot exchange data among themselves. Data exchange among the slave functional modules, if required, is always via the processor module and is totally controlled by it. Currently there are several efficient methods for data exchange between the processor and functional modules.

8.9 Summary

In this chapter, the general hardware construction of the DACU is discussed with special emphasis on the functional modules. In modern DACUs, apart from watchdog, communication modules (both serial and Ethernet) are physically integrated with the processor module. Generally, integrated processor modules have multiple communication interfaces (typically dual Ethernet interfaces and multiple serial interfaces) to support communication redundancy and/or communication with multiple external systems, as discussed in subsequent chapters.

Modern I/O modules are designed with some intelligence (microprocessor/microcontroller based), which allows them to perform some local processing and operations to reduce the routine data transfer load on the bus and the computing load on the processor. This improves the overall performance of the DACU. These issues are discussed in detail in Chapter 16.

Data Acquisition and Control Unit: Software

9

Chapter Outline

9.1 Introduction

In Chapter 8, we discussed hardware aspects of a general-purpose data acquisition control unit (DACU). Over the years, many advances have taken place in hardware technology (mainly in terms of electronics, communication, and networking). However, as explained in Appendix C, the hardware interfaces have remained virtually the same while control has become more powerful owing to processors supporting large memory and increased speed. Apart from compactness and reliability, memory and speed constraints, which were present in earlier systems, are no longer valid. Because of this, software has overtaken hardware in performing many tasks. In fact, software has eliminated the need for certain aspects of hardware interfacing electronics, and is taking care of the complete operation and control of industrial processes. Current software-based systems are more flexible and modular, allowing for modifications if required. They also provide maximum facilities, leaving the user only to customize the automation systems for particular processes. **Software in today's automation system does almost everything** to operate, analyze, monitor, and control industrial processes. This chapter discusses the application software aspects of the DACU.

Overview of Industrial Process Automation. http://dx.doi.org/10.1016/B978-0-12-805354-6.00009-8

9.2   Software Structure

The general software structure of the DACU is like that of any other real-time system. Typical layers of software in the DACU are:

- Hardware platform
- Real-time operating system
- Utility software
- Application software

Fig. 9.1 illustrates the general arrangement of these layers.

Each lower-level layer has an interface to interact with the next highest level. Here, except for the innermost or hardware layer, all layers are software. The following sections briefly explain the functions of the layers in this software structure.

9.2.1   Hardware Platform

The hardware platform layer basically consists of hardware resources such as a processor subsystem (processor, memory, and watchdog), communication subsystem (serial and Ethernet interface), and input–output (I/O) subsystem (all types of input and output) along with its interfaces to the operating system.

9.2.2   Real-Time Operating System

A real-time operating system (RTOS) is designed for real-time operations meeting response time predictability and deterministic response. Main criteria are predictability and determinism, not speed. It is also possible to have both soft and hard real-time features in the same system. Excessive overhead in operating system software can affect response time and performance. Unlike general-purpose operating systems, RTOS adds only small overhead in microseconds whereas the response requirement in automation systems is in milliseconds or above. This leaves a good portion of computing power for the execution of non–real-time tasks. The kernel of RTOS in its basic form is memory resident software that takes responsibility for the overall management

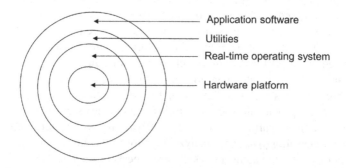

Figure 9.1 Data acquisition control unit: automation software structure.

of the real-time system by responding to time- and event-controlled tasks. The main functions are:

- *Resource management*: Sharing of resources by competing tasks as per their execution schedules. This means that tasks have the required resources allocated to them whenever they are needed;
- *Task management*: Creation of the task and its activation, running, blocking, resumption, and deactivation;
- *Task scheduling*: Scheduling multiple tasks either on a cyclic/programmed basis or on a noncyclic/preemptive basis with strict adherence to the schedule, leading to predictable or deterministic results;
- *I/O management*: Executing service tasks on either a programmed or priority basis
- *Memory management*: Allocating and deallocating the memory for tasks;
- *Intertask communication and synchronization*: Sending messages from one task to another task for their synchronization. This becomes necessary during the parallel processing of tasks, especially when execution of some tasks depends on the completion of other tasks;
- *Interrupt management*: Handling multilevel priority-based interrupts and
- *Time management*: Keeping real time and time of day, facilitating time-based execution of tasks, and monitoring elapsed time of tasks after their initiation. Timers have the highest level of interrupts and keep track of real time.

9.2.3 Utility Software

Utility software is standard and commonly used application programs developed by vendors.

9.2.4 Application Software

Application software performs automation functions (automation strategies) specifically customized for the industrial process. RTOS provides facilities with links to all resources (hardware, software, and communication). Thus the application programmer only needs to know the links to the RTOS to access all available resources.

9.2.5 Scheduling of Tasks

The DACU employs the RTOS and works with real-time task scheduling to execute automation tasks. Important real-time DACU tasks are discussed in the following sections.

9.2.5.1 Data Acquisition

This is the basic task, and the following are the most important real-time activities:

- Acquisition of changes in digital input states when they occur (e.g., the occurrence/disappearance of a process alarm or a change of event). There is a possibility of losing this information if another change takes place before the current change is acquired;
- Acquisition of the values of analog inputs with required periodicity/intervals, because their values may vary quickly (e.g., a continuous varying of a process value). There is a possibility

of missing the continuity of information if a substantial variation takes place before the current value is acquired; and

- Acquisition of the values of counters with required periodicity/intervals, because their values may get updated quickly if not acquired before the counter overflows (e.g., a continuous updating of counters with the integrated value of a process parameter).

This task is responsible to ensure that data acquisition is done within the scheduled time and no data are lost.

9.2.5.2   Data Analysis and Monitoring

Having acquired the input data, process data analysis and the monitoring task recognize the following events as early as possible to initiate the intended actions:

- Changes in the states of raw digital inputs (e.g., the occurrence of an alarm) or the derived input as a function of raw inputs (e.g., generation of a group alarm);
- Variations in the values of raw analog inputs (e.g., the occurrence of limit violations), or the derived input as a function of raw inputs (e.g., generation of composite values); and
- Changes in the raw counter values (e.g., exceeding the consumption of a process input) or the derived value as a function of raw inputs (e.g., total of several individual consumption).

This task is responsible to ensure that data analysis and monitoring is done within the scheduled time/priority.

9.2.5.3   Process Control

Having analyzed the input data, the process control step takes the following actions:

- Computing and sending variable control command (analog output) as per the required periodicity or when required by the final control element to regulate process performance (e.g., maintaining accurate flow);
- Sending discrete control command (digital output) to the final control device without delay when it becomes necessary (e.g., emergency shutdown command); and
- Sending pulsating commands (pulse output) to the final control device to regulate a process parameter (e.g., position control with a stepper motor).

This task is responsible to ensure that the issue of process commands is done at the right or scheduled time.

9.3   Application Programming

Unless programmed for specific automation functions, the DACU cannot perform the desired functions. Generally, the automation strategy is programmed by automation engineers. The first job of an automation engineer is to configure or customize the DACU for a specific process or plant. This process is called system configuration and it takes into account the number and types of process inputs and outputs required in DACU and other standard inputs. Chapter 15 discusses automation system customization in detail. The next step is to develop the specific automation program to meet the application requirements of the process. Appendix D describes the basic programming in lower-level languages (machine and assembly) and introduces higher-level language programming. This chapter explains the programming of the DACU in higher-level languages with the tools and procedures provided by DACU vendors to code the programs.

9.3.1 Higher-Level Programming

As explained in Appendix D, coding the automation strategy in assembly-level language, even though it can **exploit** the hardware features of DACU, makes the application program executable only on the specific machine or platform for which the program has been coded. Hence programming of the DACU is always done in higher-level languages so that they are easily understood by programmers and they make the program **portable** for use on other DACUs and platforms with minimum adaptation. The compiler program, which is developed by the platform vendor, converts this higher-level language program into its machine-executable equivalent and downloads it to the DACU (target machine).

The DACU is programmed for automation functions by using one or more of the following higher-level languages, which are covered by the **IEC 61131-3** standard:

- Ladder diagram (LD)
- Function block diagram (FBD)
- Structured text (ST)
- Instruction list (IL)
- Sequential function chart (SFC)

Fig. 9.2 illustrates the general structure of these higher-level automation programming languages, which are generally supported by all DACU vendors.

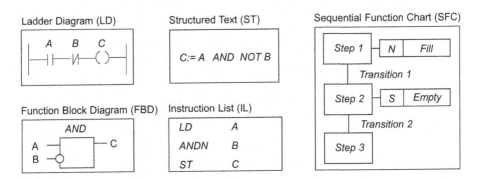

Figure 9.2 Data acquisition control unit: higher-level automation programming languages.

The IEC 61,131-3 standard allows mixing and compiling of automation programs written in all of these higher-level languages. The most popular languages used by automation engineers are LD and FBD for programming automation. They are graphic-based and are created to make program development and maintenance easier for automation and process engineers.

ST and IL are text-based languages and are similar to C/Pascal and assembly languages, respectively. Both of these languages are generally used by automation software developers. SFC is a graphic-based language with a time- and event-based interface between the DACU and the user during program development, startup, and troubleshooting.

Apart from these five, there is another additional graphical editor not yet included in the standard: **continues function chart (CFC)**. This is seen as an extension of the FBD editor. In FBD, the connections are set automatically by the operators, but in CFC they have to be drawn manually by the programmer. This also gives a free hand to the programmer because all of the boxes can be placed freely and feedback loops can be programmed without the use of interim variables.

In the following sections we discuss the features of only LD and FBD languages with examples that are commonly used in the industry.

9.3.2 Ladder Diagram

The LD originated in the graphical representation of electrical control systems using relays (relay-based logic). It is mostly used for discrete automation and is ideal for sequential control with interlocks. The specifics of LD are as follows:

- Based on the schemes/circuit diagrams of relay logic, as discussed in Appendix A
- A graphical representation of the programming elements.

The name "ladder diagram" is derived from the program's resemblance to a ladder with two vertical rails and a series of horizontal rungs between them. The rails are called "power rails" in the ladder diagram.

Fig. 9.3 illustrates a typical ladder diagram and its conventions.

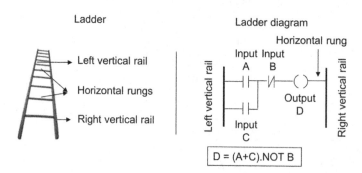

Figure 9.3 Ladder diagram: structure.

Table 9.1 illustrates the basic programming symbols in LD.
Table 9.2 illustrates some simple bit logic instructions in LD.
Table 9.3 illustrates timer and counter instructions in LD.
Table 9.4 illustrates the programming of some simple application examples in LD.

Table 9.1 Ladder diagram: basic symbols

Symbol	Functions	Operations
A ——┤├—— B	NO: normally open	Normally, O/P B is OFF. O/P B becomes on when I/P A is on.
A ——┤/├—— B	NC: normally closed	Normally, O/P B is on. O/P B becomes OFF when I/P A is on.
——(A)——	Output: multiple output can be saved in multiple parallel	Generally, this is the last instruction in the rung. Normally, O/P A is OFF. O/P A becomes on when the logic in the rung is satisfied. O/P A can also be stored as a flag in memory for subsequent use.

Table 9.2 Ladder diagram: bit logic instructions

Symbol	Functions	Operations
I/P A ——┤/├——(B)—— O/P	Inverter/NOT	O/P B is disabled (becomes OFF) when I/P A is enabled (becoming on). B = NOT A (Boolean relation)
I/P A ——┤├——┐ ——(C)—— ——┤├——┘ I/P B O/P	2 input AND	O/P C is enabled (becomes on) only if both I/P A and I/P B are enabled (becoming on). C = A·B (Boolean relation)
I/P B ——┤├——┤├——(C)—— I/P A O/P	2 input OR	O/P C is enabled (becomes on) if either I/P A or I/P B or both are enabled (becoming on) C = A + B (Boolean relation)

Table 9.3 Ladder diagram: timer/counter instructions

Symbol	Functions	Operations
A ─┤In TON├─ C B ─┤Preset	TON: on delay timer	Normally, O/P C is OFF. When I/P A is enabled (becoming on), O/P C becomes on only after the predetermined time B has elapsed.
A ─┤In TOF├─ C B ─┤Preset	TOF: off delay timer	Normally, O/P C is on. When I/P A is enabled (becoming on), O/P C becomes OFF only after the predetermined time B has elapsed.
A ─┤In O├─ D B ─┤Reset C ─┤Preset	CTU: up counter	Initially, O/P D is OFF. When each time I/P A transitions from OFF to on, the counter increments. When the counter reaches the preset value C, O/P D goes from OFF to on. I/P B resets the preset value.
A ─┤In O├─ D B ─┤Reset C ─┤Preset	CTD: Down counter	Initially, O/P D is OFF. When each time I/P A transitions from OFF to on, the counter decrements from the preset value C. When the counter becomes 0, O/P C goes from OFF to on. I/P B resets the preset value.

Table 9.4 Ladder diagram: programming of simple application examples

Scheme	Operations
Example 1: O/P latching function	
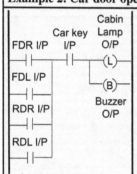	• When start I/P switch is momentarily pressed, O/P M gets enabled becomes on), sets its flag M, and gets latched (stays on) • When stop I/P switch is momentarily pressed, O/P M gets disabled (becomes OFF), resets its flag M, and gets de-latched (stays OFF). • both start and stop I/P switches are momentary types in action
Example 2: Car door open alarm generation function	
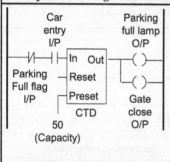	• Any one or more doors getting opened (I/Ps becoming on) and car key I/P in enabled state (on), activate cabin lamp O/P and buzzer O/P (becoming on).
Example 3: Parking full indication and closing the gate	
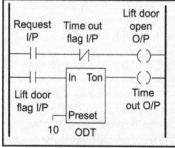	• Parking lot can hold 50 cars. The counter is preset to 50. Entry of every car decrements the counter. • As long as the number of cars in the parking lot is less than 50, the counter O/P is OFF. • Parking full flag, being NC, keeps the circuit closed and car entry input decrements the counter every time a car enters. • When the counter becomes 0, parking full O/P becomes on, gate close output also becomes on, closing the gate and making the parking full flag input open, thus opening the circuit.
Example 4: Lift door open operation	
Request I/P · Time out flag I/P · Lift door open O/P In Ton · Lift door flag I/P · Time out O/P Preset · 10 · ODT	• The time required to open the lift door upon pressing the passenger request button is 10 s. The on delay counter is programmed for 10 s. • Upon passenger registering a request, the circuit gets closed, door motor starts, and timer also starts. • Upon the timer completing 10 s, the timer holding switch opens and the circuit opens, thus stopping the motor.

9.3.2.1   Sequential Control With Interlocks

This section gives an example of programming with LD for the automation (sequential control with interlocks) of a water heater as a discrete process. Fig. 9.4 illustrates the water-heating process with instrumentation subsystems (discrete input and discrete output devices).

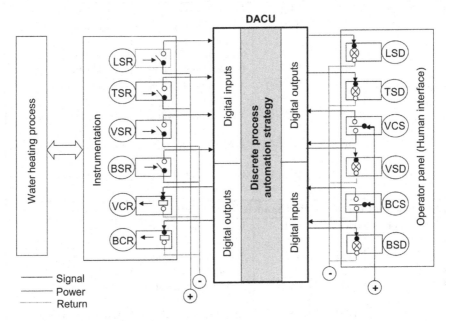

Figure 9.4 Water-heating process automation (discrete).

Table 9.5 illustrates the allocation of I/O channels to process signals.

Table 9.6 illustrates the allocation of memory locations used for specific programming purposes.

Fig. 9.5 illustrates the configuration of the DACU for automation (sequential control with interlocks) of the water-heating process.

Fig. 9.6 is a program flowchart for automation (sequential control with interlocks) of the water-heating process.

Fig. 9.7 illustrates the allocation of variables in LD programming for automation (sequential control with interlocks) of the water-heating process.

Fig. 9.8 shows the coded LD program for automation (sequential control with interlocks) of the water-heating process. The program was coded and tested using CoDe-Sys[1] software.

Today, LD supports many more programming instructions, including mathematical functions required in continuous process automation.

[1]www.3s-software.com.

Table 9.5 Ladder diagram: allocation of I/O channels to process signals (sequential control with interlocks)

From/to	I/O channel		Instrumentation device	Signal
Digital input (DI) module (eight channels: DI0–DI7)				
From process	DI0	LSR	Level switch relay	Actual level status
	DI1	TSR	Temperature switch relay	Actual temperature status
	DI2	VSR	Valve status relay	Actual valve status
	DI3	BSR	Breaker status relay	Actual breaker status
From operator panel	DI4	VCS-O	Valve control switch: open	Open valve command
	DI5	VCS-C	Valve control switch: close	Close valve command
	DI6	BCR-O	Breaker control relay: close	Close breaker command
	DI7	BCR-C	Breaker control relay: open	Open breaker
Digital output module (eight channels: DO0–DO7)				
To operator panel	DO0	LSD	Level status indication lamp	Actual level status
	DO1	TSD	Temp status indication lamp	Actual temperature status
	DO2	VSD	Valve status indication lamp	Actual valve status
	DO3	BSD	Breaker status indication lamp	Actual breaker status
To process	DO4	VCR	Valve control relay	Open/close valve
	DO5	BCR	Breaker control relay	Close/open breaker
Not used	DO6–7			

Table 9.6 Ladder diagram: allocation of memory locations for special requirements (sequential control with interlocks)

Name	Function	Purpose
FVO	Flags	To remember issue of momentary valve open command
FVC		To remember issue of momentary valve close command
FBO		To remember issue of momentary breaker open command
FBC		To remember issue of momentary breaker close command
XXX		To store intermediate results

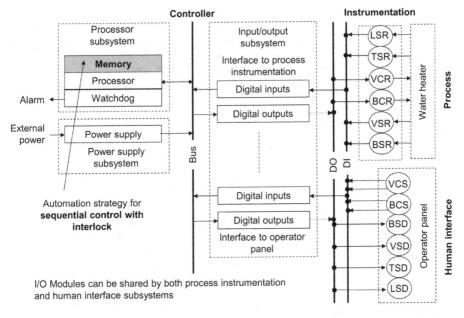

Figure 9.5 Data acquisition and control unit: configuration (discrete process automation).

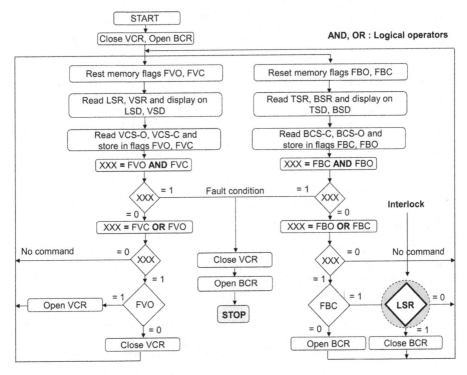

Figure 9.6 Flowchart (discrete process automation).

```
VAR_GLOBAL                              VAR_PROGRAM
    //inputs                               XXX: REAL
    LSR,TSR,VSR,BSR: BOOL;                 FVO,FVC: BOOL;
    VCS_O,VCS_C,BCS_O,BCS_C: BOOL;         FBO,FBC: BOOL;
    INIT:BOOL: = TRUE;                     XXX1,XXX2,XXX3,XXX4: BOOL;
    //outputs                           END_VAR
    LSD,TSD,VSD,BSD:BOOL;
    VCR,BCR: BOOL
END_VAR
```

Figure 9.7 Ladder diagram: allocation of variables (discrete process automation).

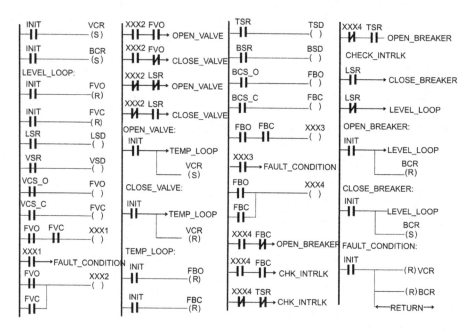

Figure 9.8 Ladder diagram: program (discrete process automation).

9.3.3   Function Block Diagram

The FBD, also a graphical representation of various mathematical and logical functions, is primarily developed for programming continuous process automation, even
though it fully supports logic operations.

Like the LD, the FBD is based on the following:

- It has graphical representations of programming elements that are modular, repeatable, and
  reusable in different parts of the program;
- The function block represents the functional relation between inputs and outputs;

- The program is constructed using function blocks that are connected together to define data exchange;
- In the programs, values flow from the inputs to the outputs through the function blocks;
- The primary concept behind FBD is data flow. The connecting lines have data types that must be compatible on both ends;
- It supports programming of binary numbers (digits 0 and 1) to deal with bit logic; and
- It supports programming of integers (single and double) and real numbers.

Fig. 9.9 illustrates a typical function block diagram and its conventions.

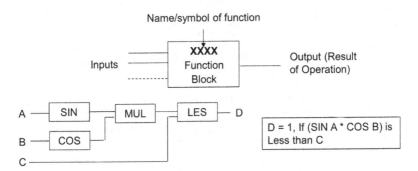

Figure 9.9 Function block diagram: structure.

The following section explains some basic and commonly used FBD instructions.

Table 9.7 illustrates bit logic instructions in FBD.

Here the bit logic instructions interpret signal states of 1 (YES) and 0 (NO) and combine them according to Boolean logic.

Table 9.8 provides compare instructions for integers (I) in FBD.

Here the inputs are integers or double integers or real (floating point), and the output is binary. If the comparison is true, the output of the function is 1 (YES). Otherwise it is 0 (NO). These instructions can are used to compare double integer and real (floating point) values as well by replacing I by D and R, respectively.

Table 9.9 lists the flip-flop instructions in FBD.

Table 9.10 gives counter [UP and down (DN)] instructions in FBD.

In an UP counter, the rising edge (change in signal state from 0 to 1) at input increments the counter and produces an output of 1 when the counter reaches the preset value. Similarly, in a DN counter, the rising edge (change in signal state from 0 to 1) at input decrements the counter from the preset value and produces an output of 1 when the counter value reaches 0.

Table 9.11 lists timer instructions in FBD.

Table 9.12 provides arithmetic instructions in FBD.

Table 9.13 shows the programming of simple application examples in FBD.

Table 9.7 Function block diagram: logic instructions

Symbol	Functions	Operations
	Insert function	
	NOT function I/P negate, O/P negate	FBD does not have a function for NOT. Inputs and outputs of function blocks can be inverted by introducing a small circle at the point of intersection.
A —[>=1]— C, B —	2 I/P OR function	C Becomes 1 when either A or B or both are 1. C = A + B
A —[&]— C, B —	2 I/P AND function	C Becomes 1 when both A and B are 1. C = A·B
A —[=]	Assign function	This is similar to O/P function in LD: produces the result of a logic operation, 1 if the conditions are satisfied, 0 if the conditions are not satisfied.

Table 9.8 Function block diagram: compare instructions

Symbol	Functions	Operations
A —[CMP ==I]— C, B —	Integer compare	If A and B are equal, then C = 1.
A —[CMP >I]— C, B —	Integer greater than	If A is greater than B, then C = 1.
A —[CMP <I]— C, B —	Integer less than	If A is less than B, then C = 1.
A —[CMP <>I]— C, B —	Integer not equal	If A is not equal to B, then C = 1.
A —[CMP >=I]— C, B —	Integer greater than or equal	If A is greater than or equal to B, then C = 1.
A —[CMP <=I]— C, B —	Integer less than or equal	If A is less than or equal to B, then C = 1.

These function blocks are for integer operands. Replacing I with D and with R refers to similar instructions for double-integer and real operands. Output C is always binary.

Table 9.9 Function block diagram: flip-flop instructions

Symbol	Functions	Operations
R — R RS S — S — C	Reset–set flip-flop	With $R=1$ and $S=0$, the flip-flop gets reset ($C=0$) With $R=0$ and $S=1$, the flip-flop gets set ($C=1$) If both $R=0$ and $S=0$, no change in C If both $R=1$ and $S=1$, set instruction dominates
S — S SR R — R — C	Set–reset flip-flop	With $S=1$ and $R=0$, the flip-flop gets set ($C=1$) With $S=0$ and $R=1$, the flip-flop gets reset ($C=0$) If both $R=0$ and $S=0$, no change in C If both $R=1$ and $S=1$, reset instruction dominates

Table 9.10 Function block diagram: counter instructions

Symbol	Functions				Operations
UP Counter CU — S — CU PV — — C R —	1	Inputs	B	CU	Increment counter
	2		B	S	Preset counter
	3		I	PV	Preset value (e.g., 0–999)
	4		B	R	Reset counter
	5	Outputs	B	C	Counter status
Counter is preset to PV with change from 0 to 1 in input S. Counter is reset to 0 with change from 0 to 1 in input R. Change in input CU from 0 to 1 increases the counter unless the counter value is already 999. Output C becomes 1 if the counter value becomes equal to PV.					
DN Counter CD — S — CD PV — — C R —	1	Inputs	B	CD	Decrease counter
	2		B	S	Preset counter
	3		I	PV	Preset value (e.g., 0–999)
	4		B	R	Reset counter
	5	Outputs	B	C	Counter status
Counter is preset to PV with change from 0 to 1 in input S. Counter is reset to 0 with change from 0 to 1 in input R. Change in input CD from 0 to 1 decreases the counter unless the counter value is already 0. Output C is 1 if the counter value becomes 0.					

B, Boolean; I, Integer.

Table 9.11 Function block diagram: timer instructions

Symbol	Functions				Operations
On Delay Timer S — ONDT TV — — C R — S ⎍ C ⎍ ⊢— TV —⊣	1 2 3 4	Inputs Outputs	B Time B B	S TV R C	Start Preset time (e.g., 0–999) Rest input Timer status

If input S changes from 0 to 1 (rising edge of start), timer starts.
If the specified time TV elapses and the state of input S is still 1, output C is 1.
If input S changes from 1 to 0, the timer stops and output C becomes 0.
If input R changes from 0 to 1 while the timer is running, the timer is restarted

Off Delay Timer S — OFDT TV — — C R — S ⎍ C ⎍ ⊢— TV —⊣	1 2 3 4	Inputs Outputs	B Time B B	S TV R C	Start Preset time (e.g., 0–999) Rest input Timer status

If input S changes from 1 to 0 (trailing edge of start), timer starts.
Output C is 1 when input S is 1 or the timer is running.
Output C becomes 0 after the elapsed time.

B, Boolean; *Time*, controller time; *TV*, timer value.

Table 9.12 Function block diagram: arithmetic instructions

Symbol	Functions	Operations
A — ADD_I B — — C	Integer add	$C = A + B$
A — SUB_I B — — C	Integer subtract	$C = A - B$
A — MUL_I B — — C	Integer multiply	$C = A \cdot B$
A — DIV_I B — — C	Integer divide	$C = A/B$

These function blocks are for integer operands. Similar instructions can be applied to double integers and real operands by replacing I by D or R. Output C is always an integer or double integer or real (as per the inputs).

Table 9.13 Function block diagram: programming of simple application examples

Example 1: Start/stop of motor

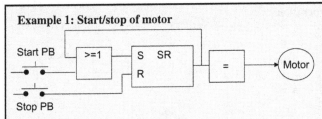

PB: Push button switch – momentary

- When start PB is pressed, input to S becomes 1. With input to R as 0, the flip-flop is set to start the motor. Feedback from FF holds the output of FF.
- When Stop PB is pressed, input to R becomes 1. With input to S as 1, the flip-flop is reset to stop the motor.
- Both of the PBs are momentary push button switches.
- SR: set Reset Flipflop

Example 2: Car cabin light/buzzer

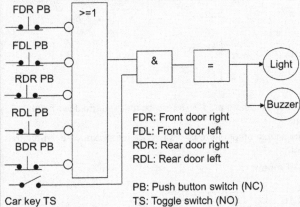

With car key TS (toggle switch) turned on and with one or more doors of FDR, FDL, RDR, RDL, PB (momentary switch) is getting opened, which activates the light and the buzzer

Example 3: Parking lot full/not full indication

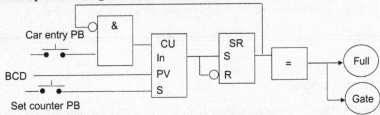

PB: Push button (NO)

When the parking lot is not full, entry of each car increments the counter. When the counter becomes full, parking lot full indication comes, gate gets closed, and further increment of the counter are blocked.

Continued

Table 9.13 Function block diagram: programming of simple application examples—cont'd

Example 4: Lift door operation

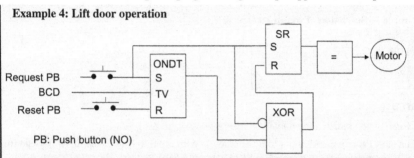

- Upon request, the flip-flop is set and the door motor and the counter start. When time is up, the flip-flop is reset and stops the motor.

Example 5: Two-step control of air conditioner

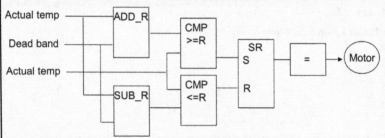

- SR flip-flop is set when the actual temperature is = (reference temperature·dead-band) starting the compressor motor.
- SR flip-flop is reset when the actual temperature is = (reference temperature − dead-band) stopping the compressor motor.

Example 6: Speed control of motor

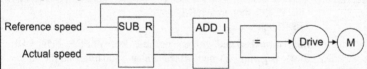

- Drive receives the control signal continuously to produce variable voltage (for DC motor) or variable frequency (for AC motor) to reduce the error between the reference and actual speeds.

9.3.3.1 Loop Control

This section illustrates a programming example with FBD for automation (loop control) of a water heater as a continuous process. Fig. 9.10 illustrates a water-heating process with its instrumentation subsystems (continuous input and continuous output devices).

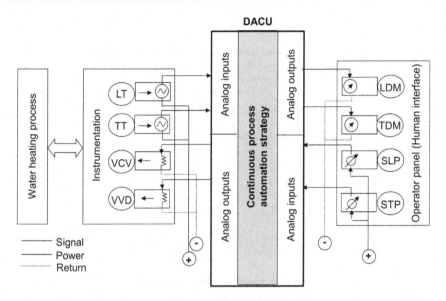

Figure 9.10 Water-heating process automation (continuous).

Fig. 9.11 illustrates the DACU configuration for the automation of a water-heating process (loop control).

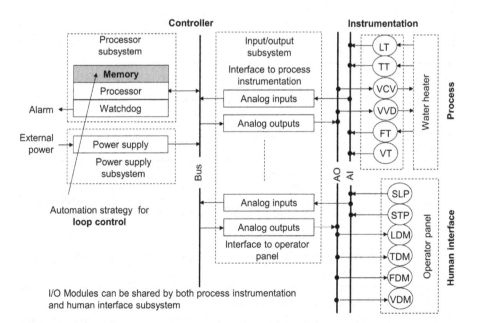

Figure 9.11 Data acquisition and control unit: configuration (continuous process automation).

Table 9.14 illustrates the allocation of I/O channels to process signals.

Table 9.14 Function block diagram: allocation of input–output (I/O) channels to process signals (continuous control)

From/to	I/O channel		Instrumentation device	Signal
Analog input (AI) module (eight channels: AI0–7)				
From process	AI0	LT	Level transmitter	Actual water level
	AI1	TT	Temperature transmitter	Actual water temperature
From operator panel	AI2	SLP	Set level pot	Set desired water l
	AI3	STP	Set temperature potentiometer	Set desired water temperature
Not used	AI4–7			
Analog output (AO) module (eight channels: AO0–7)				
To process	AO0	VCV	Variable control valve	Actual water level
	AO1	VVD	Variable voltage drive	Actual water temperature
To operator panel	AO2	LDM	Level display meter	Actual water level
	AO3	TDM	Temperature display meter	Actual water temperature
Not used	AO4–7			

Table 9.15 gives the allocation of memory locations used for specific requirements.

Table 9.15 Function block diagram: allocation of memory locations for special requirements (continuous control)

Name	Function	Purpose
XXX		To store intermediate results

Fig. 9.12 is the program flowchart for the automation (loop control) of the water-heating process.

Fig. 9.13 illustrates the allocation of variables in FBD programming for automation (loop control) of the water-heating process.

Fig. 9.14 illustrates the coded FBD program for automation (loop control) of the water-heating process.

The program was coded and tested using CoDeSys[2] software.

[2]http://www.3s-software.com.

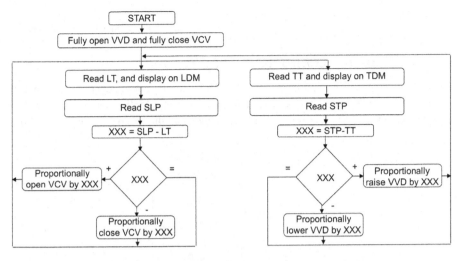

Figure 9.12 Flowchart (continuous process automation).

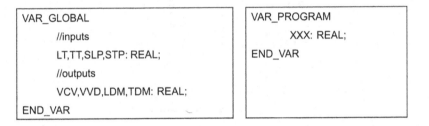

Figure 9.13 Function block diagram: allocation of variables (continuous process automation).

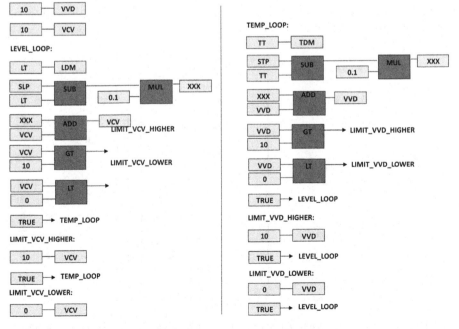

Figure 9.14 Function block diagram program (continuous process automation).

9.3.3.2   Two-Step Control With Dead-Band

This section gives a programming example with FBD for the automation (two-step control with dead-band) of a water heater as a hybrid process. Fig. 9.15 illustrates a water-heating process with instrumentation subsystems (continuous input and digital output devices).

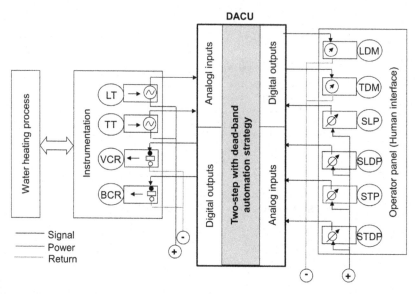

Figure 9.15 Water-heating process automation (hybrid).

Fig. 9.16 illustrates the DACU configuration for automation of a water-heating process (two-step control with dead-band).

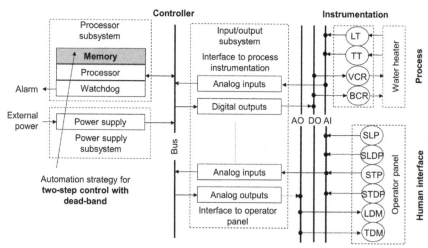

I/O Modules can be shared by both process instrumentation and human interface subsystems

Figure 9.16 Data acquisition and control unit: configuration (hybrid process automation).

Table 9.16 describes the allocation of I/O channels to process signals.

Table 9.16 Function block diagram: allocation of input–output (I/O) channels to process parameters (two-step control with dead-band)

From/to	I/O channel		Instrumentation device	Signal
Analog input (AI) module (eight channels: AI0–7)				
From	AI0	LT	Level transmitter	Actual water level
process	AI1	TT	Temperature transmitter	Actual water temperature
From	AI2	SLP	Set level potentiometer	Desired water level
operator	AI3	SLDP	Set level dead-band potentiometer	Desired water level dead-band
panel	AI4	STP	Set temperature potentiometer	Desired water temperature
	A15	STDP	Set temperature dead-band potentiometer	Desired water temperature dead-band
Not used	AI6–7			
Analog output (AO) module (four channels: AO0–3)				
To operator	AO0	LDM		Actual water level display
panel	AO1	TDM		Actual water temperature display
Not used	AO2–3			
Digital output (DO) module (eight channels: DO0–7)				
To process	DO0	VCR	Valve control relay	Open/close valve
	DO1	BCR	Breaker control relay	Close/open breaker
Not used	DO3–7			

Table 9.17 illustrates the allocation of memory locations used for specific requirements.

Table 9.17 Function block diagram: allocation of memory locations for special requirements (two-step control with dead-band)

Flags	Function
FVO	To remember issue of momentary valve open command
FVC	To remember issue of momentary valve close command
FBO	To remember issue of momentary breaker open command
FBC	To remember issue of momentary breaker close command
XXX	To store intermediate results

Fig. 9.17 illustrates the program flowchart for automation (two-step control with dead-band) of the water-heating process.

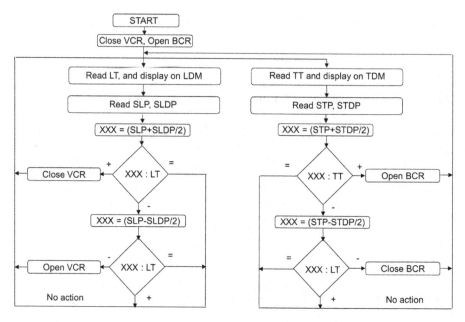

Figure 9.17 Flowchart (hybrid process automation).

Fig. 9.18 shows the allocation of variables in FBD programming for automation (two-step control with dead-band) of the water-heating process.

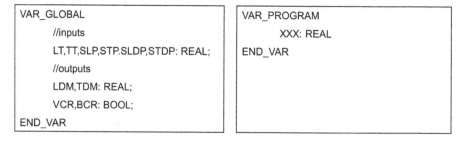

Figure 9.18 Function block diagram: allocation of variables (hybrid process automation).

Fig. 9.19 illustrates the coded FBD program for automation (two-step control with dead-band) of the water-heating process.

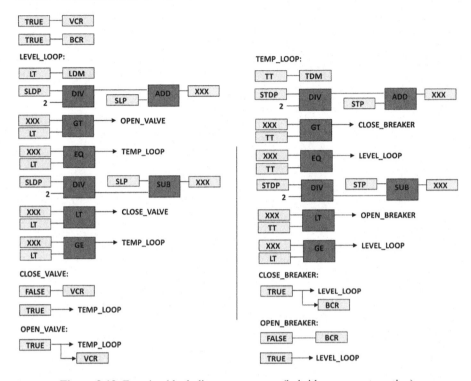

Figure 9.19 Function block diagram: program (hybrid process automation).

The program was coded and tested using CoDeSys[3] software.

9.4 Summary

In this chapter, we studied the methods for programming the DACU for customization for different applications using the IEC 61131-3 standard for programming. The entire discussion was supported by basic programming symbols and instructions in LD and FBD languages (commonly used by automation engineers) with simple application programming examples. Finally, automation programs were presented for sequential control with interlocks (discrete water-heating process), continuous control (continuous water-heating process), and two-step control with dead-band (hybrid water-heating process).

[3]http://www.3s-software.com.

Advanced Human Interface

10

Chapter Outline

10.1   Introduction

In Chapter 5, we discussed the functions of the hardware-based human interface sub-system, or operator panel. Fig. 10.1 illustrates the traditional hardwired operator panel, which is linked to data acquisition and control units (DACUs) through input–output (I/O) modules for human interface with the water-heating process.

With this approach, raw and/or derived data are displayed on lamps, meters, and counters while direct control is executed from switches and potentiometers. These active components are all mounted at appropriate locations on the passive process

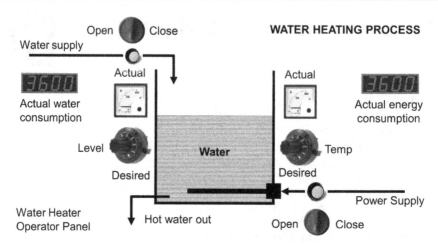

Figure 10.1 Operator panel.

diagram on the panel. Human interface functions are achieved by programming the hardwired control subsystems or DACU.

Traditional operator panels are technically simple, less expensive, still popular, and cost-effective in many small or simple applications. However they are not intelligent enough to perform anything on their own and can only send and receive **electronic signals** to and from the DACU through the I/O subsystem.

Other than sending and receiving electronic signals to and from the DACU, there is also a need to input other data into the DACU, such as date and time, so that events in the process (alarms, measurements, control actions, etc.) are recorded and stored with time tags on a real-time basis. In this chapter, the evolution, advantages, and disadvantages of advanced computer-based human interface subsystems are discussed. Advanced human interface subsystems are totally software-based and work with the latest in graphic user interface (GUI) techniques. They are called **operator stations**. The following sections discuss the evolution of human interface subsystems and transformation of the operator panel to the comprehensive operator station.

The keyboard as an inputting device has remained to date although it is slowly being replaced by touch screens. However, display technology has moved to full-color graphic display from monochrome to color and from semigraphic to full graphic.

Some common operator panels we see in our daily lives are shown in Fig. 10.2. They are also slowly being replaced by software-based human interfaces.

10.2   Operator Station

The operator station began its journey as the application of monochrome alphanumeric display and printer terminals for operator interactions with the process. These terminals were originally developed for interaction with computer systems. Although these devices differed significantly from traditional operator panels, they were

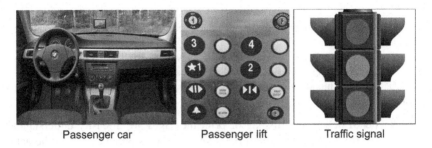

| Passenger car | Passenger lift | Traffic signal |

Figure 10.2 Examples of human interfaces.

software-based and were good enough for limited human interface functions. Apart from outputting data on the display screen or printing on paper in alphanumeric characters, they provided keyboards for inputting data into the DACU.

10.2.1 Traditional Terminals

Fig. 10.3 illustrates typical and classical display and printer terminals of earlier days that were extensively used as interactive terminals to manage computer systems.

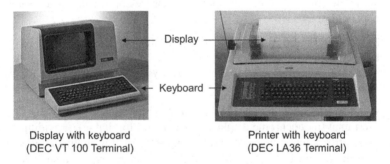

Display with keyboard   Printer with keyboard
(DEC VT 100 Terminal)   (DEC LA36 Terminal)

Figure 10.3 Classic alphanumeric terminals.

The keyboard of these terminals generated only **alphanumeric characters (ASCII coded)** and displayed or printed them using a **low-resolution 5×7 dot-matrix structure**. Display terminals typically had 24 lines with a width of 80 characters per line. Printer terminals had 80/132 characters per line. The printer terminal provided continuous printout of dialog sequences (query to and response from the DACU) whereas the display terminal rolled out the continuous display, both on a first in–first out basis. The printer terminal provided a hard copy but the display terminal did not.

The operator performed the following sequence of interactions or dialog with the DACU through the keyboard and received the alphanumeric display or print on the terminal as a response:

- Operator sends a query (inputting of a predefined text message) through the keyboard of the terminal.
- Operator receives the echoes of the query by displaying or printing it on the terminal.
- DACU accepts the query, if it is valid, and processes the query.
- DACU responds (outputting of a predefined text message) through display or printing on the terminal.

The keyboard dialog and display or print messages are in easily understandable code (in the form of text) to address the relevant process parameters and the actions required. Because it was only alphanumeric, the interactive terminal did not support graphics. These terminals were communicable but they were also known as dumb terminals because they did not have built-in intelligence and they functioned only as I/O devices (inputs from the keyboard and outputs on the display screen or printer paper) totally controlled by the program in the DACU.

These terminals were employed as a human interface subsystem in automation systems as well.

10.2.2 Intelligent Terminals

To limit the cost and overhead for simple applications, special terminals are designed on the lines of traditional terminals, but with reduced hardware and functionality. These are intelligent and communicable (microprocessor-based) and replaced traditional display terminals with:

- Limited lines and limited characters per line in the display area compared with 24 lines with 80 characters per line in a standard terminal; and
- Limited keys for operations compared with 102 keys in a standard terminal. The keys are configured for specific operations.

However, by and large, the terminal maintained the same functionality of the traditional terminal (monochrome display and keyboard for dialog), providing a simple and cost-effective solution where it was needed. Because it is intelligent, it is possible to program the terminal (display lines and the key functions) for a specific requirement. They have a monochrome display.

Fig. 10.4 illustrates the examples in the industry of intelligent terminals.

Courtesy: www.renuelectronics.com

Figure 10.4 Intelligent alphanumeric terminals.

Here, a serial interface is used not only for communication with DACU on a real-time basis but also for off-line configuration of the terminal to customize the display and the function keys (assigning functions for different keys).

10.2.3   Graphic Terminals

With advances in display technology, monochrome alphanumeric-based terminals moved first to color semigraphic and then to full-graphic terminals.

10.2.3.1   Semigraphic Terminals

Interactive terminals discussed were monochrome displays employing a dot matrix method. Subsequently, the dot matrix method with a higher number of dots was used to define and construct semigraphic symbols (drawing symbols to construct the images of various objects) apart from alphanumeric characters to print and/or display the images of process symbols. With this, semigraphic images of process elements could be constructed. To start with, this was also on a monochrome display, but later it moved to a color display. Depending on the dot matrix resolution, the display quality improved in appearance. This approach worked fairly well as far as human interaction functions were concerned.

10.2.3.2   Full-Graphic Terminals

Operator stations are continuously adapting the advanced display technologies, full-color graphics, and GUI software commonly employed in **information technology** systems. The display became closer in appearance and operation to hardware-based mimic/control panels. Finally, low-resolution semigraphic displays were replaced by high-resolution full-graphic color displays.

Fig. 10.5 summarizes the three types of displays discussed so far, for easy visualization and comparison.

5 x 7 Dot matrix alpha-numeric mono-chrome

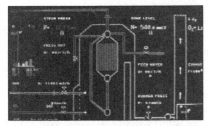
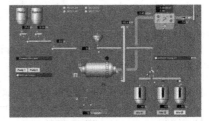

Low resolution semi-graphic colour   High resolution full-graphic colour

Figure 10.5 Different types of displays.

With full-graphic display color terminals, the mode of operator interaction with the process shifted from keyboard/coded text message display to keyboard/mouse with cursor and full-graphic display. These changes are discussed in detail in the following sections. Further discussion in this chapter is on full-graphic color-based operator stations.

10.3   Features of Operator Station

The most important and basic features of the operator station are its display screen layout and the procedure for interacting with the process for monitoring the process variables and to effect direct control of the process, all toward providing maximum convenience, effectiveness, and efficiency for the operator to manage the process. This section discusses these aspects.

To explain the concepts associated with display layout and interaction with process, let us revisit automation of the water-heating process and design a simple operator station for human interaction. Fig. 10.6 illustrates the DACU configuration for automation of water-heating process with an operator station as the human interface device.

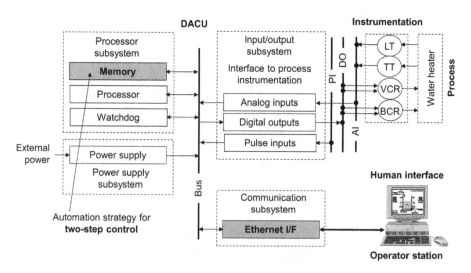

Figure 10.6 Data acquisition and control unit (DACU) configuration for two-step control.

As illustrated, the operator station is connected to the DACU over an Ethernet communication interface.

Functional requirements of the human interface that were met earlier with an operator panel are:

- Displaying valve and breaker status (digital output signals) over indication lamps
- Displaying temperature and level values (analog output signals) over meters
- Displaying power and water consumption (pulse output signals) over counters
- Setting of reference values of temperature and level (analog input signals) over potentiometers

I/O modules/signals facilitated these by interfacing the operator panel to the DACU. In this connection, readers are advised to revisit Chapter 4: Human Interface Subsystem. Now, all of these human interface functionalities need to be transferred to the operator station and eliminate I/O modules used to interface with the operator panel.

10.3.1   Display Screen Layout

Fig. 10.7 illustrates a typical display screen layout with essential features and facilities.

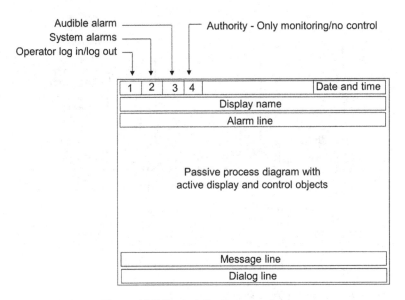

Figure 10.7 Typical display screen layout.

Explanations of various fields in the display screen are as follows:

- The operator log in–log out field displays whether the operator has logged into the operator station or logged out.
- The system alarms field displays the presence of one or more automation system alarms.
- The audible alarm field displays whether this facility is enabled or disabled.
- The authority field displays the level of authority given to the operator: full control (monitoring as well as control of process parameters) or merely monitoring.
- The date and time field displays the current date and time.
- The display name field displays the name of the specific display.
- The alarm line field displays the latest alarm.
- The passive process diagram displays the mimic of the process or process diagram, with active display and control fields/objects at appropriate places for human interaction.
- The message line field displays system messages.
- The dialog line field displays the echo of the operator dialog through the keyboard.

The display screen layout varies among systems and vendors. Modern systems provide many more facilities than are described here.

Fig. 10.8 illustrates the display screen layout based on these features for water-heating process management.

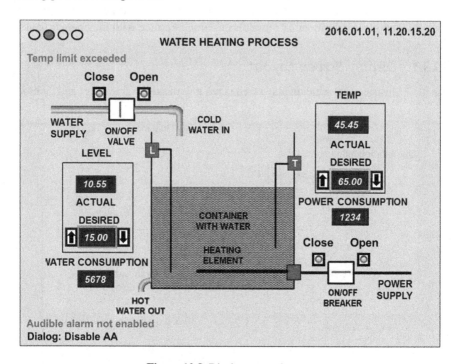

Figure 10.8 Display screen layout.

10.3.2 Interaction With Process

There are many ways to interact with the process through operator stations. Two common approaches, direct and navigated, are discussed in this section. These are applicable for both display of process parameters and control of the process.

In the water-heating process with two-step control, the following actions are taken on a continuous basis:

- The valve gets opened (to allow cold water) whenever the water level in the tank goes below the desired value; otherwise no action.
- The breaker gets opened (to disallow power) whenever the temperature of the water in the tank goes above the desired value; otherwise no action.
- The water and power consumption are measured.

The operator station is programmed for the following functions:

- Real-time display of actual and desired values of level and temperature
- Real-time display of actual consumption of water and power
- Real-time display of state (opened and closed) of the valve and breaker
- Setting (increase or decrease) and display of the desired values of level and temperature
- Direct control (open or close) of the valve and breaker as and when required

10.3.2.1   Direct Interaction

Direct interaction, a single-step operation, with the process using the operator station is done for both display of the values/states of the process parameters and for direct control of the process parameters. The sequence of operations involved is as follows:

- *Display of process values/states*: No interaction with the process is required because all the values/states of interested parameters are continuously displayed on the screen. This is seen in Fig. 10.8 for continuous display of the states of valve/breaker, values of temperature/level, and values on water/power consumption. Temperature and level values are displayed in real numbers, water and power consumption in integers, and breaker/valve status with a bar.
- *Control of process values/states*: Move the cursor and click the control buttons provided close to the device to execute the command. This is seen in Fig. 10.8 as two buttons for close and open operations provided near the valve/breaker symbol to issue the command. Similarly, up/down arrows are provided near the temperature/level display to increase or decrease reference values.

10.3.2.2   Navigated Interaction

The simple or straightforward "direct" approach is not always possible when more data need to be displayed and/or more process parameters need to be controlled with many options. This leads to overcrowding on the display area. To overcome this, a navigated interaction with the process can be employed, which is how computer systems work.

The sequence of operations adopted to control the object in a multilevel navigation, applicable for both display and control, is as follows:

- Select the desired symbol of the process parameter to be interacted by positioning the cursor on the symbol on the display screen.
- Right-click on the selected symbol to get the main menu.
- Left-click on the main menu to get secondary menu.
- Position the cursor on the selected secondary menu and left-click on the selected menu to take the next action.
- Continue until the final action, and so forth.

When this sequence of operations is complete, the intended operation is executed.

Fig. 10.9 illustrates single-, two-, and multilevel navigation steps for both display and control.

As illustrated, the navigation process can address a multiple-parameter display associated with an analog device.

Fig. 10.10 illustrates two-level interactive steps for control in the water-heating process example.

The sequence of operations to close the valve in the water-heating process is as follows:

Before interaction:

- The "bar" on the valve symbol is horizontal (valve is open)

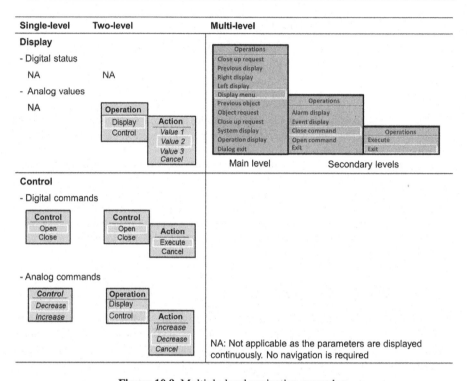

Figure 10.9 Multiple-level navigation examples.

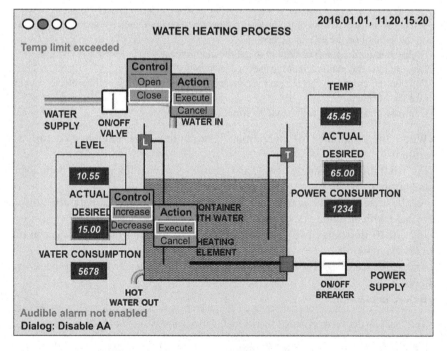

Figure 10.10 Two-level interaction with process.

During interaction:

- Position the cursor on the valve symbol on the display screen.
- Right-click on the valve symbol for main options (Close or Open).
- Move the cursor on the "Close" option.
- Right-click on the "Close" option for level 2 options (Execute or Cancel).
- Move the cursor on "Execute" option.
- Left-click to take the final action.

After interaction:

- The "bar" in the valve symbol changes from horizontal to vertical (valve closed) after the receipt of back-indication from the process.

Similarly, the sequence of operations to increase the limit by one step (5 feet) of the level is as follows:

Before interaction:

- "15.00" is displayed as the current reference value on the screen

During interaction:

- Position the cursor on "15.00".
- Right-click on "15.00" for main options (Increase or Decrease).
- Move the cursor on "Increase" option.
- Right-click on "Increase" option to get level 2 options (Execute or Cancel).
- Move the cursor on "Execute" option.
- Left-click to take the final action.

After interaction:

- The reference value "15.00" changes to "20.00" on the display.

In both direct and navigated interactions, if the operator makes a mistake (e.g., trying to change the state of objects when they are already in the intended state), the system indicates the unavailability of the option (deactivated or disabled with a faded display) and does not accept the command for further interaction. In this example, only two navigation levels are discussed. In practice, there may be many levels in the tree structure. The structure and the number of levels depend on the nature of the object and its attributes.

Fig. 10.11 illustrates an operator station display of an electrical substation with the measurements and status of the process objects continuously displayed.

In this case, a different approach is used to view the values and states of various substation parameters.

The isolators and circuit breakers (represented by I1, I2, and so on, and C1, C2, and so on) indicate their state using symbols that are either blank (device is open) or filled with color (device is closed). Similarly, values of various analog parameters (current, voltage, megawatts, megavolt amps (reactive), and tap position of transformer) are displayed next to the device. As illustrated in Fig. 10.12, graphical windows are provided for a navigation-based approach for process control.

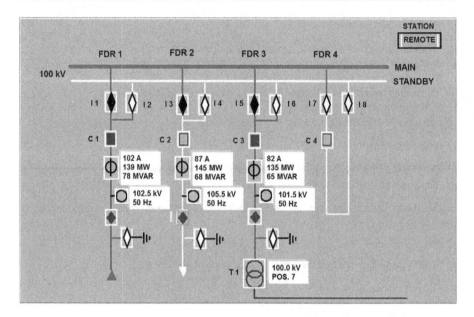

Figure 10.11 Display for electrical substation.

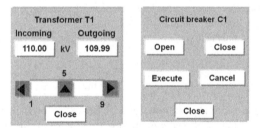

Transformer tap change   Open/Close circuit breaker

Figure 10.12 Command execution.

   Here the operator controls the devices (open/close of isolators and circuit breakers, and increase/decrease of transformer taps) through a dialog window associated with the device. By positioning the cursor on the object and right-clicking, the device is selected and a dialog box appears. To issue a command, the option within the dialog box is selected and the command is executed.

10.3.3   Other Features

Modern operator stations provide the facility to call for data related to the desired object, for viewing as well as for control. An example featuring a face plate, something similar to the face plate of equipment, is illustrated in Fig. 10.13.

   The face plate displays the status and data of the selected object and provides for the entering of data, changing of modes, etc.

Faceplate

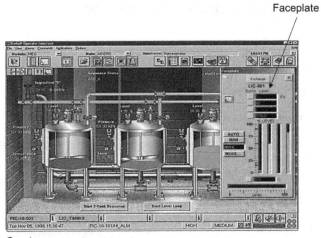

Courtesy: www.emerson.com

Figure 10.13 Face plate.

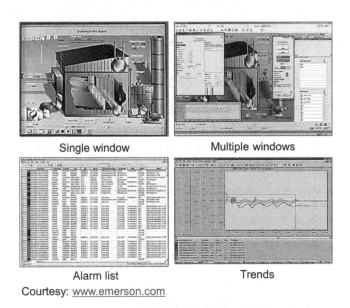

Single window   Multiple windows

Alarm list   Trends

Courtesy: www.emerson.com

Figure 10.14 Additional displays.

In addition to the facilities discussed so far, the operator station provides additional displays, as illustrated in Fig. 10.14, which are not possible in traditional mimic-based operator panels:

- Alarm/event message display
- Parameter trend display
- Multiple windows display, etc.

10.4   Comparison with Operator Panel

Table 10.1 compares the functions of the operator station with those of the traditional operator/mimic control panel.

Table 10.1 Comparison of operator/mimic panel and operator station

Functions	Operator/Mimic panel	Operator station
Appearance	Passive process diagram with active display and control components mounted at relevant locations on panel	Passive process diagram with active fields/symbols for objects placed at relevant locations on display screen
Display	On display components, such as lamps, meters, recorders, counters, etc., dedicated to process parameters on panel	Active fields near or on symbols, such as discrete and continuous objects, dedicated to process parameters on screen
Control	Commands from control components, such as switches, potentiometers, etc., mounted on panel	Commands from keyboard/mouse through control elements located on display and navigated by menus and submenus

10.4.1   Advantages of Operator Stations

Major advantages of an operator station are that it:

- Consumes less space and power;
- Eliminates I/O subsystem and associated wiring for connecting the DACU with operator panel, because the operator station is interfaced to the DACU over an Ethernet communication interface;
- Allows flexibility for easy expansion and/or modification;
- Eliminates operator movement for control execution; and
- Displays additional data, such as alarm/events, multiple windows, trends, etc.

The only disadvantage is that the operator station is that it cannot display a panoramic view of large processes in view of its limited display area.

10.5   Enhanced Operator Stations

Enhanced operator stations are employed to overcome the unavailability of the panoramic view of the process and for more displays commonly required for operation, as discussed in the following sections.

10.5.1   Multiple Monitors

In the arrangement shown in Fig. 10.15, the operator station can have more than one monitor to interact with the process, or a multiple monitor–based interaction.

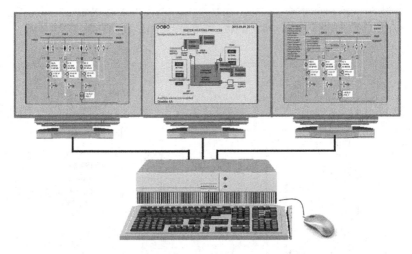

Figure 10.15 Multiple monitors with common keyboard and mouse.

For instance, one monitor may be for control and interaction with the process whereas the other monitors display information that is required most of the time. This arrangement, equipped with a special multiple-monitor graphic controller, uses a single keyboard and single mouse with a flying cursor that can jump from one screen to the other as desired.

Fig. 10.16 illustrates an example in the industry of an operator station with dual and quad monitors with a single keyboard and mouse.

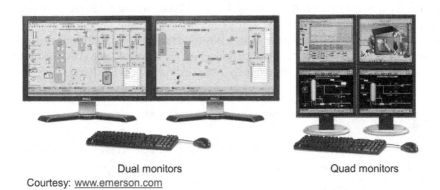

Dual monitors   Quad monitors
Courtesy: www.emerson.com

Figure 10.16 Dual and quad monitors: examples.

10.5.2 Large Screen Displays

Multiple monitors are arranged in a matrix to obtain a large screen display of the process (distributed on all of the monitors). Fig. 10.17 illustrates a display over 12 monitors (matrix of 3×4). Normally such displays are provided in the background in the control center and are not used for routine interaction with the process.

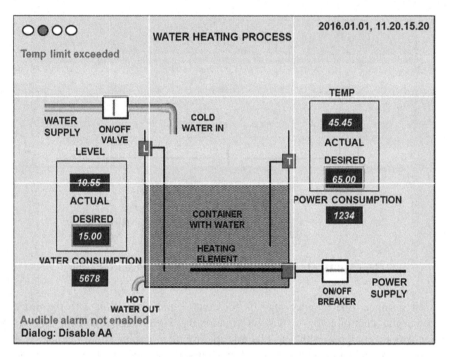

Figure 10.17 Large display with matrix of monitors.

10.5.3 Displays With Embedded Data

Current automation systems support viewing live embedded data from remote locations distributed throughout a plant, especially for surveillance for safety, security, emissions, etc., and for tracking assets, people, and so on. The data streams are integrated with the process displays on operator stations. This is possible because of the interfacing and remote controlling of networked devices installed throughout plant. An example in the industry of a display with embedded video, via remotely controlled cameras, is illustrated in Fig. 10.18.

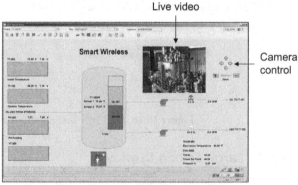

Tilt-zoom remote camera
Courtesy: www.emerson.com

Display with integrated live video

Figure 10.18 Display with embedded video.

10.5.4   Combination of Operator Panel and Operator Station

In many installations, operators need mimic panels in addition to operator stations to have a panoramic view of the process. As illustrated in Fig. 10.19, a mimic panel with only a few important active display elements is used for the panoramic view whereas normal operator stations are used for regular interaction with the process.

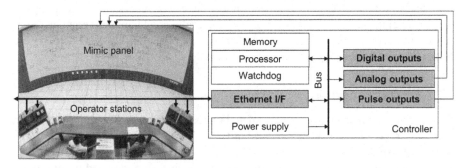

Figure 10.19 Combination of mimic panel and operator station.

10.6   Engineering and Maintenance Stations

Automation systems are designed as **platforms** with multiple features and facilities to cover a wide spectrum of applications. They need to be tailored (configured) and aligned with the process to meet the specific requirement of the process during design, installation, and commissioning. Major activities are:

- Selection, configuring, and engineering of both hardware/software and their alignment with the process
- Application software (process and other graphics, automation strategy, communication, etc.) development

Also, similar activities such as modifications and reconfiguration of the system may be required to meet changed situations during and/or after installation/commissioning.

Operator stations are designed only for interaction (monitoring and control) with the process and cannot be used for engineering activities. Hence, personnel associated with the engineering of automation systems need a different arrangement and tool to interact with the system to carry out their jobs. For this, **engineering stations** are devised as a variant of operator stations. Engineering stations are functionally and structurally similar to operator stations but with different type of displays and interaction procedures that are specific to engineering activities.

Similarly, during and after installation/commissioning, troubleshooting of faults and maintenance of hardware and software are required to keep the system up to date and running. For this, **maintenance stations** are devised as tools to carry out these activities. This is also a variant of the operator station with displays and interactions specific to maintenance activities.

Both engineering and maintenance stations share a common local area network and process data along with the operator stations. However, they generally are not located in the control room. In smaller installations, an operator station can be time shared for engineering and maintenance activities.

10.7 Logging Stations

Another important feature of the modern human interface subsystem is a logging station. A logging station is basically a printer (generally combined with graphics capabilities to support the printing of text, displays, graphs, etc.). Logging stations are generally used for data logging of different kinds.

10.7.1 Data Logging

Data logging means dumping process data in a required format. The dump can be on a printer or any other media, such as a hard disk. The following are examples of different types of data logging.

Alarm/event logging: Here, the automation system is programmed to print out a line of data with associated details whenever an alarm or an event takes place in the process/system. Operator dialog with the system is included as an event. Each logged line typically has several data fields, such as the:

- Date
- Time (up to 1-millisecond resolution)
- Type of event
- Category of the event
- Area of the event
- Name of the parameter
- State of the event, etc.

Fig. 10.20 illustrates a typical log.

Trend logging: Here, the automation system is programmed to print out continuous plots of process parameters over short intervals for close observation of the selected process parameters.

Report logging: This includes hourly, daily, and weekly reports on selected process variables (including the computed variables).

On-demand logging: The operator can request a log either on display or in print of the current display, trend plot, etc.

Date	Time	Type	Category	Area	Parameter	State
2010/01/28	20:12:19:300	Event	Process	Boiler	Temp 15	High
2010/01/28	20:12:20:300	Alarm	Process	Turbine	Vibration	High
2010/01/28	20:12:22:300	Command	User	Substation	Circuit breaker	Closed

Figure 10.20 Alarm/event log.

For logging, printers are either connected to an individual operator station (dedicated) or placed on the local area network for sharing by all operator stations. This is illustrated in Fig. 10.21.

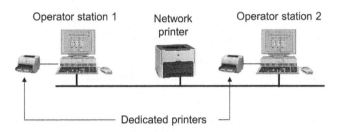

Figure 10.21 Logging stations.

10.8   Control Desk

The control desk, a set of furniture, is a physical arrangement of operator stations and other equipment, as illustrated in Fig. 10.22.

This arrangement takes care of all **ergonomic** requirements so that operators are able to interact with the process without fatigue. The control desk houses other equipment such as telephones and printers.

Courtesy: www.emerson.com

Figure 10.22 Control desk.

10.9   Modern Control Center

Integrating all of the facilities of human interfaces, control centers are established for centralized management. In control centers, whenever there is a disturbance in the process (occurrence of alarms, failures, etc.), the automation system generates an audio alarm prompting the operator to examine the details on display screens and take appropriate and immediate action. Moreover, modern control centers are designed **ergonomically** to make the control center environment fit for operator comfort, because operators have to work for long periods with repetitive activities. Modern control centers are generally, but not solely, equipped with the following:

- Operator stations for manual interaction with the process
- Telephone facilities for interacting with the plant and outside personnel
- Control desks to house operator stations and telephone equipment
- Large screen displays for real-time display of vital process information
- Printers, plotters, etc., for taking hard copy of information

Control centers are part of human interface and SCDA functionality. Fig. 10.23 illustrates examples in the industry of a modern control center.

Courtesy: www.emorson.com

Figure 10.23 Modern control center.

10.10 Summary

In this chapter, we discussed the operator station, a modern human interface subsystem based on GUI technology. Although the operator station does not offer a panoramic view of the large process, which may still be required in some cases, it provides many other benefits to compensate for this particular shortcoming. Operator stations may also have a large screen display, but the display functions remain the same. Human interfaces also include a logging facility (including storing of historical data for postanalysis) and generation and retrieval of historical data. Finally, the chapter presents control desks for housing operator stations and the modern control center for convenient, efficient, and effective operator interaction with the process.

Types of Automation Systems

11

Chapter Outline

11.1   Introduction

No one type of automation system can meet the requirements of different types of processes effectively because different types of processes have different needs. In Chapter 1, we discussed the following two types of physical processes:

- Localized processes (present over a small physical area), and
- Distributed processes (present over a large physical/geographical area).

Fig. 11.1 illustrates a typical electricity distribution process (localized as well as distributed) in a municipal area.

The overall structure of systems required for the automation of the two different types of processes differs significantly in nature. In the following sections, automation systems specially devised for these process types are discussed.

11.2   Localized Process

The localized process is present in a relatively small physical area, and the control center is physically close to the process. Fig. 11.2 illustrates the structure (logical and physical) of the automation system for the localized process.

Overview of Industrial Process Automation. http://dx.doi.org/10.1016/B978-0-12-805354-6.00011-6

Localised process   **Geographically distributed process**

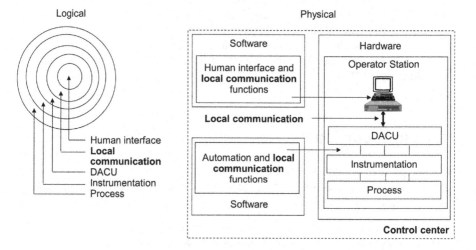

Figure 11.1 Power distribution system.

Figure 11.2 Automation system structure: localized process.

Here, we see that the data acquisition control unit (DACU) is locally connected to the operator station over the local communication line through the communication interface. Normally, an Ethernet interface is employed for communication between the DACU and the operator station. The automation systems for the localized process can be either centralized or decentralized/distributed, as explained in the following sections.

11.2.1 Centralized Control System

A centralized control system (CCS) always employs a single communicable DACU that takes the full automation load of the entire process. However, the system can have either a single operator station or multiple operator stations. In the case of CCS with multiple operator stations, multiple operators simultaneously share the load of interacting with the process. This sharing improves operator efficiency, especially when the process is large. The sharing is generally either on a functional basis or on an area basis.

Let us use the example of a power plant automation system. A power plant typically has three major functional subprocesses (boiler, turbine/generation, and auxiliaries) that produce electricity from coal and water, as illustrated in Fig. 11.3.

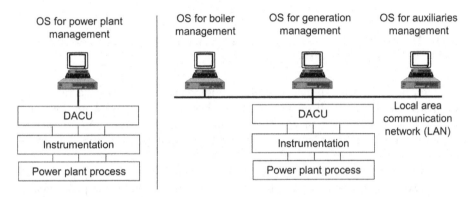

Figure 11.3 Centralized control system with single data acquisition control unit (DACU) and single/multiple operator stations.

With a single large DACU, the entire plant automation, with single or multiple operator stations, can be used for operator interaction.

Important advantages and disadvantages of CCS are as follows.

Advantages:

- Ideal for small and less spread-out processes because of minimum field cabling between the instrumentation/process points and the DACU
- Technically simple
- Less expensive

Disadvantages:

- Not economical for large and widely spread-out processes because of extensive field cabling between the instrumentation/process points and the DACU
- Failure of the DACU results in making the complete plant automation facility unavailable
- Handling, troubleshooting, maintenance, etc., of a large DACU are unwieldy

11.2.2 Decentralized/Distributed Control System

To overcome the problems associated with CCS for large or widely distributed processes, a network of many distributed communicable DACUs is formed. These DACUs share the full automation load of the entire process, as shown in Fig. 11.4.

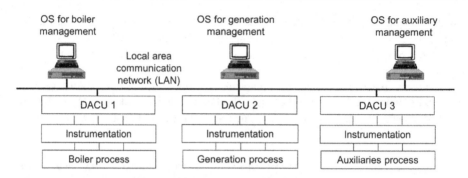

Figure 11.4 Decentralized or distributed control system with multiple data acquisition control units (DACU) and operator stations.

Here, both the DACUs and operator stations share the load on a functional or area basis. Because all of the DACUs and operator stations are on the same network, the information in one DACU can be shared by other DACUs and/or operator stations. This arrangement is called a decentralized or distributed control system (DCS).

A DCS is a local area network of DACUs and operator stations. The concept of a DCS is best explained using the example of the automation of the power plant process again with its three functional subprocesses: boiler, generation, and auxiliaries. The three smaller communicable DACUs for the three subprocesses control the complete process jointly. Important advantages and disadvantages of DCS are as follows.

Advantages:

- Ideal for large and widely spread-out processes (minimum cabling between process/instrumentation points to the distributed DACUs, because the DACUs can be placed closer to the subprocesses)
- Higher overall availability, because single DACU failure does not lead to a total automation system failure (affects only a function or a part of the plant)
- Each DACU is smaller in size, so handling, troubleshooting, maintenance, etc., are relatively easy

Disadvantages:

- Expensive
- Technically more complicated

11.3 Distributed Process

The distributed process, a group of interconnected localized subprocesses, is spread out over a relatively large physical (or even geographical) area, and the control center is physically away from the subprocesses. Fig. 11.5 illustrates the structure (logical and physical) of the automation system for one localized subprocess in a distributed process.

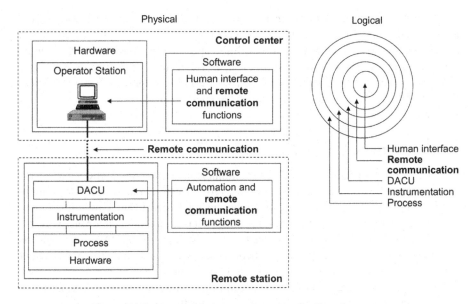

Figure 11.5 Automation system structure: distributed process.

The DACU is remotely connected to the operator station over the remote communication line through the communication interface. Normally, a serial communication interface is employed for communication between the DACU and the operator station. The automation system for the distributed processes can be a simple remote control system (RCS) or a large network control system (NCS), as explained in the following sections.

11.3.1 Remote Control System

In an RCS, as shown in Fig. 11.6, the operator station located in the control center monitors and controls the remotely located process over a communication line.

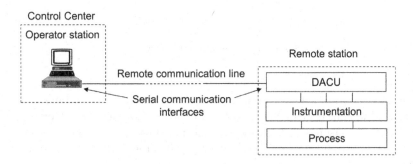

Figure 11.6 Remote control system.

The DACU, connected to the instrumentation and the process, treats the process as localized. It performs its automation functions locally without the assistance of an operator station and communicates with the operator station in the control center. The role of the operator station is only to supervise or monitor the process and issue direct control commands, if required. The RCS generally employs serial communication interfaces at both ends. The system in the control center is often called a master station.

11.3.2   Network Control System

An NCS is an extension of an RCS and simultaneously monitors and controls many geographically distributed localized processes from a central place. Multiple communicable DACUs, distributed geographically and interfaced to their own instrumentation and processes, share the common long-distance communication network for communication with the control center (operator station). Each DACU manages its own local automation independently, whereas the operator station in the control center supervises the overall automation of the entire distributed process. Furthermore, each DACU can communicate with the control center only and not with other DACUs in the network. In other words, the control center is always the master of the communication network, and any communication among DACUs is always through the control center. Fig. 11.7 illustrates the structure of an NCS.

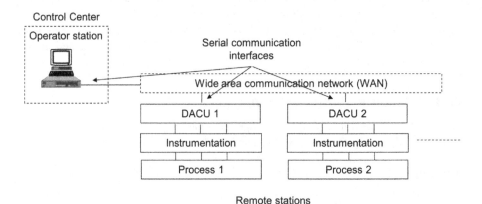

Figure 11.7 Network control system.

Furthermore, the network in an NCS refers to the network of interconnected distributed localized processes (not the communication network). For example, the electrical transmission network is a distributed process in which the localized processes are the substations, and networking is through the transmission/distribution lines.

11.3.2.1   Front-End Processor

DACUs in an NCS are loosely coupled to the control center over a wide area network (WAN). The operator station in the NCS performs two functions: human interface and remote communication. Normally, the communication links over serial interface are slow and less reliable owing to remote connectivity. Often the arrangement calls for more repetition of data exchange compared with its local communication equivalents where DACUs are tightly coupled to operator stations. Although faster and more reliable, Ethernet communication links cannot be used for long-distance remote communication.

In an NCS with many remote stations connected, the remote communication demands extensive processing time from the operator station, reducing its performance and resulting in slow response of operator interactions. To overcome this, the routine communication load (part of operator station) is off-loaded to an independent platform called a front-end processor (FEP), as illustrated in Fig. 11.8.

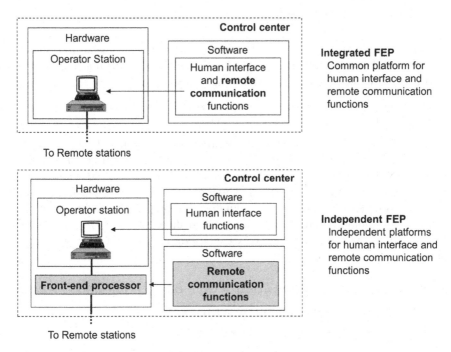

Figure 11.8 Front-end processor (FEP).

The advantage of FEPs is that they improve the performance of operator stations; disadvantages are that they are expensive and technically complex. Fig. 11.9 illustrates the implementation of an NCS for the automation of an electricity transmission network in a region employing FEPs.

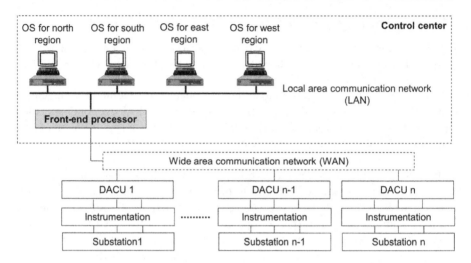

Figure 11.9 Electricity transmission network automation system.

As mentioned earlier, the physical process here is a group of geographically distributed electrical substations networked over electrical transmission/distribution lines.

A general NCS has two communication networks:

- A local area network (LAN) of operator stations and FEP within the control center
- A WAN of DACUs and FEP outside the control center

An FEP not only takes care of the routine communication load between the control center and the distributed DACUs, it also provides a buffer between high-speed LAN and low-speed WAN communication. Furthermore, depending on the processing power, an FEP can be used to preprocess the incoming data from DACUs before sending them to the operator station, and to prepare the data received from the operator station before dispatching them to the distributed DACUs. An FEP can be realized by using either the DACU platform or a general-purpose computer platform.

11.3.2.2 Data Acquisition Control Unit–based Front-End Processor

Fig. 11.10 illustrates an NCS with a DACU-based FEP.

Here, the DACU is configured with the following:

- An Ethernet communication interface for local communication between an FEP and operator stations
- A serial communication interface for remote communication between an FEP and geographically distributed DACUs
- A memory resident program for local as well as remote communication management

The advantage of the DACU-based FEP is that the system is based on a rugged hardware platform and is maintenance-free. The disadvantage is its limited

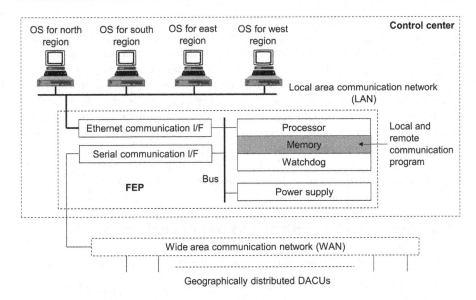

Figure 11.10 Data acquisition control unit (DACU)-based front-end processor (FEP).

memory and processing capabilities because they are specially designed for automation functions and have limited facilities for supporting communication functions.

11.3.2.3 Computer-based Front-End Processor

Fig. 11.11 illustrates a general-purpose computer-based FEP.

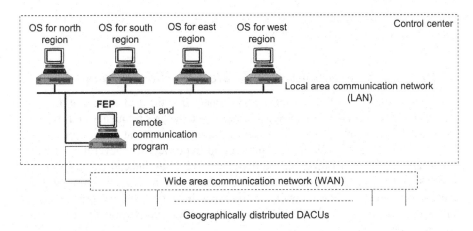

Figure 11.11 General-purpose computer-based front-end processor (FEP).

Here, the FEP is configured with a standard Ethernet communication interface of the computer for local communication with operator stations whereas the standard serial communication interface is for remote communication with geographically distributed DACUs, both with a memory resident program for local as well as remote communication in line with the DACU-based FEP. The advantage of a general-purpose computer-based FEP is that it provides an excellent platform for supporting the communication function with no limitations on memory, computing power, etc. The main disadvantages are that it is not maintenance-free and has higher overhead, which are not desirable in automation systems.

It is also possible to employ an NCS for a localized process in place of DCS, provided the functional, cost, and performance requirements are satisfied.

11.4   Supervisory Control and Data Acquisition

In the early days before the arrival of communicable DACUs, the scenario was thus:

- Only hardwired/stand-alone DACUs were employed for automation of each subprocess in a complex process.
- Each hardwired/stand-alone DACU, generally a single input/single output basis, operated independently.
- Hardwired/stand-alone DACUs, which were not intelligent and not communicable, did not support interaction among themselves for coordination.
- The computers did not enjoy the necessary confidence of the operating personnel (not meeting the acceptable level of reliability).
- Computers, as an intermediate step, were used as **supervisory** tools only to monitor, coordinate, and optimize the functions of stand-alone/hardwired DACUs, but no process control.
- The failure of the computer did not affect the process control, as the DACUs retained their set points and functioned independently of the computer.

These items are explained for the management of a group of generators in a power system to meet specific requirements.

11.4.1   Supervisory Control and Data Acquisition Example

In power system grid, all of the generators connected to the grid operate at the common grid frequency. They remain synchronized with the grid frequency and share the total load proportional to their individual capacity. Each generator takes its share of the load while complying with the common grid frequency. Whenever the consumption is less than the generation, the grid frequency increases and causes the generators to adjust automatically to the new situation while maintaining the common grid frequency. Whenever the consumption is more than the generation, the grid frequency falls and causes the generators to take the additional load proportionally at a reduced grid frequency.

As shown in Fig. 11.12, all of the generators are equipped with speed DACUs that are hardwired, noncommunicable, stand-alone, and autonomous.

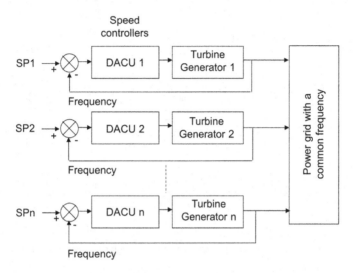

Figure 11.12 Independent operation of generators.

Each DACU, with its own set point, regulates its turbo-generator's speed/frequency. There is no coordination among the DACUs except the indirect influence by the grid frequency. Often in exceptional situations, generators are forced to take additional load beyond their share of the total.

For example during the outage of a generator, some selected generators (more efficient, less cost of production, etc.) can be forced to take the additional load by operating them at higher set points. When forced, these generators try to increase their speed/frequency. Because all of the generators have to remain synchronized with the common grid frequency, their speed/frequency cannot change. Hence, their increased frequency falls to the grid frequency by generating more power. This way, the generators can be made to take higher shares of the load but remain synchronized to the common grid frequency.

In this configuration, the supervisory computer, based on predefined criteria, computes the best set point (operating condition) for each generator to share the load and remain synchronized with the grid. Failure of the supervisory computer does not affect individual speed DACUs, because the existing set points remain unchanged. This way, the supervisory system coordinates the operation of all the speed DACUs. Supervisory control and data acquisition (SCADA) principles come in handy for the management of this situation, as illustrated in Fig. 11.13.

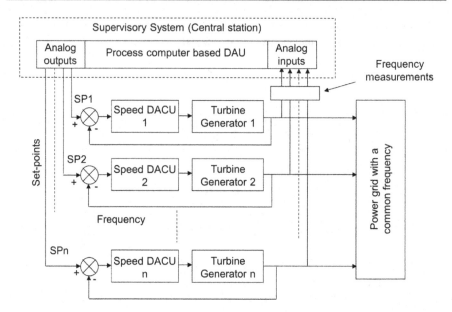

Figure 11.13 Supervisory control and data acquisition–based operation of generators.

11.4.2 Supervisory Control and Data Acquisition in Distributed and Network Control Systems

Originally, the SCADA concept was conceived to manage geographically distributed processes. Even today, an NCS, with its distributed architecture, is still referred to as an SCADA system because it resembles SCADA. This is because, whereas distributed DACUs perform the local automation functions independently, the control center (operator stations) just supervises the functioning of the distributed process and performs human interface functions. The failure of the control center and/or the remote communication network does not affect the automation functions of individual distributed remote stations. Similarly, this logic can be extended to today's DCSs.

SCADA is not a technology but an application. Today, the term "SCADA" is treated as generic because all of its features and functionalities are available in both DCS and NCS, as explained in the following sections.

Earlier, DACU itself was holding real-time process data and was making them available as and when required to other clients such as operator stations, etc. In today's automation systems, a supervisory system is called a **server** and holds all of the real-time process data for distribution to clients and all important/common functions. Furthermore, SCADA functionalities cover operator station functions as well, because the latter is meant for supervision.

Fig. 11.14 illustrates the supervisory computer in a DCS structure with common functions, coordinating the functions of the DACUs at the control center level. At the power plant level, it coordinates the functions of a boiler, generator, and auxiliary subprocesses.

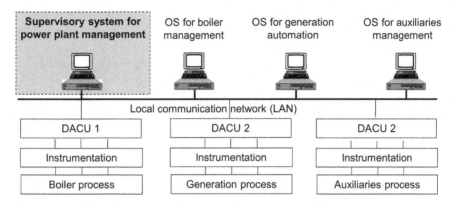

Figure 11.14 Supervisory control and data acquisition in distributed control system.

Fig. 11.15 illustrates the supervisory computer in a NCS structure with common functions, coordinating the functions of the distributed DACUs at the control center level. At the power system level, it coordinates the functions of all substations.

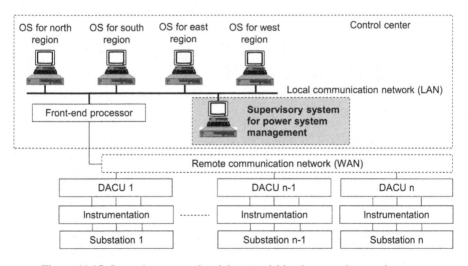

Figure 11.15 Supervisory control and data acquisition in network control system.

11.5 Summary

In this chapter, we discussed different types of automation systems for both localized and distributed processes. Centralized systems and decentralized systems were described for localized processes, RCSs, and NCSs for geographically distributed processes. Finally, SCADA functionality was discussed in both DCSs and NCSs.

Special-Purpose Data Acquisition and Control Units

12

Chapter Outline

12.1 Introduction

We discussed types of automation systems in Chapter 11. In Chapter 7, we covered the general-purpose data acquisition and control units (DACU), its application, and customization for the automation of various processes (discrete, continuous, and combination). In this chapter, we further discuss the standardization of DACUs to address the specific needs of localized/centralized and distributed/remote processes. Localized and distributed processes need functionally different and diversified types of DACUs (different hardware and/or software configurations of general-purpose DACUs) although they appear physically similar. Different processes require different types of automation strategies. Generally, discrete processes require sequential control with interlocks, whereas continuous processes require closed loop control. A hybrid process requires both. *As explained in the following sections, all special-purpose DACUs are communicable and can coexist with others over a compatible communication network.*

12.2 Localized/Centralized Processes

The automation structure of a localized/centralized process is illustrated in Fig. 12.1.

Overview of Industrial Process Automation. http://dx.doi.org/10.1016/B978-0-12-805354-6.00012-8

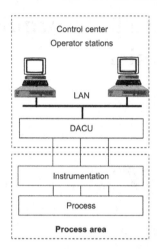

Figure 12.1 Automation system for localized process.

The emphasis here is on local automation tasks. Following are variations on general-purpose DACUs customized for localized processes:

- Programmable logic controller (PLC)
- Loop controller
- General-purpose controller (controller)

The following sections explain in detail the differences and application areas of these special-purpose DACUs.

12.2.1 Programmable Logic Controller

PLCs were developed as standalone replacements for traditional relay logic (or its solid-state equivalent, as explained in Appendix A) for discrete process automation. In principle, this is a DACU configured with only digital inputs and outputs capable of executing only logic functions to support sequential control with interlocks. The main performance requirement here is the **speed of execution** of the automation strategy (control logic). The hardware and software are specifically designed to address this requirement.

The current PLC is designed with high-performance processors, which also optionally supports analog data handling and exploit the available/built-in computing power of the processor. All general features of the PLC, including its name, have remained unchanged over time, and the term "PLC" has become generic. Today's PLC supports limited continuous process automation while according highest priority and speed to its basic functions (logic processing for discrete process automation). Also, the modern PLC is communicable and can become part of a network of automation equipment.

Fig. 12.2 illustrates examples in the industry of PLCs with extendable input–output (I/O).

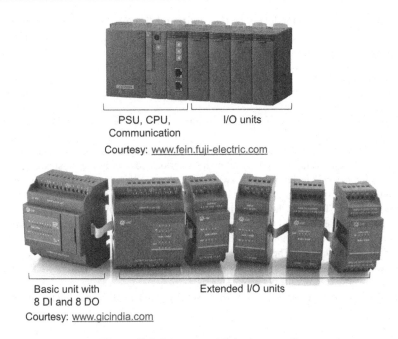

PSU, CPU,   I/O units
Communication
Courtesy: www.fein.fuji-electric.com

Basic unit with   Extended I/O units
8 DI and 8 DO
Courtesy: www.gicindia.com

Figure 12.2 Programmable logic controller.

Fig. 12.3 shows a simple PLC-based water level indicator and controller for application in domestic buildings and apartments.

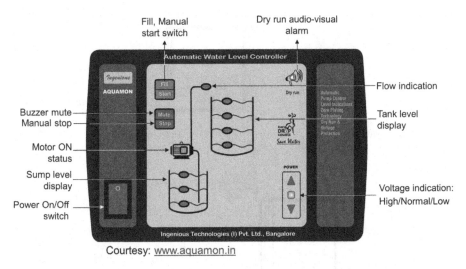

Courtesy: www.aquamon.in

Figure 12.3 Programmable logic controller–based water level indicator and controller.

Depending on the water levels in the sump and the overhead tank, a continuous supply of water is ensured to residents automatically, subject to safety conditions.

Functions of the equipment are as follows:

- *Display on operator panel*: Flow indication, level indications in sump and overhead tank, supply voltage status indications (high, low, and normal), and motor status indication (manual, auto, and dry run)
- *Direct control on operator panel*: Auto/manual selection, buzzer enable/disable (mute) selection, and fill operation
- *Interlock conditions*: The motor starts only when the water level in the overhead tank goes below the lowest level, provided the following interlock conditions are satisfied:
- *General interlocks*: The auto/manual switch is in auto mode, supply voltage to the motor is within the limits or normal, and water level in the sump is above the lowest level.
- *Sump interlock*: To avoid the possibility of the motor switching on and off continuously (hunting) until the overhead tank is above a low level, the motor is made to start only once the water reaches the second level in the sump.
- *Overhead tank interlock*: The motor can also start if the sump reaches the full level and the tank level is below the high level (100%). The motor does not start again for this condition until the overhead tank falls below 75%.
- *Actions*: The motor automatically stops whenever any of the following conditions are detected:
    - Overhead tank full
    - Sump empty
    - Motor dry run
    - Abnormal voltage
    - Both overhead tank and sump are empty

This is an example of sequential control with interlocks implemented with PLC. Implementation of the application of this product is illustrated in Fig. 12.4.

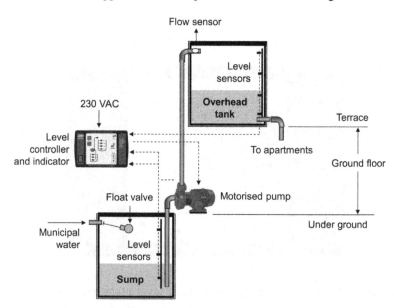

Figure 12.4 Water level indicator and controller setup.

12.2.2 Loop Controller

This DACU is configured with analog inputs and analog outputs as a replacement to the earlier standalone solid-state loop controllers to support continuous process automation (closed loop control). Unlike the traditional PLC, which can handle only logic functions, the loop controller is designed to handle arithmetic and mathematical functions in the context of continuous control. An important performance requirement of the loop controller is its ability to handle analog data with **higher accuracy**. Because analog process parameters vary relatively slowly in real time, speed of execution is not an important criterion. Hence hardware and software are specifically designed to address the accuracy requirement.

Because they are processor-based, modern loop controllers also provide a few (but limited) digital I/Os that are functionally related to the basic analog I/O. Apart from their general applications, loop controllers are also used as **backup systems** for critical loops in continuous process automation in case of the failure of the distributed control system (DCS) (main/integrated control system) in a process plant. Modern loop controllers are communicable, have their own built-in human interfaces, can be a part of a larger network of automation equipment, and can support advanced automation strategies such as proportional, integral, derivative control.

Fig. 12.5 shows examples of a single loop controller with an integrated operator interface.

Courtesy: www.yokogawa.com   Courtesy: www.fein.fuji-electric.com

Figure 12.5 Single loop controllers.

Fig. 12.6 shows the simple application of a single loop controller regulating liquid flow in accordance with the set-point value.

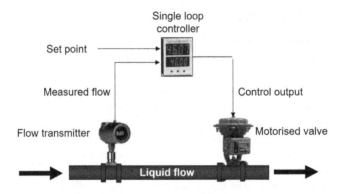

Figure 12.6 Liquid flow regulation.

The controller receives analog inputs (continuous actual flow from the flow transmitter and desired flow or set-point from the operator panel), and generates a continuous control signal to operate the motorized valve to regulate the liquid flow (output following the reference value or reducing the deviation). The industry also provides for dual-loop and multiple-loop controllers for specific applications.

12.2.3 Controller

The general-purpose controller, or **controller**, is technically a functional combination of a PLC and multiple-loop controller that implements both discrete control (sequential control with interlocks) and continuous control [multiple-input/multiple-output (MIMO) control] in process automation. In continuous processes, although the main requirement is continuous process automation, there are also discrete subprocesses that need automation. In view of this, the controller supports both digital and analog I/O. **However, this is a general-purpose controller with more emphasis on continuous process automation (MIMO control) and higher accuracy in handling analog data. Speed is not an important requirement here.**

Fig. 12.7 illustrates examples in the industry of a controller.

Figure 12.7 Controller.

Fig. 12.8 shows an application of controllers in a DCS.

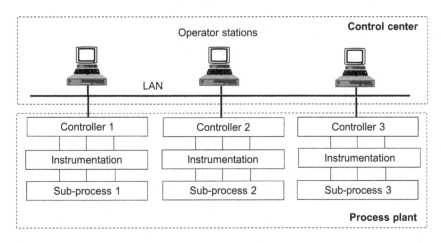

Figure 12.8 Controllers in a distributed control system.

Fig. 12.9 illustrates the application of controllers in a DCS for power plant automation.

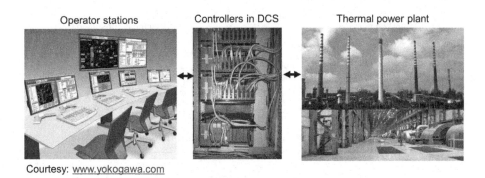

Courtesy: www.yokogawa.com

Figure 12.9 Application of controllers in a distributed control system (DCS) in power plant automation.

Here all of the subprocesses in the power plant are managed by a set of controllers and operator stations on a local area network (LAN).

12.3   Remote/Distributed Processes

The automation structure of remote processes is illustrated in Fig. 12.10.

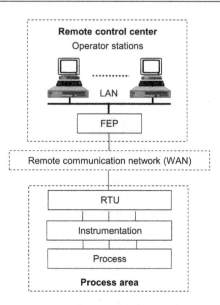

Figure 12.10 Remote control system.

There was a need to for manage remote processes from a central location. Hardwired systems were employed for this purpose. The schematic shown in Fig. 12.11 is applicable to an electric substation.

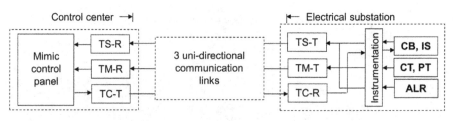

TS: Tele-signaling, TM: Tele-metering, TC-Tele-command, T: Transmitter, R: Receiver

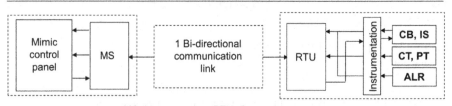

MS: Master station, RTU: Remote terminal unit

CB: Circuit breaker, IS, Isolator, CT: Current transformer, PT: Potential transformer, ALR: Alarms

Figure 12.11 Approaches to remote monitoring and control.

Data acquisition from a remote station was called telemetering (analog measurands such as current, and voltage) and telesignaling (digital indications such as switchgear status). Control of the remote station was called telecontrol (digital devices such as switchgears). Here the term "tele" means performing an operation from a distance. This can be seen in words such as "telephone," "television," "teleprinter," and so on. Separate/independent unidirectional systems were used for remote data acquisition and control.

The control center, known as the master station, had facilities to display the status of switchgears (on–off) on indication lamps and values of electrical parameters (voltage, current, etc.) on meters. Control switches were provided on the mimic panel to execute commands (open–close) of switchgears in the other direction. This arrangement required three communication links between each remote station and the master station for telemetering, telesignaling, and telecontrol, as shown in Fig. 12.11.

Each link had a transmitter at one end and a receiver at the other, performing the following operations.

From the remote station to master station:

- Acquisition of values of continuous parameters locally from the substation and transmission of these over the communication link to the master station for display on the mimic panel
- Acquisition of states of the discrete parameters locally from the process and transmission of these over the communication link for display on the mimic control panel

From the master station to the remote station:

- Sending control commands issued through control switches on the mimic panel for discrete parameters of the process over the communication link and passing these on locally to the process

12.3.1 Remote Terminal Unit

The telemetering, telesignaling, and telecontrol equipment were hardware-based and did not support built-in intelligence or capabilities. These were subsequently replaced by a communicable DACU. The name for this special-purpose DACU is **remote terminal unit (RTU)**. The emphasis here is on data communication and exchange between the RTU and the master station; traditionally it designed to perform telemetering, telesignaling, and telecontrol functions.

Early RTUs, which were designed with low computing power, mainly supported remote data acquisition and transfer to the master station in one direction and receipt of control command from the master station and transfer to process. Modern RTUs, built with powerful microprocessors, support intelligence functions such as local execution of automation functions. Fig. 12.12 illustrates the appearance of typical examples in the industry of RTUs.

Courtesy: www.siemens.com   Courtesy: www.yokogawa.com

Figure 12.12 Remote terminal unit.

Fig. 12.13 shows a typical application of an RTU for remote monitoring and control of an electrical substation.

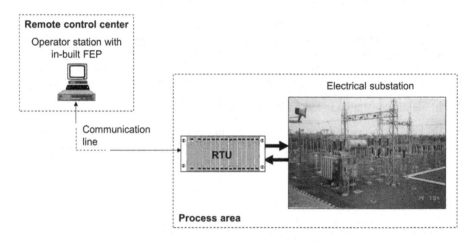

Figure 12.13 Remote management of electrical substation.

Remote data acquisition and control concepts originated in electrical transmission and distribution systems because these processes are large and geographically distributed, and there was a need to manage the distributed process from a central location. To start with, power lines were employed for communication in the form of power line carrier communication (PLCC). The communication medium for data was the same as that for speech and protection in electrical transmission and distribution companies. The application later migrated to an NCS-based approach. Currently, the NCS approach is also applied to other geographically distributed processes, such as oil/gas transmission systems and water transmission and distribution, as illustrated in Fig. 12.14.

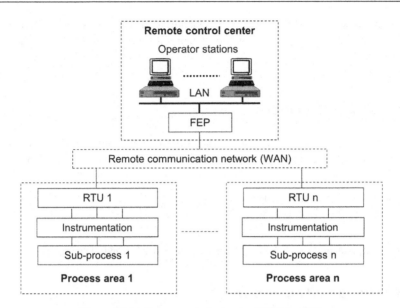

Figure 12.14 Remote terminal units (RTU) in NCS.

Fig. 12.15 illustrates the application of RTUS in hierarchical NCS for power system management.

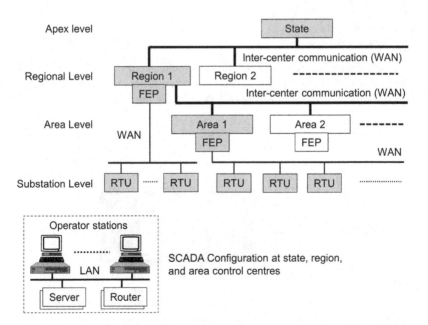

Figure 12.15 Hierarchical NCS configuration for power system management.

The special-purpose DACUs discussed so far are based on process computers. However, there are other players and variants in the category of special-purpose DACUs that compete with the PLCs and controllers. These are discussed briefly in the following sections.

12.4   Other Variants

Some other players and variants in the category of special-purpose DACUs, specialized and optimized for different requirements are:

- PC-based controllers
- Programmable automation controllers
- FPGA/FPAA-based controllers
- Embedded controllers

These special-purpose DACUs are discussed in the following sections.

12.4.1   PC-Based Controller

The combination of ever-declining prices, the increasing reliability of commercial personal computers (PCs), and the increasing reliability of operating systems has led to **soft-logic** control applications resulting in the development of PC-based controllers. The benefits of PC-based controller are excellent for:

- Cost-effectiveness
- Alternatives for many applications
- Networking capabilities
- Programming environment

PC-based controllers traditionally link an adapter card on a PC to I/O modules, and they work with customized applications written for control and communication. Fig. 12.16 illustrates the architecture of a PC-based controller.

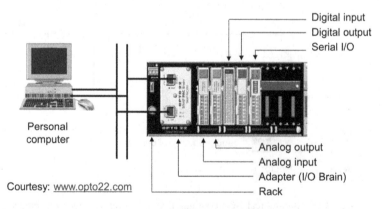

Figure 12.16 PC-based controller.

The adapter shown in the figure is the **logical interface** between the PC and the I/O modules over a dual Ethernet interface (see Chapter 14). The communication link can also be a serial interface. Fig. 12.17 illustrates various functional components associated with the PC-based controller.

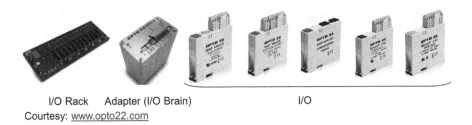

I/O Rack   Adapter (I/O Brain)   I/O
Courtesy: www.opto22.com

Figure 12.17 Components of PC-based controller.

12.4.2 Programmable Automation Controller

A programmable automation controller (PAC) combines the features and capabilities of a PC-based controller and PLC. Functionally, the PAC and PLC serve the same purpose, because they are primarily used to perform automation tasks. Compared with PLCs, PACs are highly open and modular in their architecture. Hence, hardware selection can be made from the open market with no worry about compatibility.

The following are some important features of PACs compared with PLCs:

- Provide the best of both the PLC and PC worlds
- Programmable using more generic software tools, whereas PLCs are programmed in ladder logic diagrams, function block diagrams, etc.
- Combine high performance and I/O capabilities of PLCs with the flexible configuration and integration strengths of PCs
- Have a common hardware platform for wide-ranging purposes

Fig. 12.18 illustrates an example in the industry of a PAC.

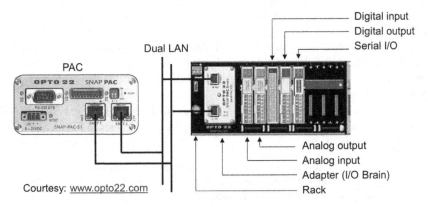

Figure 12.18 Programmable automation controller.

Here, the PC in the PC-based controller is replaced by a PAC. Also, the adapter (I/O brain) is the logical interface between the processor and the I/O modules. Fig. 12.19 illustrates various functional components associated with a PAC.

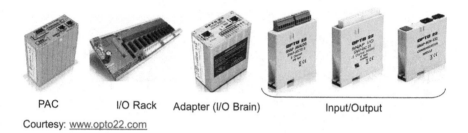

PAC   I/O Rack   Adapter (I/O Brain)   Input/Output

Courtesy: www.opto22.com

Figure 12.19 Components of programmable automation controller (PAC).

In relation to the industry examples previously discussed, the adapter (I/O brain) in a PC-based controller or in a PAC does the following:

- Serves as the interface between a PC/PAC and its I/O
- Manages I/O modules for data acquisition and control
- Can work over an Ethernet interface (I/F) or serial I/F (with or without redundancy) for data exchange

The major difference between a PC-based controller and a PAC is that the PAC employs a **hardened industrial PC** (specially designed PC with high reliability to work with harsh industrial environments).

12.4.3 Field-Programmable Gate Array/Field-Programmable Analog Array-based Controller

As a natural extension of integrated circuit (IC) technology, field programmable gate arrays (FPGAs) were designed to realize discrete process control and field programmable analog arrays (FPAAs) for continuous process control, or a combination hybrid control. FPGA contained configurable digital blocks whereas FPAA contained configurable analog blocks. FPGAs and FPAAs are reprogrammable silicon chips. Using prebuilt function blocks and programmable routing resources, one can configure these chips to implement custom hardware functionality without adding up individual circuits.

The programmability of FPGA/FPAA-based controllers fills the gap between hardwired and soft-wired (processor-based) technologies. Advantages of FPGA and FPAA technology are that they:

- Do not have a fixed hardware structure like microprocessors or microcontrollers
- Have high performance, take less time to implement and market, are cost-effective, have high reliability, and have long-term maintenance

- Are flexible toward modifications/extensions unlike fixed hardware for fixed application and are easily upgradeable or do not call for a completely new system
- Have faster execution owing to lesser overhead in applications in which real-time response is considered the main feature
- Have reusable, reprogrammable elements or lower nonrecurring engineering cost for redesigning
- Are good for prototyping and economical for small volume before going in for the soft-wired version

Although technically FPGAs/FPAAs have all of the possibilities, there are a few disadvantages, such as that they:

- Are not suitable for low-cost systems as microcontrollers are cheap
- Need special skills for programming, and also the development systems are expensive

Fig. 12.20 illustrates a typical FPGA-based controller module.

FPGA Chip

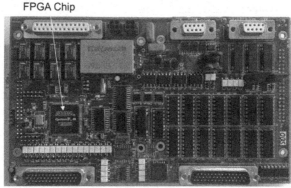

Courtesy: www.sunluxtech.com

Figure 12.20 Typical field-programmable gate array (FPGA)-based controller module.

12.4.4 Embedded Controller

An embedded system is frequently a computer that is implemented for a particular purpose. Contrary to general-purpose computers, embedded systems usually have only a single task or a small number of related tasks that they are programmed to perform. An embedded system can also be defined as a single-purpose computer, typically microcontroller-based.

In the automation context, an embedded system is:

- Designed to perform a special task in a most efficient way, interacting with physical elements in the real world (driving a motor, sensing a temperature, etc.)
- A combination of both hardware and software that performs a specific task
- Built to perform its duty, completely or partially independent of human intervention

Most embedded systems are time-critical applications, meaning that they work in an environment in which timing is important. The results of an operation are relevant only if they take place in a specific time frame.

Fig. 12.21 illustrates the development steps of an embedded solution.

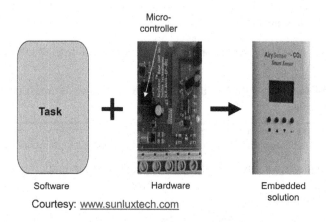

Micro-
controller

Task

Software   Hardware   Embedded
Courtesy: www.sunluxtech.com   solution

Figure 12.21 Embedded controller.

The biggest advantage is that they can be made cheaper for a particular application. They also support communication ports such as UART, USB, RS 232c, and RS 485, Wi-Fi, Bluetooth, and even Ethernet, etc. However, unlike PLCs, embedded systems are not easily field programmable. They need special skills and tools.

12.5 Summary

In this chapter, we discussed the need for special-purpose controllers, namely PLCs, loop controllers, and general-purpose controllers (controllers). Although it is technically possible to employ a general-purpose controller for all applications, its features are not required in all applications. Thus the design and development of special-purpose controllers to meet specific application areas has led to performance improvement, optimization, and cost-effectiveness. Today's PLCs, controllers, and RTUs, which do not differ much in their hardware and software structure, have all the features of general-purpose controllers. They can technically be employed to execute automation and data communication functions for all types of processes. However, their performance and cost may vary. Special points to be noted are as follows:

- PLCs are primarily designed and optimized for executing logic functions (sequential control with interlocks) in discrete process automation with emphasis on **speed.** They also support continuous process automation in a limited way.
- Loop controllers are primarily designed and optimized to execute arithmetic/mathematical functions (single and multiple-loop control) in continuous process automation with an emphasis on **accuracy.**

- Controllers, technically a combination of both PLC and multiple-loop controller functions, are more oriented and optimized for continuous process automation.
- RTUs are primarily designed and optimized for data acquisition locally from the process/ transmission to the remote control center and receipt of command from the remote control center/transfer locally to the process. They also support local automation in a limited way.

With their built-in differences, PLCs, controllers, and RTUs support all automation functions (data acquisition, data processing, control, and communication). Only the priorities of the functions are different. In practice, it is possible to employ a PLC or controller as an RTU and vice versa, provided the functional, performance, and cost requirements are met. Finally, this chapter ends with a brief discussion on PC-based controllers, programmable automation controllers, FPGA/FPAA-based controllers, and embedded system controllers.

System Availability Enhancements

Chapter Outline

13.1   Introduction

Any interconnected hardware and software in an automation system, however reliable, are susceptible to failures for a variety of reasons, and they can cause partial or total failure of the automation functions. When a functional unit fails, corresponding automation functionality becomes unavailable, affecting the overall performance of

Overview of Industrial Process Automation. http://dx.doi.org/10.1016/B978-0-12-805354-6.00013-X

the industrial process. In some industrial processes, partial or total failure of the automation system may not be a serious issue. In other cases, especially in continuous processes, the automation system needs to function continuously with the utmost reliability and availability. Functioning of the automation system can be a 24/7 job. This chapter addresses the issue of how to design an automation system with 100% availability when required.

We will discuss availability analysis followed by provisions for enhancing availability, beginning with some basics on availability enhancement through standby or redundancy schemes.

13.2   Standby Schemes

The following sections discuss various standby schemes, including the no-standby scheme.

13.2.1   No Standby

In the case of the failure of a unit [a controller in a distributed control system (DCS) or front-end processor (FEP)/remote terminal unit (RTU) in a network control system (NCS)], its functionality is unavailable for the entire duration of the following sequence:

1. Occurrence of fault
2. Noticing of fault
3. Diagnosing of fault
4. Rectification of fault
5. Reinstallation of rectified unit
6. Recommission of rectified unit
7. Resumption of operation

This sequence is a time-consuming process and the unit's functionality is unavailable until the full sequence of actions is completed. Resuming the operation as quickly as possible depends on how well one can reduce the time taken for the sequence.

In some cases in which the continuous operation of a unit is neither warranted nor demanded, one can wait for the unit to go through the previous sequence of actions. However, this calls for an efficient service setup.

13.2.2   Cold Standby

An extra unit called **cold standby**, which is physically and functionally identical to the working unit, is kept ready for immediate one-to-one replacement if the working unit fails. This way, the failed unit can be repaired without great urgency. Although it provides better availability, this approach leaves a break in the operation of the automation system, and it cannot continue from the state where the working unit had stopped. In other words, the previous data or history stored in the failed unit is not available to the cold-standby unit to continue the function. However, this arrangement is better than no standby, and it is acceptable in many situations; it is also economical and technically simple.

13.2.3   Hot Standby

The drawback to the cold-standby arrangement (the inability to continue the function from where the operation had stopped) is overcome by **hot standby**, as illustrated in Fig. 13.1.

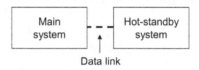

Figure 13.1 Hot-standby with data link.

This arrangement provides for a link between the main and the hot-standby unit for data exchange.

Normally, the hot-standby unit is kept ready to take over from the failed unit automatically. There are many approaches to the takeover scheme, such as the ones described:

- Both the main and hot-standby units receive the data and update their own databases. Only the designated main unit takes the control actions. Because the hot-standby unit is in continuous dialog with the main/working unit, upon recognition of any failure, the standby unit automatically takes over the control without a break.
- The hot-standby unit periodically initiates a dialog with the main/working unit, gets the latest data, and updates its own database (data shadowing). In case the main unit does not respond, the standby unit considers the main unit to be faulty and starts functioning from its latest available data. The maximum loss of data is limited to the time between the two consecutive dialogs (switchover time).

The following sections analyze availability issues in DCS and NCS and the application of the hot-standby technique to them.

One cause of a fatal failure is the failure of the power supply unit in the controller. This can be overcome by having two power supplies in parallel inside the rack, as shown in Fig. 13.2.

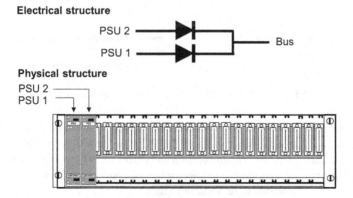

Figure 13.2 Paralleling of power supply units.

Normally, both power supplies share the load, and upon failure of one power supply unit, the other, healthy one takes the full load. Hence, in the following discussions, power supply units are not considered in the availability analysis and enhancement.

13.3   Distributed Control System

A DCS has several controllers and operator stations on its local area network (LAN). The analysis is made for one link with critical items in the data flow. This is applicable to other links as well.

13.3.1   Availability Analysis

An availability analysis in DCS can be made with the information flow from the process to the instrumentation subsystem (level 1), to the control subsystem (level 2), to the communication subsystem (level 3), and to the human interface subsystem (level 4). This is illustrated in Fig. 13.3.

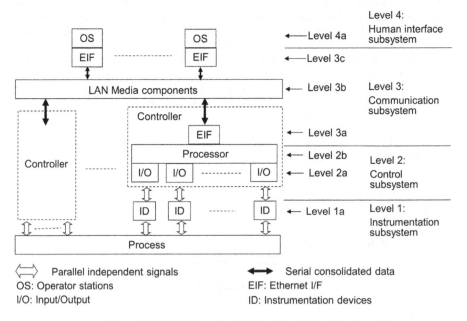

Figure 13.3 Data flow in a distributed control system.

Any failure (full or part of a subsystem) can be fatal or nonfatal. A fatal failure is one that affects the overall functioning of the DCS. In the following discussions, the failure of a controller or its communication interface, for whatever reasons, leads to the unavailability of automation functions for the subprocess to which the controller is assigned. This is considered to be a major failure.

With this information flow, the following sections explain the implications of the failure of a particular subsystem or one of its modules at various levels on the availability of automation functions for a subprocess.

13.3.1.1 Level 1: Instrumentation Subsystem

Level 1 has an instrumentation subsystem with the information flow in parallel. It is connected to parallel input/output (I/O) channels of the I/O modules in the controller. In other words, one device for one process parameter is connected to one I/O channel in one I/O module (see Section 13.5 for I/O redundancy).

13.3.1.1.1 Level 1a: Instrumentation Devices
The failure of any instrumentation device affects only the data acquisition or control function of the particular process parameter connected to the failed device. This is a nonfatal failure, because it does not totally affect the automation functions of the subprocess.

13.3.1.2 Level 2: Controllers

Level 2 has a control subsystem with two functional modules in series in the information flow, namely, I/O modules and the processor module in the controller.

13.3.1.2.1 Level 2a: Input–Output Modules
Because all I/O modules work in parallel, the failure of any I/O module affects only the data acquisition or control function of the particular process parameters connected to the failed I/O module. This can also be treated as a nonfatal failure, because it does not affect the total automation functions of the subprocess.

13.3.1.2.2 Level 2b: Processor Module
The processor module can fail owing to either a hardware failure (power supply, processor, and memory) or a software failure. Processor module failure affects the total controller. This is a fatal failure, because it totally affects the automation functions of the subprocess leading to the unavailability of the following functional subsystems (cascading effect):

- Operator stations dedicated to the failed controller, although they may be healthy
- Other operator stations in the DCS that need interaction with the failed controller
- Other controllers in the DCS that need interaction with the failed controller

13.3.1.3 Level 3: Local Area Network

Level 3 has a communication subsystem with three functional modules in series in the information flow, namely the Ethernet interface (I/F) in the controller, local communication network, and Ethernet I/F in the operator station.

13.3.1.3.1 Level 3a: Ethernet Interface in Controller
With the failure of its Ethernet I/F, the controller cannot communicate with the subsystems at the same level (other controllers) and at the higher level (operator stations). This makes the operator interaction totally unavailable (similar to the failure of the processor module).

However, with a healthy processor module and I/O modules in the controller as well as healthy instrumentation devices, execution of the automation functions of the assigned subprocess is not affected, except where the controller depends on the operator stations and other controllers. Hence, this failure can be treated as nonfatal, because it does not affect the automation functions of the subprocess.

13.3.1.3.2   Level 3b: Local Area Network Media Components
Because the LAN media components (cables, connectors, and hub) are passive, they generally do not contribute to failures of the network. Therefore, this is not considered in the availability analysis.

13.3.1.3.3   Level 3c: Ethernet Interface Module in Operator Station
This failure makes the concerned operator station totally unavailable (cannot interact with the controllers) even though the other functional parts of the operator station may be healthy. This amounts to the failure of the operator station discussed in the next section. Here also, the failure is nonfatal, because we miss only the human interface functions assigned to the failed operator station, which does not affect the functioning of the automation of the subprocess.

13.3.1.4   Level 4: Operator Stations

Level 4 at the apex has a computer system functioning as the operator station and is built of hardware (power supply, processor, and memory) and software (human inter-face functions).

13.3.1.4.1   Level 4a: Operator Station
The failure of an operator station can result from the failure of its computer system (hardware and/or software) or of its Ethernet I/F module (operator station becoming totally ineffective). However, whether the failure is caused by its Ethernet I/F or the computer system, the human interface functions of the failed operator station can be moved to one of the healthy operator stations. Hence, this is also a nonfatal failure, because it does not affect the automation of the concerned subprocess. However, the only minor drawback of this approach is the extra loading of one of the operators of the healthy operator stations.

13.3.2   Availability Enhancement

Next, we will discuss availability enhancement in DCS. From the preceding sections, we can conclude that the most critical modules in the information flow whose individual failure contributes to the fatal failure in a DCS are the:

- Processor in the controller
- Ethernet I/F in the controller
- LAN

With the assumption that only one critical module in the information flow can fail at any given time, a way to increase the availability of overall DCS functioning is to provide for a redundancy, or hot standby, to these critical modules.

13.3.2.1 Processor in Controller

Fig. 13.4 illustrates the standby arrangement for the processor module.

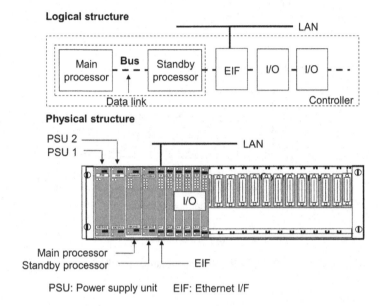

Figure 13.4 Standby for processor in controller.

The hot-standby processor automatically takes over the function of the main processor whenever it fails.

13.3.2.2 Ethernet Interface in Controller

Failure of the controller's Ethernet I/F can lead to the isolation of the controller with other subsystems (other controllers, operator stations, etc.) on the LAN. This fatal failure can be overcome by providing standby/redundancy to the Ethernet I/F, as shown in Fig. 13.5, with the Ethernet I/F integrated with the processor module.

Upon recognition of the failure of the Ethernet I/F in a processor module, the other processor module with a healthy Ethernet I/F takes over and continues the communication over the LAN.

13.3.2.3 Local Area Network

To further increase the availability of local communication systems, one can consider the physical redundancy for LAN components, as shown in Fig. 13.6.

This arrangement also addresses the possible failure of any LAN component.

In this arrangement, one Ethernet I/F is connected to LAN 1, whereas the other is connected to LAN 2. Simultaneous failure of Ethernet I/F 1 and LAN 2 or Ethernet I/F 2 and LAN 1 leads to total failure irrespective of duplication of the processor and Ethernet I/F in the controller and the LAN.

Logical structure

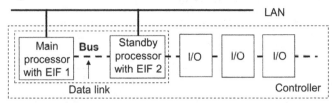

Physical structure

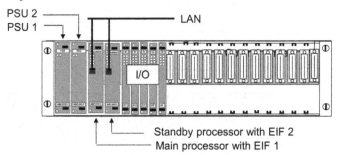

PSU: Power supply unit   EIF: Ethernet I/F

Figure 13.5 Standby for processor and Ethernet interface in controller.

Logical structure

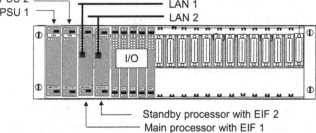

Physical structure

PSU: Power supply unit   EIF: Ethernet I/F

Figure 13.6 Standby for local area network (LAN) media components.

To overcome this drawback, the arrangement employs main and hot-standby processors with two Ethernet interfaces each. These are connected to both of the LANs. Any simultaneous failure of Ethernet I/F 1 and LAN 2 or Ethernet I/F 2 and LAN 1 does not lead to the total failure of the local communication. This arrangement is illustrated in Fig. 13.7.

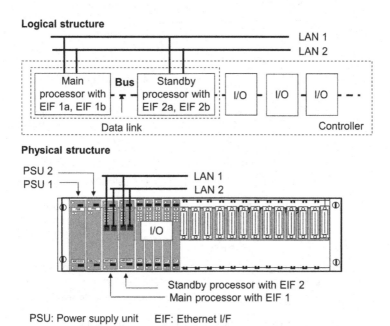

Figure 13.7 Dual local area network (LAN) with dual Ethernet interface in controller.

13.4   Network Control System

Most of the discussion about DCS holds true for NCS as well, especially at the top three levels (operating system, LAN, and FEP). The following sections discuss the availability analysis and enhancement at other levels.

13.4.1   Availability Analysis

Availability analysis in NCS can be made with the information flow in a distributed process from process to instrumentation subsystem (level 1), to control subsystem (level 2), to remote communication subsystem (level 3), to FEP (level 4), to local communication subsystem (level 5), and to human interface subsystem (level 6). This is shown in Fig. 13.8.

As in the case of DCS, any failure (full or part of a subsystem) in information flow can be fatal or nonfatal. A fatal failure is the one that affects the overall functioning of the NCS. In the following discussions, the failure of an RTU or its communication interface, for whatever reasons, leads to the unavailability of automation functions for the subprocess to which the RTU is assigned. This is generally considered acceptable.

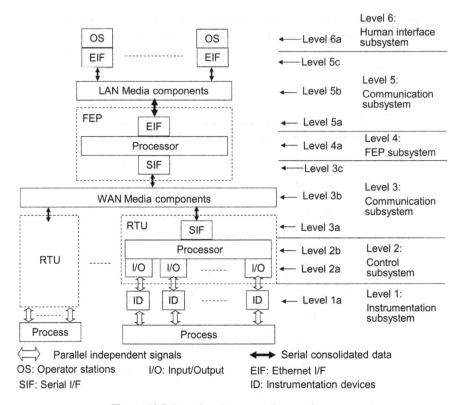

Figure 13.8 Data flow in a network control system.

Using this information flow, the following sections illustrate the implications for automation functions of the failure of a particular subsystem or one of its modules at various levels.

13.4.1.1   Level 1: Instrumentation Subsystem

This is identical to level 1 in DCS with RTU in place of the controller.

13.4.1.2   Level 2: Remote Terminal Units

This is also identical to Level 2 (controller) in DCS with RTU in place of the controller.

13.4.1.3   Level 3: Wide Area Network

In this case, depending on the topology of the wide area network (WAN), failure of WAN components may affect a few RTUs. Other aspects are similar to level 3 (LAN) in DCS with a serial I/F in place of an Ethernet I/F.

13.4.1.4   Level 4: Front-End Processor

This case is similar to level 2 (controller) in DCS with FEP in place of the controller.

13.4.1.5   Level 5: Local Area Network

This case is also similar to the local communication subsystem (LAN) in DCS.

13.4.1.6   Level 6: Operator Stations

This case is similar to level 4 (human interface subsystem) in DCS.

13.4.2   Availability Enhancement

Availability enhancement in NCS is similar to that of DCS. Hence, we can conclude that the most critical modules in the information flow whose individual failure contributes to a fatal failure in NCS are the:

- Serial I/F and processor in RTU
- WAN
- Serial I/F, processor, and Ethernet I/F in FEP
- LAN

Once again, the assumption here is that only one critical module in the information flow can fail at any given time. The approach is to increase the availability of the overall functioning of the NCS to provide for a redundancy or standby to the critical module.

13.4.2.1   Serial Interface and Processor in Remote Terminal Unit

Fig. 13.9 illustrates the redundancy arrangement in an RTU for a serial I/F and processor.

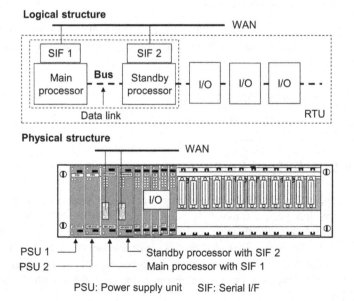

Figure 13.9 Standby for processor and serial interface in a remote terminal unit (RTU).

Here, the hot-standby processor takes over the functions of the failed main processor with data updating over the RTU bus.

13.4.2.2 Wide Area Network

Fig. 13.10 illustrates the redundancy for a WAN.

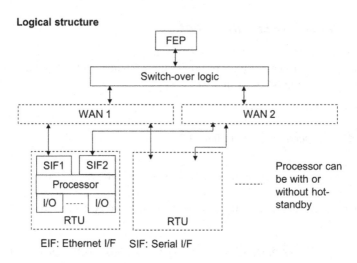

Figure 13.10 Standby for wide area network (WAN).

13.4.2.3 Serial Interface, Processor, and Ethernet Interface in Front-End Processor

Fig. 13.11 illustrates the provision of standby at an Ethernet I/F, processor, and serial I/F in FEP. LAN itself is used as the data link between the main processor and the hot-standby processor.

13.4.2.4 Local Area Network

LAN redundancy is identical to LAN redundancy in DCS.

13.5 Input–Output Redundancy

So far, the application of standby/redundancy concepts for vital components in automation systems has been discussed. Improving availability in the automation system is meaningful provided there is adequate redundancy in the process plant for its vital equipment. This is not discussed here because it is out of scope for the chapter. However, to support redundant equipment in the process plant, an automation system can support I/O redundancy for the redundant plant equipment with its associated instrumentation.

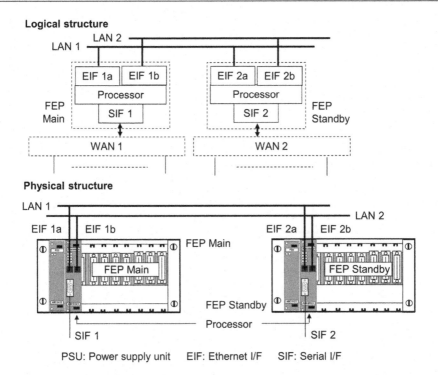

Figure 13.11 Standby for front-end processor (FEP) processor and communication interfaces.

13.6   Summary

In this chapter, we discussed issues related to the availability of DCS and NCS when a specific critical module fails in the information flow. Upon recognition of the failure of the main unit, several types of standby schemes may be employed. Hot standby is used to continue the operation from the point where the main unit failed. However, because this approach is complex and expensive, it is implemented only when absolutely required.

Common Configurations

14

14.1   Introduction

In Chapter 13, we discussed how to provide hot standby for various critical components in the data flow link so that the total operation of the automation system, whether it is a distributed control system (DCS) or network control system (NCS), is always available. In this chapter, using the approaches available for enhancing system availability, some common configurations are discussed for increasing the functionality of both DCS and NCS.

14.2   Distributed Control System Configurations

In the following sections, some typical configurations of the DCS above the controller level are discussed, assuming that the controllers are already provided with hot standby.

14.2.1   With Operator Stations

Fig. 14.1 illustrates the configuration of controllers with hot standby.

This is the basic configuration in which one or more operator stations with a specific area or function assignment carry out the operator functions. Normally any operator station can be used to monitor the process variables assigned to the other operator stations. However, the issue of commands to the process is restricted only to the authorized operator station. If one of the operator stations fails, another healthy operator station, under authorization, can perform its full functions. The drawback of this approach is that the operator station to which the functions of the failed operator station are transferred becomes overloaded.

Overview of Industrial Process Automation. http://dx.doi.org/10.1016/B978-0-12-805354-6.00014-1

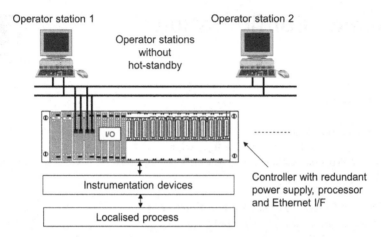

Figure 14.1 Distributed control system configuration with operator station.

14.2.2 With Supervisory Stations

In many applications, there are some common or supervisory functions for coordinating the functions of the controllers in DCS. It is generally not preferable to share one of the controllers for this function, because the performance of the controllers becomes reduced. If we incorporate supervisory functions in any operator station, it becomes unavailable if that particular operator station fails. To overcome these problems, the supervisory functions are provided on an independent supervisory station, as illustrated in Fig. 14.2.

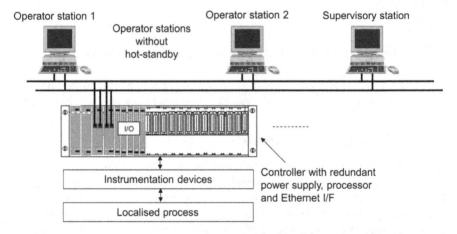

Figure 14.2 Distributed control system configuration with supervisory station without standby.

The independent station for common supervisory functions is called a server, and modern servers hold complete process data unlike controllers of the past. If the supervisory functions are critical, the obvious next step is to provide a hot standby for the supervisory station, as illustrated in Fig. 14.3.

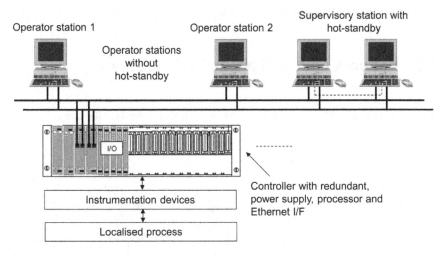

Figure 14.3 Distributed control system configuration with supervisory station with standby.

14.2.3   *With Application Stations*

In some special cases, there is a need to execute complex real-time application functions with heavy computing, and these must be performed within a fixed time. Combining complex application functions with supervisory functions may lead to the performance degradation of both functions. Once again, to overcome this, an independent station (with or without hot standby) can be employed exclusively for the execution of the complex application programs. Fig. 14.4 illustrates the configuration of an independent application station with hot standby.

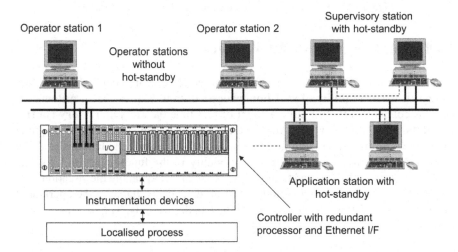

Figure 14.4 Distributed control system configuration with application station with standby.

These independent stations are generally called application servers.

14.3   Network Control System Configurations

All previous discussions pertaining to DCS are equally applicable to NCS. As shown in Fig. 14.5, the only difference is that front-end processors replace the controller level in NCS.

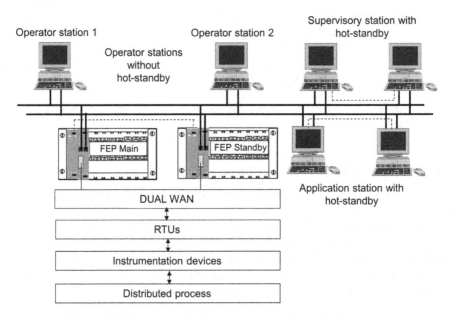

Figure 14.5 Network control system configuration with all features.

14.4   Real-Time Data Availability

In all of these configurations, one important aspect of computerized automation systems is the **availability of real-time process data** in control center. Automation systems collect and store a large amount of real-time data in **supervisory control and data acquisition servers**. All of the data received and stored are not required for normal operations. Only limited data in pockets are required to handle the object of interest. However, in case of any abnormality in the functioning of a specific object, a few more data related to the object are required to handle the situation. The power of the computer-based system is that all data are available to the operator at his fingertips so that any contingency can be handled quickly. This is illustrated in Fig. 14.6.

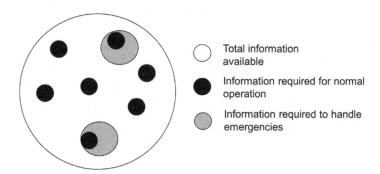

Figure 14.6 Data availability and use in automation system.

14.5 Summary

In this chapter, we discussed a few typical configurations of DCS and NCS, with only operator stations for basic monitoring and control of the localized and distributed processes. Extension of this configuration to support supervisory station and application stations for the execution of complex functions is discussed. The technical complexity and cost of the configuration increases with the addition of more functionalities.

Customization

15

Chapter Outline

Overview of Industrial Process Automation. http://dx.doi.org/10.1016/B978-0-12-805354-6.00015-3

15.1   Introduction

Automation systems are not designed or developed for each process or application. They are built as a common platform and are configured or customized (for hardware, software, and applications) to meet specific physical and functional requirements of the process/application. **In this chapter, only the basic configuration of automation is considered to explain the customization. The customization of all extensions to the basic system (discussed in forthcoming chapters) is not considered here.** General automation system hardware and software structures of the **basic** distributed control system (DCS) and network control system (NCS) are illustrated in Fig. 15.1.

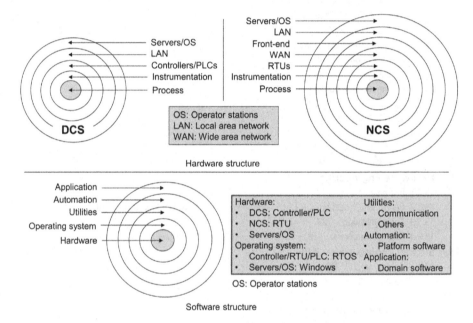

Figure 15.1 Data acquisition and control unit: automation system hardware–software structure.

The structure is employed here to discuss the configuration (customization) of the platform to meet the specific requirement of applications. For customization, the first and foremost thing is to understand the application of the process, its environment, its specific requirements, and so forth, to customize each and every subsystem of automation and finally integrating them. Basically the automation system consists of two parts, hardware and software, although they work together as an integral part. Each part within hardware and software is to be individually selected, configured, and integrated to form an automation system for the specific application, as discussed subsequently.

15.2   Hardware Selection

Hardware can be broadly divided into automation hardware, computer hardware, communication hardware, and support hardware. These are discussed in the following sections.

15.2.1   Automation Hardware

This includes instrumentation, control, communication, front-end processor, and human interface which are of industrial grade.

15.2.1.1   Instrumentation

Generally in continuous processes (process plants), the focus is more on analog parameters whereas digital parameters are used in discrete processes (factories). Following are important parameters for selection:

- Type of device: input or output, analog or digital
- Number of devices and their types
- Signal type: 4–20 mA DC or other for analog; 0/24 VDC or other for digital
- Application environment: safe or hazardous
- Construction: food-grade (food and beverage), hazard-resistant (petrochemical), corrosion-resistant (chemical), high voltage–protected (electrical), etc.
- Tag details: analog input–output (I/O) (range, span, engineering unit, accuracy, limits, redundancy, etc.); and for digital I/O (*NO/NC, dry/wet contact, interrogation voltage, redundancy, etc.*).

15.2.1.2   Control

Control units are programmable logic controllers (PLCs), controllers, and remote terminal units (RTUs). For all of these units, common hardware parameters to be selected are:

- Units:
  - Centralized or distributed; if distributed, the number of units and their distribution (area-wise or function-wise)
  - Operating power: 24 VDC or 230 VAC or other (normally 24 VDC for remote installations)
  - Redundant or nonredundant
  - Sizing of each unit based on its I/O count
  - Processor: Speed, memory capacity, watchdog, integrated Ethernet and serial ports, rack extension (serial or parallel), maximum I/O capacity and rack capacity, provision for redundant processor, power supply, communication modules, etc.
- I/O:
  - Count for each unit
  - Analog signals: 4–20 mA or other, ADC/DAC (resolution and accuracy), isolated or nonisolated, etc.
  - Digital signals: 0/24 VDC or others, isolated or nonisolated, etc.

15.2.1.3 Communication

The communication units under consideration are:

- Local area network (LAN) interfaces: Number of ports, redundant or nonredundant
- Serial interfaces: Number of ports, redundant or nonredundant

15.2.1.4 Front-end Processor

Front-end processors are employed in NCS in control centers. There are two types of platforms: automation hardware-based and computer hardware-based. The former is discussed here; important parameters for their selection are:

- Redundant or nonredundant
- Number of ports supported

15.2.1.5 Human Interface

The human interfaces under consideration are:

- Operator panel for small processes (both display and control)
- Mimic diagram board for medium processes (both display and control)
- Large display board for large processes (only display salient information)

15.2.2 Computer Hardware

Computer-based hardware includes SCADA and application servers, operator stations, engineering and maintenance management stations, management information system (MIS) and remote terminals, computer-based front-end processors, printers, etc.

15.2.2.1 Common

Common to all computer-like equipment are:

- Industrial or commercial grade
- Open or proprietary
- Redundant or nonredundant
- Reliability and availability
- Processing power, memory (cache, primary, and secondary)
- LAN and serial interfaces (redundant or nonredundant).

15.2.2.2 Specific

Specific to different types of computer-like equipment are:

- Human interface: Thin or thick clients, display area, resolution, interaction (touch or nontouch), single or multiple monitors, number, and their distribution.
- Front-end processors: Number of ports
- Printers: Dedicated or shared (networked)

15.2.3   Communication Hardware

These is general-purpose communication hardware that is adopted by automation systems.

15.2.3.1   Common

Common to all communication equipment are:

*   Grade: Industrial or commercial
*   LAN interfaces: Redundant or nonredundant, hubs and switches, Ethernet cables (Cat5), etc.
*   Serial interfaces: Redundant or nonredundant, routers, serial cables (RS232c), etc.
*   Network: Topology, redundancy, response, etc.
*   Environment: Hostile, distance, etc.
*   Media: Copper, optical fiber, wireless, etc.
*   Components: Gateways, bridges, modems, etc.

15.2.3.2   Specific

Selection of the data communication network is based on the type of automation system:

*   **LAN interfaces**: control network of (a) PLCs/controllers, servers, and operator stations in DCS and (b) front-end processors, servers, and network of operator stations in NCS
*   **WAN interface**: remote network of RTUs and front-end processors in NCS.

15.2.4   Support Hardware

Support hardware includes power supply systems, control desks, and hardware panels (operator panel, mimic panel, large screen display, etc.)

15.2.4.1   Power Supply

The power supply equipment under consideration are:

*   Bulk 24 VDC supply: Auxiliary powering of 4–20 mA for analog devices and for isolation/level conversion for digital I/O signals
*   Power supply system: Float-cum-boost charger or Uninterruptible power supply system (UPSS): sizing, standby battery capacity, redundancy, diesel generator support, etc.

15.2.4.2   Others

The control room should take account of the workplace, monitoring and visual effects, cable management, lighting, and acoustics. Above all, the most important consideration is ergonomics for operator comfort and convenience.

Control desks: The design of the control desk depends on the work space required for housing:

*   Number of operator stations
*   Number of printers (if housed on the control desk)
*   Other instruments such as telephones

Hardwired panels: Operator panels, mimic panels, large screen displays, etc.

15.3 Hardware Engineering

Hardware engineering involves the physical integration of all of these items to form the hardware configuration of the automation system.

15.3.1 Instrumentation

Instrumentation includes the instrument list, cable schedule, junction box schedule, hookup diagrams, loop drawings, panel drawings, panel wiring drawings, interconnection cable schedules, and channel base layouts, etc.

15.3.2 Controllers, Programmable Logic Controllers, and Remote Terminal Units

This includes the calculation of heat dissipation, processor loading, and bus loading controllers, PLCs, and RTUs.

15.3.3 Communication

Communication includes specifications, communication mapping lists, switch architecture, routers, and so on.

15.3.4 Support Hardware

The equipment under consideration are:

- **Power supply**: Power distribution schemes, power cable sizing, isolation coordination [molded case circuit breakers (MCCBs)/miniature circuit breakers (MCBs)/fuses], MCCB/MCB/fuse sizing and rating, power distribution drawings, and earthing/grounding schemes.
- **Control desk**: Ergonomic design
- **Operator panels and mimic panels**: Layout, wiring
- **Large display screens**: Ergonomic design

15.4 System Engineering

System engineering is the basic activity of building the architecture interconnecting various hardware and software components specific to the project or plant. Application engineering is building the interface with the plant process.

15.4.1 Configuration

Configuration is part of system engineering. The configurator is a tool supplied by the vendor to assist application engineers in establishing a link between automation system hardware, software, and physical parameters of the process. The tool:

- Is an application-independent facility to customize a general automation platform to specific requirements
- Provides an interactive process through which the customer inputs project-specific data into a general-purpose automation system
- Software that runs normally on a host system, produces data, and downloads them to the target system

Configuring is a procedure to:

- Prepare an I/O list and link it to the process signals and database
- Define signal particulars such as input or output, analog or digital, alarm or event, high/low limit settings, etc.
- Prepare basic human interface layouts on the screen such as process diagrams, alarm and event lists, trend lists, etc.

The configuration of individual systems is discussed next.

15.4.1.1 Controllers, Programmable Logic Controllers, and Remote Terminal Units

The activities involved are detailed below.

- Define signal particulars such as input or output, analog or digital, alarm or event, high/low limit settings, etc.
- Configure station/node numbers, configure domains and IP addresses, configure the number of function blocks/registers to be used, configure alarms, messages, and trends buffer, and so on.

15.4.1.2 Computer Systems

Computer systems require general configuration such as IP addresses, switches, etc.

15.4.1.3 Human Interface

Human interface layouts (graphics) on the screen are done for:

- Operator stations: Process diagrams, alarms, event and trend lists, controller pages, face plates, etc.
- Asset management stations: Diagnostics, etc.
- MIS and remote terminals

15.4.1.4 Communication system

Communication system require general configuration such as node addressing, routing, etc.

15.5   Application Programming

Application programming involves customizing the control units with an appropriate automation strategy to manage the process. Based on Standard IEC 61131-1, following programming languages are employed for this:

1. Instruction list
2. Ladder logic diagram (LLD)
3. Function block diagram (FBD)
4. Structured text
5. Sequential function chart

Applications engineers commonly use LLD and FBD whereas other languages are used for application software development. Application programming for PLCs, controllers, and RTUs are discussed next.

15.5.1   Programmable Logic Controllers and Controllers

The requirements and the activities are:

- Need extensive programming of local automation functions
- Employ programming languages of IEC 61,131-1 standard (LLD and FBD)
- FBD is employed for continuous process whereas LLD is used for discrete process
- Control schemes for simple/complex loops, cascade/split range, sequence logic, communication interface, etc.

15.5.2   Remote Terminal Units

A local automation facility is generally is not used in RTUs, even though modern RTUs provide for them. However, if this functionality is needed, programming is done employing the IEC 61,131-1 standard.

15.6   Power Supply Systems

The following highlights the importance of a reliable power source for an automation system. All subsystems of the automation system (instrumentation, control, human interface, etc.) should be available as long as the process is in operation and producing the result. In Chapter 13, we discussed the need to provide standby or redundancy to vital components so that the failure of any one does not affect functioning of the automation system. However, any failure of the external power supply leads to total unavailability of the automation system even though the latter may be healthy. This section discusses how to overcome this problem and how to customize it.

15.6.1   Float-Cum-Boost Charger System

Normally, float-cum-boost chargers with battery backup are provided to the automation equipment where the power requirement is relatively low (controllers/RTUs installed in remote places requiring no-break power). A cost-effective way is to provide no-break DC power. Fig. 15.2 illustrates the two configurations for this arrangement.

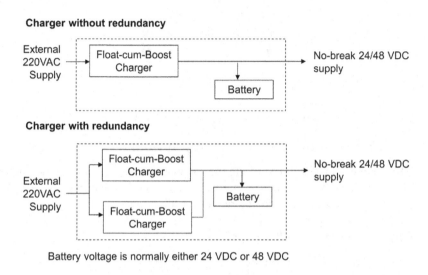

Figure 15.2 Charger configurations.

In the configuration without redundancy for the charger, during normal operation (when the external supply is present), the charger not only charges the battery (trickle charging) but also feeds the power to the controller/RTU. In other words, during normal operation the battery gets trickle charged and floats. When the external input power supply fails, the battery takes over to feed the controller/RTU (bump-less switchover). In the redundant charger configuration, each charger is individually designed to take the full load upon failure of the other. However, in a normal operation, both chargers share the total load (trickle charging and feeding controller/RTU). This arrangement provides time to have the faulty charger repaired and reinstalled. If necessary, there can be a standby for the battery as well.

Trickle charging compensates for the loss of power in the battery while in floating mode. Boost charging charges the battery in offline mode when the battery is fully discharged.

Sizing of the float-cum-boost charger is as follows:

- Charger sizing is done on the basis of the normal load (trickle charging of battery and feeding controller/RTU).
- Battery sizing (ampere hours) is done on the basis of the duration for which the battery support to controller/RTU is required on external power supply failure or charger failure.
- Redundancy to the charger and/or battery is provided, if the situation demands, to increase availability.

Fig. 15.3 illustrates the industry examples of float-cum-boost charger.

Courtesy: www.aplab.com

Figure 15.3 Charger examples from the industry.

15.6.2   *Uninterruptible Power Supply System*

The UPSS approach is employed where the automation equipment is installed in the control center (controllers, operator stations, servers, etc.), requiring no-break AC power. Fig. 15.4 illustrates two configurations for this arrangement.

UPSS Without redundancy for rectifier/inverter

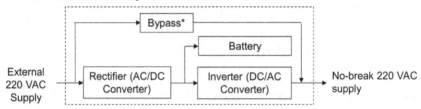

USS With redundancy for rectifier/inverter

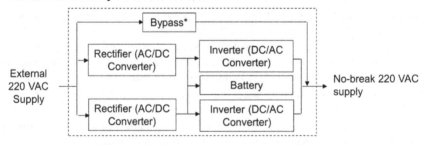

*Bypass can be manual or automatic (static)

Battery voltage is normally either 110 VDC or 220 VDC

Figure 15.4 Uninterruptible power supply system: configurations.

In the configuration without redundancy for the rectifier/inverter, during normal operation (when the external supply is present), the rectifier simultaneously charges the battery (trickle charging) and feeds the inverter. The inverter, in turn, converts DC power to AC power and feeds the automation equipment. In other words, during normal operation, the battery floats. When the external power supply fails, the battery takes over to feed the inverter, which then feeds the automation equipment (bump-less switchover). In the second configuration (with redundancy for the charger/inverter with each rectifier/inverter individually equipped to take the full load), both of the rectifiers/inverters share the total load (trickle charging and feeding the automation equipment) in normal operation. When any one of the rectifiers or inverters fails, the healthy rectifier or inverter takes the full load. Here also, there can be a standby battery if the situation warrants. Additional availability can be provided with a bypass to switch the external power supply to the automation equipment when the entire UPSS fails. Bypass is a switch to connect to the output mains if the UPSS fails, and it can be either manual or automatic (static).

Sizing of the UPSS is as follows:

- Rectifier/inverter sizing is done on the basis of the normal load (trickle charging of battery and automation equipment in the control center).
- Battery sizing (ampere hours) is done on the basis of the duration for which battery support to automation equipment is required upon external power supply failure.
- Redundancy to various subsystems in the UPSS is provided, if the situation demands, to increase availability.

In addition to these, to improve availability, where grid supply is not reliable, one can use diesel generators of adequate capacity and rating. Fig. 15.5 illustrates examples in the industry of UPSSs.

Courtesy: www.aplab.com

Figure 15.5 Uninterruptible power supply system: examples in the industry.

15.6.3 Battery Bank

The battery cell is specified in terms of its rated voltage and rated ampere hours, which means how much current (amperes) can be drawn from the battery at the rated voltage, and how long. The more current (amperes) is drawn, the shorter is its duration of support. To obtain the specified voltage (typically 24 or 48V in the case of a charger and 110 or 220V in the case of a UPSS), the required number of battery cells is stacked up and connected serially to form a battery bank or stack. Normally the battery floats and draws power from the charger/rectifier only for trickle charging to compensate for its no-load loss. The batteries need to be boost charged after sustained use to bring them to a normal level. Fig. 15.6 illustrates an example of an industrial battery and a battery bank.

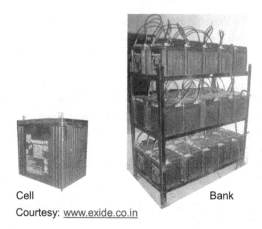

Cell   Bank
Courtesy: www.exide.co.in

Figure 15.6 Battery cell/bank.

As already indicated above, battery sizing (Ampere Hours) is done on the basis of the duration for which the battery support to the automation equipment is required on external power supply failure.

15.6.4   Power Distribution

For a good installation, it is not enough to have a charger or UPSS for power backup. Apart from the automation equipment, there may be other equipment that needs no-break power. To facilitate this and to provide proper distribution of the load for all equipment that needs no-break power, a distribution panel or board is normally employed, as illustrated in Fig. 15.7.

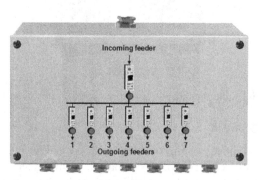

Courtesy: www.fabionix.co.in

Figure 15.7 Power distribution panel.

As seen here, the incoming feeder (no-break supply from the charger or UPSS) and the outgoing feeders to all of the equipment (including the automation equipment) are provided with adequate protection and display facilities (an MCB and indication lamp).

15.7 Summary

Because automation systems are not designed for each application, it is necessary to customize a general-purpose platform. An automation system has two parts: hardware and software. Hardware customization involves selecting and engineering each subsystem, such as instrumentation, control, communication, and human interface. Software customization needs system engineering and configuring of each subsystem and application programming of control units. Finally, the customized subsystems are integrated to meet the application requirement. The chapter ends with a discussion of a power system that supports the work of an automation system.

Data Communication and Networking

16

Chapter Outline

Overview of Industrial Process Automation. http://dx.doi.org/10.1016/B978-0-12-805354-6.00016-5

16.1   Introduction

Data communication is the transfer and exchange of digital information or intelligence between two points over a data network. For communication to happen, a path or medium (wired or wireless) must exist for data to travel from one end to the other. Equipment must exist at the sending end of the path to condition the message and place it on the medium in an acceptable form. Similarly, the equipment must also exist at the receiving end of the path to extract the message from the medium and understand its meaning.

In this chapter, we will discuss digital or data communication for the transfer of digital data from one point to another over a data network.

16.2   Communication Networks

In this section, some basics on data communication networks and their structure, operation, etc., are discussed in the context of automation.

16.2.1   Network Media

The communication path or media can be broadly divided into:

- guided (wired)
  - open wire
  - telephone cables
  - power line carrier
  - coaxial cable
  - optical fiber, etc.
- unguided (wireless)
  - multiple access radio
  - microwave
  - satellite communications
  - global system for mobile communication, etc.

Transmission on all of these communication media, except for optical fiber, is inherently analog. Over these media, digital data cannot be transmitted because they lose their characteristics over a long distance. This can be overcome by transmitting them over analog media after converting (modulating) data into their suitable analog form. However, transmission over optical fiber media is generally in digital form.

16.2.2   Network Topologies

There are different types of network topologies to facilitate data exchange among partners. A few basic types are discussed in the following sections. Some common network topologies are star, multidrop/bus, ring, and mesh, as shown in Fig. 16.1.

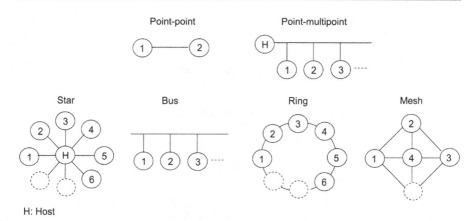

Figure 16.1 Network topologies.

Features of these network topologies are:

- The star network is the simplest topology with a dedicated link between two nodes. This network performs better (faster), the sent signal reaches only the intended node, failure of one node does not affect other nodes (high availability), it has centralized management, and it is easy to troubleshoot and maintain. However, it is expensive and depends on centralized management failure, which affects the entire network.
- The bus/multidrop network allows many participants to share a common medium. This network is less expensive because each node has equal access to the medium, it is good for local area networks (LANs), and it is easy to set up and extend. However, it has a limited number of nodes, which reduces performance with an increase in the number of nodes.
- The ring networks are highly organized. Performance is better than for bus topology, node connectivity is ensured, and each node has equal access to the medium. However, failure of one node affects the network, and network components are expensive.
- The mesh network provides connectivity among all nodes. If the direct path is not available, alternate paths are available via other nodes. This is normally used in wireless networks.

Data communication networks are functionally divided into:

- Local Area Network (LAN): A network of computers/devices that spans short distances in a relatively small area. A LAN is normally confined to a single room, a building, or a group of buildings. The LAN is logically a bus network. However, it can be arranged as a physical star network, which offers higher availability.
- Wide Area Network (WAN): A network of computers/devices (or LANs) that extends long distance in a geographic area. A WAN connects computers or LANs located over different cities. A WAN is generally a combination of different types of topologies networked.

Media used in a LAN is either an unshielded twisted pair or a shielded twisted pair, whereas media for a WAN is generally a shared wideband network (coaxial cable, optical fiber, microwave, etc.). Fig. 16.2 illustrates the logical schematics of LAN and WAN networks.

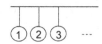

Local area network (LAN)

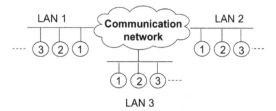

Wide area network (WAN)

Figure 16.2 Local area and wide area networks.

16.2.3 Network Components

Apart from the communication medium, common hardware (network) devices employed in the networks are:

- Network interface: An active/intelligent device that interfaces or links the computer to the communication network.
- Repeater: An active device that repeats (powering and reshaping) data signals for longer distances and has no logical function.
- Hub: A passive device that links the trunk cable to the spurs, making a physical star network.
- Switch: An active/intelligent device similar to the hub that routes or switches data to its destination.
- Bridge: An active/intelligent device that links two networks of different characteristics such as speed, message format, etc.
- Router: An active/intelligent device that links different networks and routes data around the network.
- Gateway: An active/intelligent device that links one network to another with remote networks of different or dissimilar communications protocols (see Section 16.5).

Fig. 16.3 shows schematics of these network devices.

16.3   Signal Transmission

This section explains how digital signals are transmitted on analog media so that the sent signal reaches the end point reliably and efficiently.

16.3.1   Encoding

Encoding is a process specially formatting the bit sequence of data for efficient transmission and storage (for example, increasing the bandwidth of the medium by reducing the number of transitions). Some general methods of encoding are presented in Fig. 16.4.

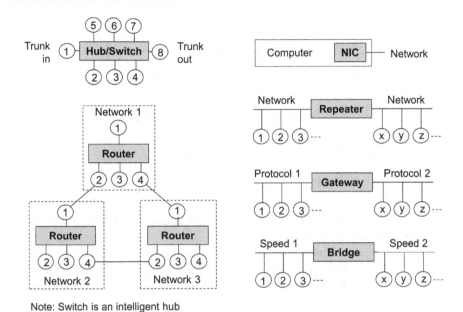

Figure 16.3 Common network devices.

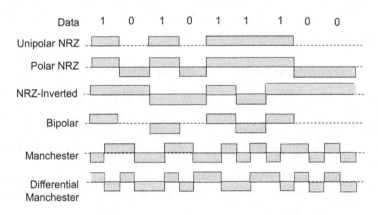

Figure 16.4 Data communication: encoding.

At the sending end, encoding is done on the original bit stream. At the receiving end, decoding is done on the received bit stream to retrieve the original bit stream.

16.3.2 Modulation

As mentioned, digital data cannot be transmitted in digital form over a long distance on the analog channel unless they are suitably converted into their analog form. Modulation is the process of superimposing digital data with a carrier signal for analog

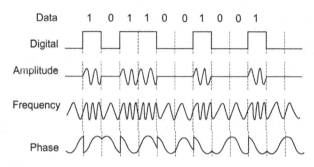

Figure 16.5 Data communication: modulation.

transmission (facilitates digital data transmission on analog media). Three basic modulation methods are presented in Fig. 16.5.

At the sending end, modulation of the original bit stream to its analog equivalent is performed; at the receiving end, demodulation of the received analog signal is done to retrieve the original bit stream.

16.3.3 Multiplexing

Multiplexing is a process in which several low-bandwidth channels can share a single high-bandwidth channel for better bandwidth use. There are two ways of multiplexing, frequency division multiplexing (FDM) and time division multiplexing (TDM), which are explained in this section.

16.3.3.1 Frequency Division Multiplexing

FDM applies only to analog channels. Many low-bandwidth analog channels (or digital channels modulated to their analog equivalent) share a high-bandwidth analog channel on a frequency division basis. Here, the individual low-bandwidth analog channels with their own carrier frequencies are superimposed on the carrier frequency of a high-bandwidth channel. The concept is illustrated in Fig. 16.6.

FDM is similar to several slow-moving vehicles on different roads converging onto a highway with many lanes. Incoming vehicles take their allotted lane to move in parallel in the same speed.

16.3.3.2 Time Division Multiplexing

TDM applies to digital channels. Here, the individual and low–bit rate digital channels share slots in a high–bit rate digital channel, on a time division basis. In other words, every incoming channel is given a fixed time sequentially to place part of its data in the time slot of the high-speed channel. The concept is illustrated in Fig. 16.7.

TDM is similar to items on several slow-moving conveyer belts shifting to a fast-moving conveyer belt sequentially within the allotted time to move all items serially with higher speed.

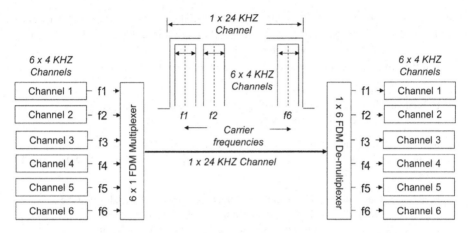

Figure 16.6 Data communication: frequency division multiplexing.

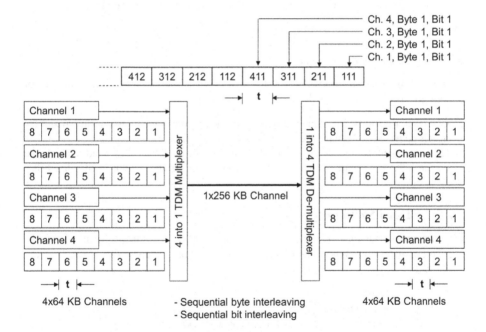

Figure 16.7 Data communication: time division multiplexing.

16.4 Data Transmission

The basic unit of data transmission is either a bit or a byte (group of 8 bits). The transmission of digital data can be done either in parallel (1 byte at a time) or in serial (1 bit at a time). Parallel transmission (as employed in data acquisition control units for communication between a processor and functional modules) calls for eight parallel paths. Serial transmission, on the contrary, calls for one communication path. However, serial transmission takes more time or is slower (parallel to serial conversion of a

unit of data at the sending end and its serial to parallel conversion at the receiving end)
for transmission of the same amount of data. Although it is slower, serial communication is economical for long-distance data transmission.

Information transfer can be one of the following over the medium:

- Simplex: One direction, requires one path
- Half duplex: Both directions but one direction at a time; requires one path
- Duplex: Both directions but simultaneously; requires two paths

Most commonly used LAN networks work in half-duplex mode.

Furthermore, data transmission can be either synchronous or asynchronous. In
other words:

- In asynchronous transmission, each byte is preceded by a start bit (indicating the arrival of
  a new byte) and ending with a stop bit (completion of byte transfer). Characters/bytes may
  follow random intervals. This mode is slower and is used for low-volume data transfer.
- In synchronous transmission, there is a constant time between successive bits/bytes. Timing
  is achieved by a common clock at both ends. This mode is faster and is used for high-volume
  data transfer.

16.4.1   Data Connection

Data communication between nodes can be either connection-oriented or
connection-less:

- Connection-oriented: The connection is established before any data is transferred and the
  data is delivered in the same order in which they were sent.
- Connection-less communication: Each data unit is individually addressed and routed based
  on the information carried in each unit instead of a prearranged setup with a fixed data
  channel.

16.4.2   Data Switching

Data transmission activity follows after establishing network connections called switching. To connect two points in the communication network, three types of switching are
employed:

- Circuit switching
  - Connection-oriented transaction
  - A dedicated connection (similar to telephone or telex)
  - Connection remains until both parties decide to quit
  - Interactive
  - No delay
  - Minimum overhead (header information is sent only once)
  - Not efficient

This practice is almost obsolete now and is given here only for comparison.

- Message switching or store-forward
  - Connection-less transaction

- Message sent to switching center
- Switching center forwards the message to intermediate/final switching center close to the destination when facilities become available. This process continues until the message is delivered at the destination
- Noninteractive
- Delay for long messages
- Minimum overhead (header information is sent only once)
- More efficient over circuit switching
- Packet switching
  - Connection-less transaction
  - Improved way of message switching
  - Long messages are broken into small/optimum-sized packets
  - Extra overhead (header information is repeated for all packets in the same message)

16.4.3   Data Error Detection and Correction

To ensure that data received are the same as data transmitted, check bits are introduced in the message, encoded, modulated, and placed on the media. Upon receipt of the message (after demodulating and decoding), the check bits are regenerated and compared with the received check bits. Any discrepancy indicates that the message is corrupted. Typical check bits are check sum, vertical redundancy check, longitudinal redundancy check, and cyclic redundancy check. These check bits increase the overhead of the data transmission because they are introduced before sending and are stripped off upon receiving. Normally data retransmission is employed if the message is detected with an error. Data corrections methods are employed in special situations. Fig. 16.8 illustrates the generation of check bits/bytes.

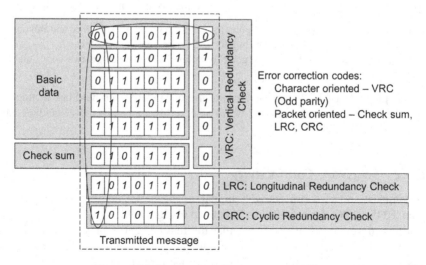

Figure 16.8 Data communication: error checking.

16.5   Data Communication Protocol

In general, communication protocol is an agreed formal set of rules and conventions for transacting data exchange between the sender and receiver. Typically, the protocol enables the sender to send all prerequisites that are agreed upon with the receiver, such as:

- Who the sender is?
- How the sender initiates the request for transaction?
- How the receiver conveys the agreement for initiating the transaction?
- What to do with the data that are being exchanged?

This agreement between the sender and receiver is formulated in the form of rules for data exchange between the sender and the receiver and is called a **communication protocol**.

To facilitate this, the communication protocol governs the hardware (mechanical and electrical) and software (message format and its exchange) connections among two or more digital systems. It ensures:

- Hardware connectivity: Mechanical and electrical compatibility (connectors, cables, levels, and timing for voltage and current signals, etc.).
- Software connectivity: Data segmentation and reassembly, encapsulation, connection control, ordered delivery, flow control, error control, synchronization, addressing, multiplexing, other transmission services, etc.

Together, hardware and software connectivity form the communication protocol.

16.5.1   Hardware Connectivity

Typical and commonly employed hardware connectivity standards are:

- Slow-speed data transfer: RS 232c (point to point, star topology, 3/4 wires, and full duplex) and RS485 (point to multipoint, bus topology, 2/3 wires, and half duplex).
- High-speed data transfer: Ethernet (point to multipoint, bus topology, two wires, and half duplex).

As explained in this chapter, apart from these standards, the field bus technology in the automation domain employs several other hardware connectivity standards specific to its applications.

16.5.2   Software Connectivity

The simplest software connectivity, similar to the postal delivery system, can be explained through a message format illustrated in Fig. 16.9.

In this message format, it is seen that the real content is only the data field to be delivered. The rest of the fields are simply the added overhead to ensure that the message is delivered reliably. In practice, message formats are different for different protocols, depending on their design and structure.

Postal delivery system

Sender's address	Receiver's address	Service type	Content	Confidentiality

Sender's address	: On left lower corner of envelope
Receiver's address	: On right centre of the envelope
Service type	: Certificate of posting, registered, ack. due, etc.
Content	: Letter to be delivered
Confidentiality	: Open, sealed, unsealed, etc.

Data communication system

Sender's address	Receiver's address	Function	Data	Error checking

Sender's address	: Sender's node number
Receiver's address	: Receiver's node number
Function	: Actions to be performed
Data	: Content to be delivered
Error checking	: Message security

Figure 16.9 Software connectivity: message format.

Although the hardware connectivity part of the protocol is standardized and accepted, that has not been the case in the software connectivity part of the protocol. Hence, protocol generally means only software connectivity. Henceforth in this chapter, protocol refers only to software connectivity. Communication protocols can be either proprietary or open.

Proprietary protocols have:

- Advantages: They are efficient because they fully exploit the vendor's platform features.
- Disadvantages: Development is expensive and time-consuming because it is designed for only for the vendor's platform (it cannot be used by other vendors), the customer depends on the vendor because it runs only on the vendor's platform, it is expensive because the base population is relatively small, etc.

Open protocols have:

- Advantages: The structure of the protocol is either standardized or is made publicly available; productivity and efficiency are increased because of the large scale and low-cost production; small firms can compete; technology transfer and dissemination of information are facilitated; and resources are conserved because only one item of equipment is needed where many were needed before.
- Disadvantages: Standards tend to freeze technology and take a long time to be certified, so that technology might become obsolete before it is adapted; many different standards bodies produce many different standards for the same piece of equipment; manufacturers tend to supply standards plus a little more to retain an edge over others; they are not efficient in performance owing to overheads from excessive standardization.

Open protocols are preferred to make multiple-vendor supplied systems, equipment, and devices:

- Compatible (ability to coexist)
- Interoperable (ability to be operated together)
- Interchangeable (ability to be replaced)

Open protocols make customers independent of vendors for their choice of equipment.

16.5.3   Protocol Standards

In view of the advantages of open protocols, standards are developed supporting:

- Interoperability to allow products from several manufacturers to be connected with their own and those from other manufacturers, exchange data, and use them in their own functions, working equally well for all systems.
- Standard and free configuration of different philosophies for communicating with their own products and with other products, allowing free allocation of functions.
- Long-term stability and usability (future-proof) in communication in an environment of continuously evolving and fast-growing communication, networking, and information technology.

The basic philosophy in a communication protocol is a **command-response (hand-shaking)** procedure that has been in use to transact data exchange. The basic sequence of the command-response procedure is simply:

- Sender sends a command to the receiver
- Receiver recognizes the command
- Receiver deciphers the command and responds

This fundamental command-response protocol becomes complex when rules are added to:

- Establish the connection between sender and receiver
- Define the meaning of the command
- Decide whether to accept the command or reject

These rules at various stages of data exchange, creating a state machine, are called a **communication stack**. There are communication stacks that are basically executables in the devices as per predefined standards. Generally, communication models are designed based on such communication stacks. One such standard is the seven-layer Open Systems Interconnection (OSI 7) model recommended by the International Standards Organization (ISO). This is called the **ISO/OSI 7 Standard** for intersystem communications. Objectives of this standard are to develop a modular, uniform, future-proof, and transparent protocol structure for data communication networks. The seven layers are physical, data link, network, transport, session, presentation, and application, as illustrated in Fig. 16.10.

Each layer in the structure deals with one aspect of networking and communicates with the adjacent layers in both directions and formats the data. Protocols enable an entity in one host to interact with a corresponding entity at the same layer in another host. Fig. 16.11 illustrates the physical and virtual data paths in an ISO/OSI 7 model.

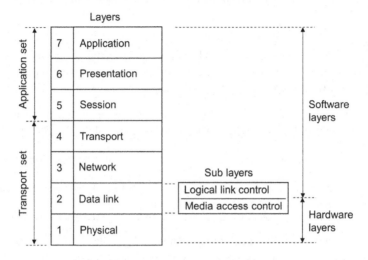

Figure 16.10 ISO/OSI 7 model.

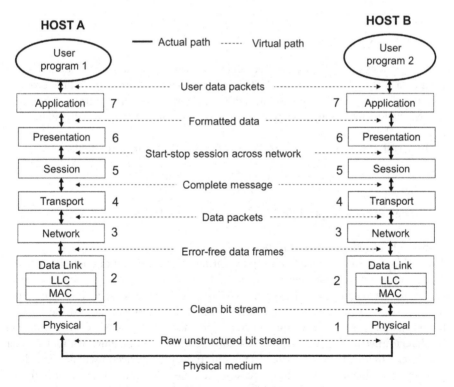

Figure 16.11 Data path in ISO/OSI 7 model.

The next sections briefly explain the roles and the responsibilities of each layer.

16.5.3.1 Physical

This layer defines the mechanical and electrical characteristics necessary for an interface to establish, maintain, and terminate the link between two physical devices. Characteristics of this layer are that it:

- Defines transmission media, an actual set of wires, connectors with pin assignments, and voltage/current levels of connecting (sending and receiving) devices, signaling, encoding, and modulation.
- Handles only the bits and physical transferring of bit streams (messages) between the two nodes.
- Delivers bit-by-bit, and synchronizes bits for synchronous serial communications and start–stop signaling for asynchronous serial communication.
- Ensures a 1 is received when a 1 is transmitted and a 0 is received when a 0 is transmitted.

The sender accepts bit streams from the data link layer and delivers an encoded and modulated bit stream (raw/unstructured) over the physical medium while the receiver accepts the modulated and encoded raw/unstructured bit stream from the physical medium, and demodulates, decodes, and delivers the clean bit stream to the data link layer.

16.5.3.2 Data Link

This layer is responsible for error-free data transfer from a node to the other nodes that need data within the same network. This layer manages the priority and the order of data transfer requests, as well as data, address, priority, medium control, error checking, etc., all related to message transfer, and is concerned with physically addressing devices and network topology.

To perform these functions, this layer is further divided into two sublayers:

- Lower sublayer: Media access control (MAC): A hardware layer
- Upper sublayer: Logical link control (LLC): A software layer

The next sections explain the functions of these sublayers.

16.5.3.2.1 Media Access Control

This sublayer, which depends on a physical layer, is a procedure controlled by the sender, in which several devices (on multidrop or bus networks) compete to gain access to the medium for data transfer. This sublayer decides, among competing devices, on who should gain the access to the media and send data. All devices on the bus gain access to the medium in some specified order. The MAC procedure can be **deterministic,** in which all devices gain access to the medium without fail, or **nondeterministic,** in which there is a possibility that the device(s) will not be able to gain access to the medium within a reasonable time. The receiver has no role in this sublayer irrespective of the network topology.

Fig. 16.12 illustrates some commonly used MAC procedures.

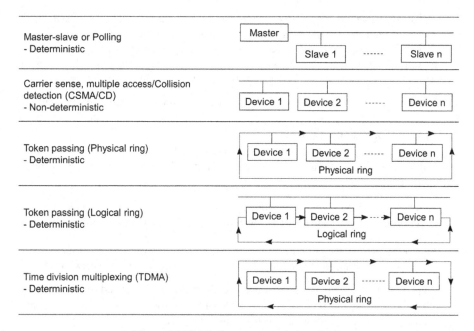

Figure 16.12 Media access control methods.

Popular MAC procedures are:

- Master–slave: Here, the master controls the bus and polls each slave sequentially to receive data or send data. The polling is cyclic, meaning the master starts polling slave 1 and continues until all slaves are polled, ensuring all slaves get the chance to gain access to the medium for data transfer before returning to slave 1. This is a deterministic procedure because all slaves are served by the master. Slave-to-slave communication is via the master only. Slaves can be polled based on their physical or logical address. This procedure is also called polling.
- Carrier sense, multiple access/collision detection: Here, all devices have equal right to gain access to the medium whenever they want. The competing device, before attempting data transfer, checks whether the bus is free. If the bus is free, the device takes control of the bus and proceeds with data transfer to other device(s). If the bus is not free, the device drops out and tries after some random time. If more than one device simultaneously competes for the bus, a collision takes place. Sensing the collision, competing devices drop out and each will try independently after some random interval. This procedure goes on until one of the competing devices gains access to the bus. This is a nondeterministic procedure because there is the probability that a device may not gain access to the media at all within a reasonable time. The fewer devices there are on the bus, the better the success rate will be.
- Token passing: Here, a device on the network holding the token becomes the bus master, transfers the data, and passes the token on to the next device in sequence, after completing the data transfer. This sequence goes on until all of the devices are served and the cycle repeats. This is a deterministic procedure because all of the devices have the chance to become the bus master once during each cycle. The sequence can be either physical (token ring) or logical (token bus). This procedure is also known as a round robin.

- Time division multiple access: Here, each device becomes the bus master for a fixed time to conduct data transfer and the control goes to the next device in sequence after a fixed duration whether or not data transfer is completed. This sequence goes on until all devices are serviced and the cycle repeats. This is also a deterministic procedure because all devices are served even though the devices may complete data transfer in more than one cycle.

There are many derivatives and variations to these methods to suit different applications.

16.5.3.2.2   Logical Link Control

This sublayer, independent of physical layer, establishes the link, and decides how much data are to be sent and their error detection and possible correction. In other words, in this layer:

- The sender accepts data from the network layer, breaks them into manageable units (frames), adds the header and trailer to the frame, disassembles the frame into bit streams, and delivers it to the physical layer.
- The receiver accepts the clean bit stream from the physical layer, assembles the bits into frame, checks the frame for errors, and delivers the error-free frame to the network layer.

A typical message format, called a protocol data unit, is illustrated in Fig. 16.13.

Subsequent layers in ISO/OSI standard do not really address data delivery; instead they address rules to be followed to bring data to the destination.

16.5.3.3   Network

This layer provides a data routing path and physical address mapping for network-to-network communication. If the data link layer defines the boundaries of a network, the network layer defines how internetworks (interconnected networks) function. The network layer is concerned with actually getting data from one computer to another even if it is on a remote network; in contrast, the data link layer deals only with devices that are local to each other.

A typical example is Internet protocol (IP). This protocol provides the rules for defining a number that uniquely identify the logical addresses of senders and receivers in a network (say, IPv4). Once the address is assigned to the node, which is different

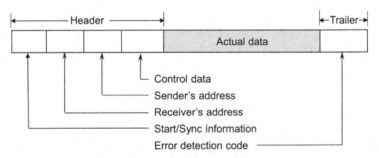

Start/Sync information indicates the arrival of new packet
Header and trailer are data transmission overhead

Figure 16.13 Typical message format: protocol data unit.

from a physical address (say, an MAC address in a physical layer), the IP packet contains this unique IP address along with other information pertaining to the other layers. The data can be sent with an IP address in one shot or it may contain several fragments. The IP protocol defines at a very high level:

- Rules for extracting the address, type of service, and information to other layers in the model
- How to get fragmented data together upon receiving them at the receiver
- Error handling, etc.

A typical example is address resolution protocol (ARP). This protocol is used for the resolution of a network layer address (say, IPv4) into a data link layer address (MAC address), which is a critical function in multiple-access networks. In other words, ARP turns the IP address on a local network into its MAC address.

16.5.3.4   Transport

This layer is responsible for end-to-end communication over a network providing logical communication among application processes running on different hosts. This layer is also responsible for managing error correction, providing quality and reliability to the end user, and enables the host to send and receive error corrected data, packets, or messages over a network.

A typical example is transmission control protocol (TCP). This protocol establishes and maintains a reliable connection between the sender and receiver and controls the data flow, considering the hardware and software limitations of the sender and receiver. A typical rule for transmission control is to establish a connection by both the sender and receiver agreeing that the resources associated with them are ready for data exchange. A TCP employs a **three-way handshake** procedure to establish the connection:

- The sender node sends an SYN (synchronisation) data packet over an IP network to a receiver node on it or on an external network. The objective of this packet is to ask whether the receiver is open for a new connection.
- The target receiver node, with an open port, accepts and initiates a new connection when it receives the SYN packet from the sender node. The receiver node responds and returns a confirmation receipt: an ACK (acknowledgment) packet or SYN/ACK packet.
- The sender node receives the SYN/ACK from the receiver node and responds with an ACK packet.

16.5.3.5   Session

This layer controls connections among multiple computers and tracks dialogs among computers or sessions. The session is a logical link between two parties involved in exchanging data without bothering with the delivery of data. This layer establishes, controls, and ends sessions between local and remote applications. This layer basically establishes a long-term dialog between the sender and receiver and connects to presentation/application layers. In the case of an application layer, the connection is via application program interfaces.

A typical example is a remote procedure call. This protocol facilitates information exchange between the sender and receiver to continue in dialog without worrying about the underlying physical connection. This is a protocol one program can use to request a service from another program located in another computer in a network without having to understand network details. This employs client/server architecture (see Section 16.6.1).

16.5.3.6 Presentation

This layer is used to present data to the application layer in an accurate, well-defined, and standardized format; it is also sometimes called a syntax layer. This layer is responsible for data encryption/decryption, character/string conversion, data compression, and graphic handling. This layer assembles information between the sender and receiver to perform additional tasks: for example, the sender may want to send confidential data intended for a specific receiver whereas all other receivers, whoever hears the data, may not decipher the information.

A typical example is secure sockets layer (SSL) protocol. This protocol is a cryptographic protocol designed to provide communications security over a computer network. Several versions of this protocol are in widespread use in applications such as Web browsing, email, instant messaging, and voice-over-IP. Major websites use SSL to secure all communications between their servers and Web browsers. The primary goal of the SSL protocol is to provide privacy and data integrity between two communicating computer applications.

16.5.3.7 Application

The application layer is used when the network meets end-user programs or serves as the window for users and application processes to access network services. This layer interfaces with the operating system/other application and standardizes the data. This layer is responsible for process-to-process communication across a network and provides a firm communication interface and end-user services. This is the only layer that directly interacts with the end user and provides services such as simple mail transfer protocol, file transfer, Web surfing, etc.

This layer, which is closest to the user, is responsible for:

- Handling user data
- Providing a link between the user and the network
- Providing services that directly support user applications
- Enabling communication between two user applications
- Providing the interface between software running in a device and the network

This is really not the user application but communication software supporting user application programs.

Protocols developed **before** ISO/OSI 7 recommendation are not modular and may not have all the functionality and the layers. Typical examples are field bus protocols.

16.6   Interprocess Communication

Applications or processes, whether they reside within a computer or device or are distributed across many computers or devices in the network, require data to be transferred from one application to another (sharing data among themselves). There are several architectures to accomplish this data transfer between two partners: client/server, report distribution, publisher/subscriber, producer/consumer, etc. The most commonly used method is client/server architecture. This is briefly described in the next section in an automation context.

16.6.1 Client/Server Model

Participants in this model could be computers or application processes (software) in computers. In case of processes, both can reside in the same computer or distributed in other computers elsewhere in the network. In this method, a process on the network is either a client or a server. The server is a provider of services (resources) whereas the client is the requester of services. This arrangement facilitates clients sharing resources available in servers irrespective of where they are located in the network.

Here a client requests the server for a resource and the server responds and delivers the resource. This method is connection-oriented, used for queued, unscheduled, user-initiated, confirmed, and one-to-one communication. Here queued means the messages are sent and received in the order submitted for transmission according to their priority and without overwriting previous messages. In the automation context, this is used for applications such as operator-initiated requests for set-point changes, tuning parameter access/change, alarm acknowledge, access display views, remote diagnostics, etc.

16.6.2 Interprocess Connectivity

Many times, applications or processes (software packages) on the network require data from other applications hosted by different vendors, and are normally inaccessible without authorization. Fig. 16.14 illustrates the traditional way of sharing the data or interprocess communication.

In the traditional approach, special efforts are required to make applications communicate with other application programs. It is expensive and time-consuming to

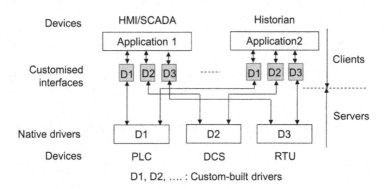

Figure 16.14 Interprocess communication: traditional.

develop such customized interfaces for each case. These interfaces are proprietary but do not employ widely used standards for interface development.

In this direction, **Dynamic Data Exchange (DDE)** was first introduced by Microsoft for interprocess connectivity allowing one program to subscribe to data made available by another program, and to be notified whenever that item changes. DDE was superseded by more sophisticated object linking and embedding technology, built on top of Microsoft's Component Object Model (COM). COM technology in networked systems is known as Distributed COM (DCOM). DDE is also based on the client/server method of data transfer.

An application of COM/DCOM technology, called **open platform communications (OPC)**[1,2] **(Object Linking and Embedding for Process Control)** technology, was developed as an open standard for interprocess communication/data sharing with the following important aspects in mind:

- Focus on automation
- Interoperability
- Client/server architecture
- Scalable solution
- Secure, flexible, and easy for end users
- Standard and common interface for software-to-software data sharing

Two types of OPC devices, namely, OPC server and OPC client, serve this purpose.

An OPC server is a program that fetches data from native protocols/drivers (device-dependent) and provides them to OPC clients. Similarly, an OPC client, a program that communicates with an OPC server through a strictly defined interface, receives data from an OPC server for further processing. An OPC server does not eliminate the native device protocol but provides an interface with it, addressing issues involved with multiple proprietary protocols, customized drivers, complex integration, etc. An OPC server can be conceptually treated as a standardized driver. An OPC works as an interface layer between the data source and data sink, allowing them to exchange data without each knowing anything about the other.

Fig. 16.15 illustrates the software structure of the OPC way of interprocess communication.

In the following sections, we will briefly discuss OPC Classic [OPC Data Access (OPC DA), OPC Alarms and Events (OPC AE), and OPC Historical Data Access (OPC HDA)] and OPC Unified Access (OPC UA).

16.6.2.1   OPC Classic

OPC Classic provides the interface between automation system software and other application packages. In the context of automation, OPC provides the following specifications:

- OPC DA: This specification defines the OPC DA server transferring of real-time data to OPC clients without each having to know the other's native protocol. In the context of

---

[1] https://opcfoundation.org/
[2] www.matrikonOPC.com

automation, a typical OPC-compliant server is a programmable logic controller whereas a typical OPC-compliant client is a human machine interface device using a common communication interface. OPC DA deals only with real-time data.

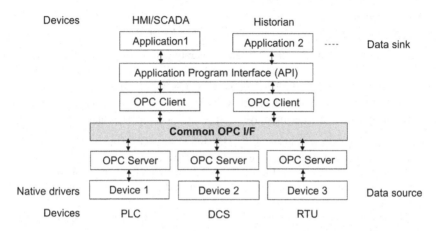

Figure 16.15 Interprocess communication: open platform communication (OPC).

An extension of OPC DA is OPC XML Data Access (OPC DA XML). XML-based languages use Web services and provide flexible exchange of structured information between collaborating applications. Further, XML technology is available across a wide range of platforms and XML DA interface simplifies sharing and exchange of OPC data among the various levels of the plant hierarchy (from low-level devices to enterprise systems, and beyond) via the Internet. This is more suitable for complex data exchange and is commonly used as a bridge between a control network and enterprise level.

- OPC Alarms and Events (OPC AE): This specification defines the OPC AE server capturing of alarms and events automatically from OPC-compliant automation system and makes them available to any interested OPC-compliant client application. The OPC AE server only reports alarms and events using a common communication interface previously defined in the automation system. Typical OPC clients are alarm and event logging.
- OPC Historical Data Access (OPC HDA): This specification defines the OPC HDA server to retrieve archived process data, say from an OPC-compliant historian. Normal OPC-compliant clients are trending applications, spreadsheet applications, etc. This is in contrast to an OPC DA specification that deals with real-time data.
- OPC Data Exchange (OPC DX): OPC DA, OPC AE, and OPC HDA servers acquire data from the plant floor and move it vertically into the enterprise system. In contrast, OPC DX is designed to move plant floor data horizontally among OPC DA servers.

Each of these protocols has its own commands for read, write, etc. The most commonly used OPC specifications are OPC DA, OPC AE, and OPC HDA.

16.6.2.2 OPC Unified Access

As mentioned earlier, OPC Classic was developed on the COM/DCOM model of Microsoft Windows technology. On the contrary, OPC UA was based on service-oriented architecture, which is vendor, product, and technology neutral. OPC UA employs the following protocol standards:

- UA binary (best performance and less overhead)
- UA Web service [Simple Object Access Protocol (SOAP)]
- A combination where the code is binary but the transport layer is SOAP: A compromise between efficient binary coding and firewall-friendly transmission

These allow programs to run on different operating systems and communicate using Hypertext Transfer Protocol and its Extensible Markup Language. Furthermore, OPC UA integrates the functionality of all individual OPC Classic specifications into one framework extending interoperability standards. OPC UA standardizes not only the interface but also the transmitted data. This multilayered approach accomplishes and supports:

- Functionality equivalent of all classic specifications (mapped to UA)
- Universality across all platforms developed by Microsoft, Apple, Android, Linux, etc.
- Scalability-wide infrastructure (from embedded microcontrollers to high-end servers)
- Security with encryption, authentication, and auditing; it is also firewall-friendly (see Section 16.7)
- Interoperability with extensibility for the addition of new features without affecting existing applications, and it is future-proof

OPC UA is the communication technology built specifically for data to pass through firewalls, specialized platforms, and security barriers to arrive at a place where they can be turned into information. OPC UA is designed to connect databases, analytic tools, and enterprise resource planning systems, and systems with real-world data from low-end controllers, sensors, actuators, and monitoring devices that interact with process plants generating real-world data. OPC UA is now IEC Standard 62,541 and is called an **Open Platform Communication**.

16.7   Cyber Security

Today's industrial automation systems have taken advantage of advanced general-purpose information technology (IT). However, the same technology has made today's industrial automation systems increasingly vulnerable to security intrusions (malicious or otherwise) from within the plant and outside.

Automation systems [say, distributed control systems (DCS) or network control systems (NCS)] are traditionally built on proprietary technology that provided a reasonable level of security from unauthorized access owing to its closed architecture. Also, automation systems were not generally connected to the outside world. Hence, they were considered secured and their security was not an issue for a long time. However, this is no longer the case because all subsystems in automation systems now are IT-like and network based. Other reasons are that:

- Proprietary systems are becoming a legacy and are paving the way for open and connected systems.
- Automation systems implementation continue to move toward the use of off-the-shelf technologies such as Windows operating systems, Ethernet communications, etc., for

reducing costs, improving performance, enabling interoperability, and adding new capabilities.

• Because people are data driven, being connected to the external world is no longer an option but a necessity for successful business operations.

With all of these factors, open technologies expose automation system to the same types of security issues as those of IT systems. They have become prone to internal and external security threats commonly experienced in any networked IT system.

Today, even the stand-alone automation system has very high network content, be it DCS or NCS. This is because of its large spread in the plant or process (locally in DCS and geographically in NCS) with many weak spots in its networks. The problem becomes even more complicated with the presence of wireless networks. An anonymous hacker, anywhere in the virtual space, can throw a wrench in the working of the automation system and create severe damage. The problem becomes even more serious when the system becomes connected to the external world.

Cyber security is about how to secure automation systems from being hacked while meeting operational needs in terms of confidentiality, availability, and integrity. This section addresses security issues that arise in stand-alone (not connected) automation systems. Issues related to the connected (to the external world) system will be discussed in Chapter 20.

16.7.1 Security Deployment

The basic requirement is to build a dynamic model for end-to-end security in automation systems, considering the technological and organizational issues, and to embed a robust security deployment model to:

• Establish a chain of trust among devices, data, and systems regardless of the application.
• Authenticate and validate everything at every point within the trusted system to ensure trusted interoperability and integrity.

These should be an implementable security model to work not only with new systems but also with legacy systems to protect the investment already made. In addition, because security is a dynamically evolving process (continuously changing in needs, threats, policies, and threat detection methods), the solution must be flexible enough to adapt and update continuously to new situations. This embedded security deployment model must establish and ensure trusted interoperability essential for interconnectivity within the automation system. Core requirements of the model are:

16.7.1.1 Hardened Devices

This is the beginning of the trust chain that validates the identity of the device. This method employs embedded tamper-proof microprocessor-based hardware supporting data encryption to determine and block the software application from its execution, if it is found to have been tampered with.

16.7.1.2   Secured Communications

Application of trusted transactions allows communication within the authorized zones and with the devices ensuring the trust and integrity of data within each zone. Embedded solutions facilitating this are:

- Intelligent gateways: They physically isolate the legacy systems from the outside world, limiting the attack surface of an automation system by providing an interface between the legacy system and the new system. This method can secure the legacy devices without any modification.
- Trusted execution environments: They provide security and privacy anywhere and enhance security by preventing any device from executing malicious code. They use virtualization and encryption technologies to create secure containers for applications and data that are accessible only to the approved devices. These environments are secure, trusted zones that ensure tamper-proof protection of data and make data and applications invisible to unauthorized persons who may transport, store, and process sensitive information.

16.7.2   Security Management

Security monitoring is a prerequisite for effective security management. This calls for setting up a centrally managed system for end-to-end security alerts of the entire system. Features are:

- Gathering, consolidating, correlating, assessing, and prioritizing security events from all managed devices that affect the automation system for quick and easy visualization.
- Bringing together situational and contextual awareness of all events through a process of baseline trending, anomaly detection, and alerting.
- Differentiating among normal and abnormal operational patterns and refining policies to minimize false alerts and responses.
- Conducting forensics to gain greater insight in case of a security incident or device failure.

   The standard for the cyber security in automation systems is covered by ISA 99 or IEC 62443.

16.8   Safe and Redundant Networks

The network needs to be robust against unintentional faults as against the intentional attacks for which the security measures are already in place. Two approaches are:

- Safe network: This improves the reliability of successful communication over gray channels. This technique applies a cyclic redundancy check with sufficient length encapsulated along with the data packets to verify the integrity of the data packet. The data are sent along with their error detection and correction codes. Detection of these codes at the receiving end is used to correct the changed bit, if any. The safe network has different levels of failure probability of, for example, one in a million bits to detect an error in the transmitted data.
- Redundant network: They carry the same data on both main and redundant channels. The receiver accesses the data from both main and redundant channels. Upon detection of a fault in one of the channels, data from other channel are used by the application.

Many other techniques can be employed to make the communication more robust and against unintentional faults over the network owing to harsh environments. Depending on the critical nature of the application, the integrity levels are defined and the corresponding safe communication system is employed to meet the end-user demand.

16.9 Summary

This chapter started with a basic introduction to data communication and networking and briefly covered network media, topologies, components, and signal and data transmission aspects. Subsequently, the chapter covered in detail the data communication protocol, hardware and software connectivity, and protocol standards followed by interprocess communication with introduction to OPC and OPC UA. The chapter ended with a brief description of cyber security in automation systems.

Fieldbus Technology

17

Chapter Outline

17.1 Introduction

In large plants, instrumentation devices are installed and linked close to process equipment to reduce the length of control cables for technical and commercial reasons. Instrumentation devices are linked to the controller in the control center with signal cabling. Normally the process points or devices are widely distributed in the plant and the control center is away from them. This leads to long and extensive cabling between the control center and process points. Apart from technical disadvantages, this approach is expensive owing to the cost of cables, cabling, and associated maintenance in the long run. This is the starting point that drove the development of fieldbus technology. This chapter discusses not only this technology but also several subsequent extensions of this technology.

Overview of Industrial Process Automation. http://dx.doi.org/10.1016/B978-0-12-805354-6.00017-7

17.2   Centralized Input–Output

In Chapter 8, we discussed the construction of general-purpose data acquisition and control units with multiple communication interfaces that are primarily for data exchange with compatible systems.

In all of the discussions so far, we selected a configuration of the controller in which the input–output (I/O) modules are placed in the same rack as the processor module. In case of more I/O modules (which are project dependent), additional or supplementary racks are employed with the bus physically extended from the main rack to the supplementary racks. Here the I/O modules are close to the processor physically and logically, and they adopt parallel data transfer with the processor over the bus. This I/O arrangement is called centralized I/O (CIO). A schematic of CIO and physical interconnections among the process, instrumentation devices, and the CIO is illustrated in Fig. 17.1.

In CIO, the I/O is physically and logically an integral part of the controller and the instrumentation devices are interfaced (hardwired) to the I/O channels on a one-to-one basis. Instrumentation devices are not intelligent (dumb), meaning that they just convert physical signals into electronic signals and vice versa, and can be either analog or digital. Each process signal requires one exclusive I/O channel for data transfer in one direction, depending on whether the signal is an input or an output. Furthermore, the number of instrumentation devices required equals the number of process inputs and outputs. The data transferred between the controller and the instrumentation devices is in the form of an **electronic signal** representing the physical signal.

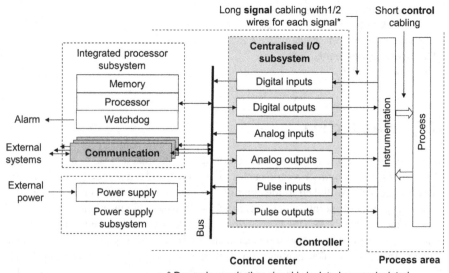

Figure 17.1 Centralized input–output schematic.

17.2.1   Intelligent Centralized Input–Output

Traditionally, CIO modules were not intelligent and were instructed to perform their functions by the processor. The processor had to scan the channels in all input modules for new data, processing, diagnostics, etc. This resulted in too much computational burden on the processor, leading to a reduction in the controller's overall performance. Making the I/O modules intelligent (microprocessor-based) was not feasible earlier in view of cost considerations. Cost-effective microprocessors and microcontrollers have now made intelligent I/O a reality. Physically there is no difference between not-intelligent and intelligent I/O modules, and they are one-to-one replaceable, as illustrated in Fig. 17.2.

Changes are required only in the software. However, the intelligent I/O contributes to the overall improvement in the performance of the controller by taking away the routine I/O data-processing load from the processor. The performance improvement generally comes from the following:

*   Preprocessing of the acquired data from the process in the input module itself before sending it to the processor (analog inputs: filtering, linearization, offset correction, limit checking, etc.; digital inputs: antibouncing, time tagging, sequencing, etc.)
*   Postprocessing of commands received from the processor in the output module itself before sending it to the process (analog outputs: set-point manipulation, etc.; digital outputs: command security checking, command execution monitoring, etc.)
*   Execution of diagnostic program of the I/O module in the I/O module itself and reporting any faults to the processor

Fig. 17.3 illustrates the physical structure and an example in the industry of CIO.

As a part of the controller, the CIO is installed within the control room and is generally away from the instrumentation devices and the process. Fig. 17.4 illustrates the field engineering associated with CIO.

The number of wires in the signal cabling depends on whether the signals are isolated.

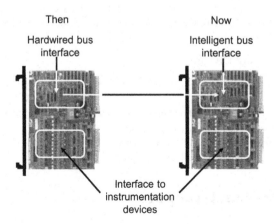

Figure 17.2 Centralized input–output: transition to intelligent input–output.

Schematic   **Industry example**

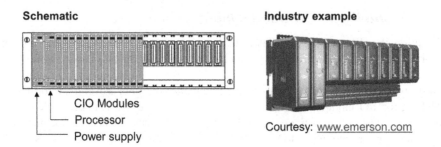

CIO Modules
Processor
Power supply

Courtesy: www.emerson.com

Figure 17.3 Centralized input–output: physical structure.

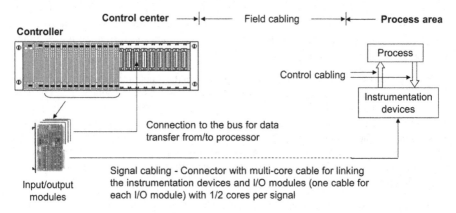

Connection to the bus for data
transfer from/to processor

Signal cabling - Connector with multi-core cable for linking
the instrumentation devices and I/O modules (one cable for
each I/O module) with 1/2 cores per signal

Input/output
modules

Figure 17.4 Centralized input–output: field engineering.

17.2.2   Advantages and Disadvantages

The advantages of CIO are that:

- It is ideal for small and less–spread-out processes.
- The controller is less expensive because I/O modules are placed close to the processor.
- There is fast data transfer between the processor and I/O modules owing to parallel communication over the bus.

  The disadvantages of CIO are that:

- The controller is bulky, occupies more space, and consumes more power because the I/O is concentrated.
- There is a high cost of signal cabling between the instrumentation devices and the controller (including installation and maintenance).
- There is **signal transfer** between the instrumentation devices and the controller (possibility of data quality degradation and/or loss during data transfer).

17.3   Remote Input–Output

The biggest disadvantages of CIO, cables, and cabling cost led to the evolution of remote I/O (RIO), a special-purpose controller that exploited the communicability

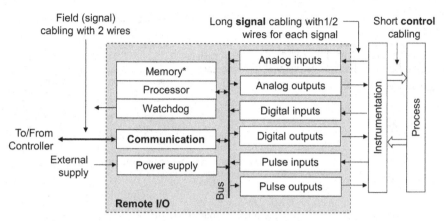

Figure 17.5 Remote input–output: schematic.

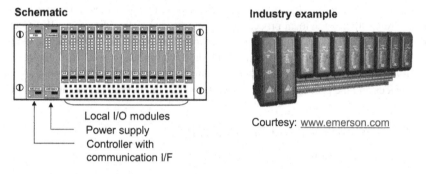

Figure 17.6 Remote input–output (RIO): physical structure.

feature of the processor. It moved I/O closer to the process, reducing the cabling cost to a minimum. The schematic RIO and interconnections among process, instrumentation devices, and controller are illustrated in Fig. 17.5.

As seen here, the RIO is not physically an integral part of the controller, and interfacing between the I/O and the controller is over a pair of wires with the communication module in controller. Features of RIO are:

- Communication cable with two cores for bidirectional data transfer of many process parameters
- **Data transfer** (not signal) between the controller and RIO (no possibility of data degradation and/or loss during data transfer)

Fig. 17.6 illustrates the structure and an example in the industry of RIO.

RIO is equipped with a microprocessor and software to handle only data acquisition from process/transfer to the controller in one direction and data receipt from the controller/transfer to process in the other direction.

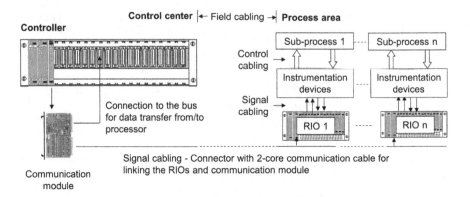

Figure 17.7 Remote input–output (RIO): field engineering.

The I/O that existed earlier in the control center as an integral part of the controller is relocated to the process area to become an integral part of RIO, reducing the cabling cost drastically. Physically, RIO resembles the controller, but it has a specific function. RIO just acquires the data from the process through its local input modules and dispatches it to the controller after some local processing. Similarly, in the other direction, RIO receives the data from the controller and dispatches it through its local output modules to the process after some local processing. Local I/O modules within RIO can be intelligent as well.

RIO is installed outside the control room and is generally closer to the instrumentation devices and the process. Fig. 17.7 illustrates field engineering associated with RIO.

17.3.1   Advantages and Disadvantages

The advantages of RIO are that:

- It is not an integral part of controller; moves I/O closer to process
- It is interfaced with the controller on a one-to-many basis; one serial interface channel for many RIOs
- It is suitable for large and physically spread-out processes
- Controller in control center is slim and power-efficient because the I/O is distributed
- Much less cabling between process instruments and controller (two wires linking all RIOs) results in less installation and maintenance cost
- Data transfer (not signal) occurs between the RIO and controller (no signal quality degradation)

Disadvantages of RIOs are as follows:

- Expensive because they are intelligent and communicable
- Slow data transfer between the controller and RIO modules owing to serial communication with the controller; not as much of an issue now considering the faster response in modern electronics and communication

For commercial reasons, some vendors have discontinued CIO and employ RIO for all applications.

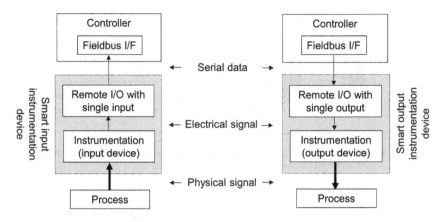

Figure 17.8 Smart input–output (I/O): schematic.

17.4   Fieldbus Input–Output

The obvious next step is to merge the RIO and the traditional instrumentation device to form an intelligent and communicable instrumentation device or **Self-Monitoring, Analysis, and Reporting Technology (SMART)**[1] device. The schematic of SMART I/O is illustrated in Fig. 17.8.

Fig. 17.9 illustrates the construction and the industry examples of a SMART instrumentation device.

Fieldbus devices in the process industry (temperature, level, pressure, etc.) physically look like their nonintelligent counterparts even though they are totally different functionally.

A fieldbus is a serial bus that runs throughout the process plant, linking all of the SMART I/O devices with the controller. The communication module at the controller is called the fieldbus interface (I/F), the I/O system is called the fieldbus I/O (FIO), and the SMART device is called the fieldbus device.

The conventional way has been hardwiring (cabling) to carry a single process variable (input or output) either in discrete (24 VDC) or analog (4–20 mA) form. On the contrary, the fieldbus is a digital, bidirectional, multidrop communication system for linking multiple intelligent field devices and other compatible automation equipment. In simple terms, a fieldbus is a two-wire single cable network system that can replace conventional ways of transporting data between the field devices and the controller.

Fieldbus is one of the best examples of the application of information, communication, and networking technologies in the automation domain and in an industrial network system for real-time distributed control (control on wire) supporting an open, digital, multidrop communications network for intelligent or smart field devices through data communication and networking for plant data acquisition and control.

Fig. 17.10 illustrates the logical and physical layout of the fieldbus system.

[1]InTech, ISA, March 2006.

Schematic

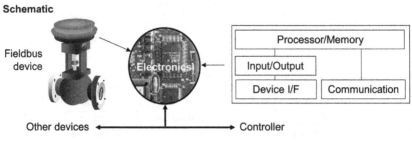

Data acquisition, control, and communication unit interfaced to instrumentation device -
All integrated and embedded

Industry examples

Figure 17.9 Smart input–output (I/O): physical structure.

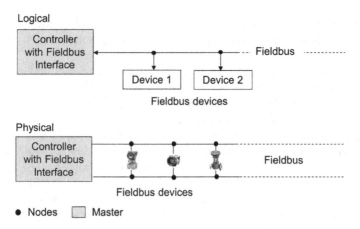

Figure 17.10 Fieldbus input–output: logical and physical layout.

FIO, like RIO, is also installed outside the control room and is generally closer to the process. Fig. 17.11 illustrates field engineering associated with FIO.

Fig. 17.12 illustrates an example of an intelligent and communicable energy meter employed in the electrical industry to measure all electrical parameters of an outgoing feeder.

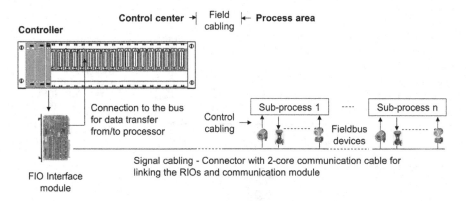

Figure 17.11 Fieldbus input–output (FIO): field engineering.

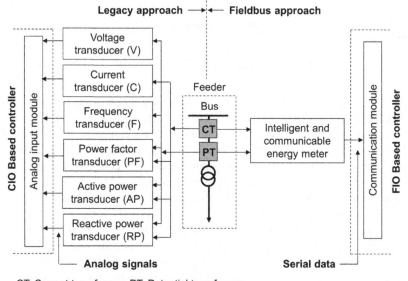

CT: Current transformer, PT: Potential transformer

Figure 17.12 Electrical substation instrumentation.

This is an example of how one intelligent and communicable device can send multiple parameters to the control center over fieldbus, replacing several independent transducers for the same job.

In this case, the traditional approach is to have independent transducers for each power system parameter, which requires feeder voltage input, feeder current input, or both. The intelligent energy meter takes feeder voltage and feeder current inputs and computes all of the parameters, and then sends them to the controller over the serial link.

Schematic

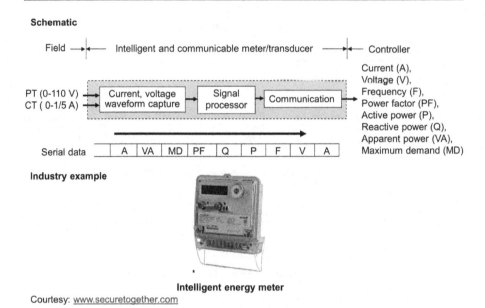

Industry example

Intelligent energy meter
Courtesy: www.securetogether.com

Figure 17.13 Intelligent and communicable energy meter.

Fig. 17.13 shows an example and function of an intelligent energy meter.

Here the waveforms of voltage and current inputs are continuously captured, and the instantaneous values of all of the electrical parameters are computed as functions of the input voltage and current and sent to the controller on a serial link as digital data. The energy meter, which is intelligent and communicable, can compute lots of data, such as energy and maximum demand, and send them to the controller. In this way, a single intelligent and communicable device can replace many nonintelligent devices.

Fig. 17.14 summarizes the differences between FIO and CIO.

An intrinsically safe barrier, a device to protect equipment and people from possible occurrence of high-voltage surges on the field cable, will be discussed in Chapter 18.

In practice, the green field projects can go directly to FIO. However, in case of expansion or retrofitting of a plant, existing or legacy instrumentation devices can be made part of the FIO network using RIO to support the legacy devices. This approach protects the investment already made on legacy devices. Here, both RIO and FIO are required to support the common communication protocol.

17.4.1   Advantages and Disadvantages

Apart from all of the advantages of RIO, because they are intelligent, fieldbus devices can:

• Diagnose their own health independently and send the diagnostic data to the controller for predictive maintenance.

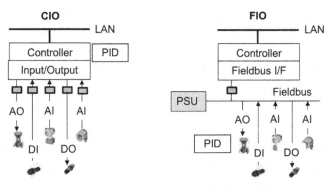

Non-intelligent devices
- One variable, one direction
- Centralised I/O and control

Fieldbus devices
- Multiple variable, both directions
- Decentralised I/O and control

PID : Proportional, Integral, Derivative control (Loop controller)
☐   : Intrinsically safe barrier (ISB)
PSU: Power supply unit

Figure 17.14 Fieldbus input–output (FIO): comparison with centralized input–output (CIO).

- Perform/implement local data processing, thereby reducing the burden on the controller, hitherto centrally executed in the controller. A notable example is single loop control (proportional, integral, derivative). This not only improves the performance of the controller but also eliminates the timing constraints or conflicts in the execution of loop control functions with different priorities.
- Provide almost unlimited data (diagnostics, status, etc.) about themselves to the controller and to all the other members on the network, unlike traditional instrumentation devices, which provide one type of data per device.
- Facilitate bidirectionality, remote calibration, field-based control, and plugs-and-plays (shorter engineering cycle); they are smaller and have high-speed connection.
- Support data integrity to send/receive only good data, so no data are lost. The controller or a fieldbus device can request retransmission in case bad data are received.
- Enable seamless integration of data right from the field level to the management level.
- Carry, along with data, the power required to power field devices for their operation. This is an important requirement in hazardous areas to conform to safety requirements.

Apart from the disadvantages of RIO, in FIO:

- Devices are relatively expensive, complex, and require training.
- There are mixed signals on the same bus.
- Architecture, configuration, monitoring, diagnostics, and maintenance tools are highly complicated.
- Device manufacturers have to offer different versions of devices owing to different fieldbus standards.
- Standards may predominate or become obsolete, increasing the investment risk.

The trend is more toward FIO and the CIO is on the decline, especially for large plants. However, RIO is expected to remain for some time to protect legacy I/O devices. The cost savings mainly comes from the reduction in:

- Cables, cable laying, commissioning, diagnostics, maintenance, etc.
- Marshaling cabinets
- Intrinsically safe barriers
- I/O modules
- Power supply units
- Control room size

The overall tangible savings is substantial (more than 40%). This is apart from the intangible benefits of fieldbus technology.

Fig. 17.15 illustrates the transition of the I/O system from CIO to RIO to FIO.

Fieldbus technology facilitates the decentralization of intelligence in the automation system through a network of controller, RIOs, and fieldbus devices.

17.5 Fieldbus Communication

Contrary to normal data communication networks, fieldbus networks call for specific requirements such as:

- Constant network use because data acquisition and control is on a continuous basis
- Relatively short response time to match process requirements
- Deterministic communication so that no device on the network either misses or delays sending/receiving data

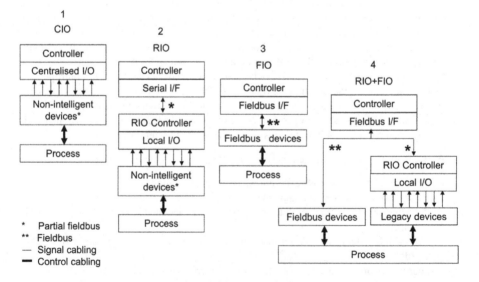

Figure 17.15 Transition in input–output subsystem.

Most of the discussion presented on data communication and networking in Chapter 15 applies to the fieldbus networks as well.

17.5.1   Fieldbus Reference Model

Contrary to the standard ISO/OSI seven-layer reference model discussed in Chapter 15, most fieldbus protocols have a simplified structure with only three layers: physical, data link, and application Fig. 17.16. Some fieldbuses use two more layers: transport and network. The simplified fieldbus model is illustrated in Fig. 17.17.

A fieldbus protocol stack is an optimized combination of data link and application layers and is specific to the fieldbus.

An additional layer, called a user layer, is introduced at the top of fieldbus protocol stack for easy configuration and programming of the control strategy based on IEC 61,131-3 [function block (FB)] programming language. This is not part of the ISO/OSI standard and is specific to the fieldbus model. This was **first** introduced by Foundation Fieldbus[2] to provide an interface with process (sensor/control elements) and user interaction with the control system specifically for process control applications. Common control schemes are standardized and made part of the user layer in the protocol, and are open. This facilitates configuration of the control strategy using the function blocks and avoids customized programming.

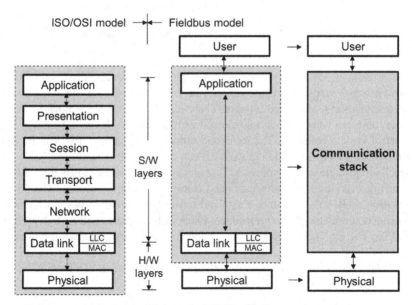

Sub layers: LLC - Logical link control, MAC - Media access control,
User: User defined/configured function block application (control) program

Figure 17.16 ISO/OSI and fieldbus protocol models.

[2]www.fieldbus.org.

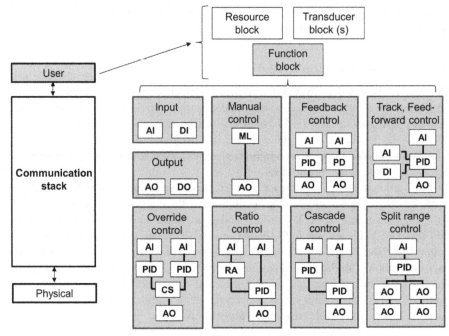

Typical control function blocks

Figure 17.17 User layer structure (Fieldbus Foundation).

The user layer has three blocks: resource block, transducer block, and FB. The resource block describes the characteristics of the fieldbus device, such as the device name, manufacturer, and serial number. There is only one resource block in a device whereas there can be more than one transducer block in one device. The FB provides the control system behavior. The input and output parameters of FBs can be linked over the fieldbus. The execution of each FB can be precisely scheduled. There can be many FBs in a single user application. Transducer blocks decouple FBs from the local I/O functions required to read sensors and command output hardware. They contain information such as the calibration date and sensor type. Fieldbus devices are configured using resource blocks and transducer blocks whereas the control strategy is built using FBs. The structure of the user layer and typical control schemes are illustrated in Fig. 17.17.

Fig. 17.18 illustrates the data flow (both physical and virtual) in a fieldbus protocol between two partners (Host A and Host B).

17.5.2 Fieldbus Networks

No single protocol meets the requirement of all applications. Hence, different protocols are designed for discrete and continuous processes, and, within these, for different applications. As mentioned earlier, discrete processes call for speed, whereas continuous processes call for accuracy.

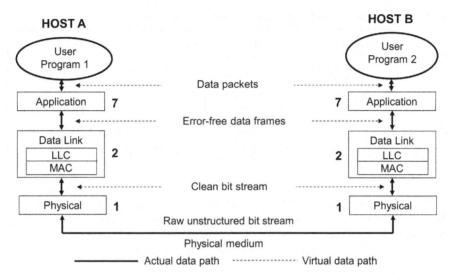

Figure 17.18 Data flow.

Networks in the automation system are:

- Field networks (Sensor/actuator and device)
- Control networks (Backbone)

The following sections explain commonly used fieldbus networks.

17.5.2.1 Field Networks

This is a low-speed network at the lowest level in the hierarchy that includes field instrumentation devices such as sensors, actuators, transmitters, control valves, etc., and which:

- Replaces legacy structures of 4–20 mA and 24 VDC technology
- Handles small data with temporal consistency
- Transfers data serially from field devices to the controller and vice versa
- Offers diagnostic and configuration capabilities
- Offers local processing power and intelligence

17.5.2.1.1 Sensor/Actuator

These are generally for discrete process automation in which:

- Networks are basically designed to support discrete I/O
- Domain activities are fundamentally discrete and binary, replacing discrete I/O (digital I/O modules)
- Protocols are simplest and focus principally on supporting binary inputs and outputs (discrete sensors and actuators such as push buttons, limit switches, solenoid valves, motor starters, etc.)
- Protocols are designed for very fast cycle times
- The cost of a network node is relatively low

Table 17.1 Fieldbus examples of discrete process automation

Fieldbus		Masters	Nodes	Reference
CAN	CANopen	Multiple	64	www.can-cia.org
LON	LonWorks	Multiple	32K	www.lonmark.org
Interbus-S		Single	256	http://www.interbus.de
AS-i	ASi	Single	31	www.as-interface.net

Application areas are factory automation (assembly, packaging, material handling, pick and place, etc.) and building automation (heating, venting, and cooling control).

Table 17.1 provides some typical protocols in this category.

17.5.2.1.2 Device

Devices, also low-speed networks, are generally for continuous process automation in which in:

- Networks are intended more to support process automation and more complex transmitters and valve actuators
- Domain activities are fundamentally continuous and analog
- Typical devices are transmitters (pressure, temperature, flow, level, etc.) and actuators (motorized valves, pneumatic positioners, etc.)
- Protocols do not need fast cycle times
- The cost of a network node is relatively higher because more sophisticated functions are built-in devices

Application areas are process automation (chemical, steel, cement, power generation, pulp and paper, etc.).

Fig. 17.19 illustrates a typical field-level network.

Table 17.2 provides some typical protocols in this category.

17.5.2.2 Control Network

This is a high-speed network, also known as a backbone network, integrating all other systems, which:

- Provides temporal data consistency and event order.
- Facilitates information flow primarily for the loading of programs and configuration of devices.
- Connects controllers, operator stations, remote I/Os, and linking devices (gateway for low-speed field networks), scanners, recorders, drives, etc.
- Provides real-time and high-throughput capabilities.
- Supports deterministic and repeatable transfers of all critical control data in addition to transfers of non–time-critical data.
- Supports critical uninterrupted operation.
- Handles large amounts of data and nodes.

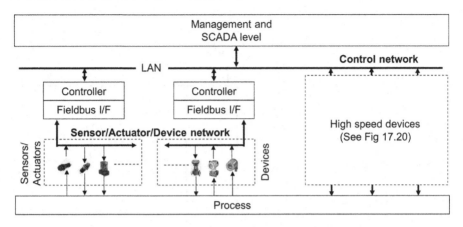

Figure 17.19 Fieldbus network: field level.

Table 17.2 Fieldbus examples of continuous process automation

Fieldbus		Masters	Nodes	Reference
Fieldbus Foundation H1	*Fieldbus*	Single	64	www.fieldbus.org
HART	HART	Multiple	32	http://en.hartcomm.org
Profibus PA		Single	32	www.profibus.com

The application is mainly in backbone networking.

Fig. 17.20 illustrates a typical control level network.

Table 17.3 provides some typical protocols in this category.

17.5.3 Fieldbus Interface

The serial I/F employed in FIO is called a **fieldbus I/F**. So far, we have been talking about communication programs or software that facilitates data transfer between any two intelligent devices with the help of I/Fs present at both ends and supported by an appropriate communication program. Participating devices are the controller and RIO/fieldbus devices. Communication between any two intelligent devices can take place only if they are compatible in connectivity in terms of both hardware and software. As dealt with in Chapter 15, this compatibility in connectivity is called **communication protocol**.

17.5.3.1 Traditional

Fig. 17.21 illustrates the traditional way of communicating using a nonintelligent communication interface (communication module with simple serial interfacing).

The communication protocol, which uses special software, resides in the controller memory and is executed by the controller (in addition to the execution of

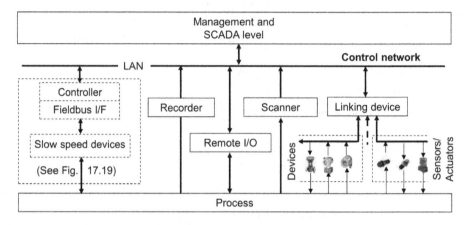

Figure 17.20 Fieldbus network: control level.

Table 17.3 Examples of a control network

Fieldbus		Masters	Nodes	Reference
Fieldbus Foundation HSE	Fieldbus	Multiple	—	www.fieldbus.org
MODBUS	Modbus	Single	247	ww.modbus.org
Profibus DP	PROFIBUS	Multiple	127	www.profibus.com
ControlNet[4]	ControlNet	Multiple	99	www.odva.org

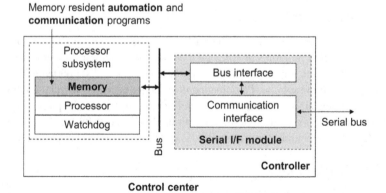

Figure 17.21 General serial interface.

automation functions). A conventional nonintelligent serial interface module facilitates the following:

- Establishes **hardware connectivity** between partners
- Assembles serial information received from the sender into parallel information (serial to parallel conversion) and transfers this to the processor over the bus
- Disassembles parallel information transferred from the processor over the bus into serial information (parallel to serial conversion) and sends this to the receiver over the medium

17.5.3.2 Intelligent

Because of its high overhead, execution of the communication protocol software demands considerable processor time from the controller. This communication is beyond just the normal data transfer between the two partners and is extensive, continuous, and repetitive. This can drastically reduce the performance of the controller.

Shortcomings in the traditional nonintelligent serial I/F led to the development of the intelligent (microprocessor-based) serial I/F, which decentralizes the execution of automation and communication functions, as illustrated in Fig. 17.22.

Here the local processor in the intelligent serial I/F module takes the communication load from the controller. This module can be personalized to different protocols by changing the protocol software in its program memory.

17.5.4 Fieldbus Protocol Standards

The most commonly used hardware connectivity standards in fieldbus are **RS-232c** and **RS-485** for serial interfaces and **RJ-45** for Ethernet interfaces, established by standards organizations. In addition to these, for technical reasons, many fieldbus protocols have their own standards for hardware connectivity: for example, IEC 61,158-2 in Foundation fieldbus.

Today, the communication protocol refers more to software connectivity, because hardware connectivity standards are few and standardized. Similarly, there are many industry standards for fieldbus **software connectivity**. This communication protocol

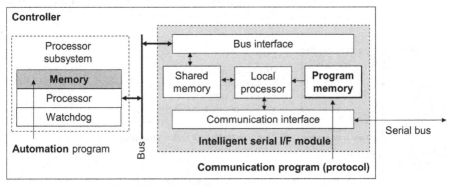

Figure 17.22 Intelligent serial interface.

software resides in all partners for data transfer. The standard, specified by IEC 61,158, includes several established protocols.

The original standard for eight types of fieldbus protocols was released in 1999 under IEC 61,158[3] (digital data communications for measurement and control; Fieldbus for use in industrial control systems). Subsequently, in 2008, the number of fieldbus protocols in the standard was enhanced to 17. In this, the list was reorganized as **Communication Profile Families.** Additional technologies have been added and some existing technologies have been removed.

17.6   Fieldbus Device Integration

Field devices need to be integrated with the automation system to operationalized their work. Because field devices are intelligent and have a number of configuration options, they need to be customized for individual use (configuration and control). For this, there is a need for software tools for each type of device. To avoid writing new software tool for each new device type, common integration tools are developed to support all types of devices. For vendors, the creation of a common tool is less expensive, less time-consuming, and less effort, rather than writing an entirely new software tool for each type of device. For end users, this means that they can choose the field devices they prefer from any vendor and still gain full access to the device's capabilities from a single configuration tool.

The first step was to define a common approach to create a device description (DD), a formal description of the data and operating procedures for a field device, including its commands, menus, display formats, etc., describing exactly what can be done with the particular device.

A DD is a piece of software that resides in the host or controller and operates the field device. This is similar to printer drivers that are used to operate different printers and other similar devices connected to the computer. Fig. 17.23 illustrates the integration of a printer with a computer.

As in case of printers, device vendors supply the DD on a CD along with the device or make it available on the Web for downloading. Any host or controller can operate the device if it has the device's DD.

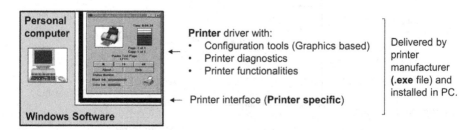

Figure 17.23 Printer integration with computer.

[3]www.iec.ch.

The DD is written in a standardized programming language known as device description language (DDL), similar to the programming language C in structure. It was developed such that, through the interpretation of a DD, a host could configure and control many different devices.

A PC-based tool called the **Tokenizer** converts the DD source code files as output files by replacing key words and standard strings in the source file with fixed **tokens**. With this, the DD provides the information needed for the host or controller to understand the meaning of the device data and its functions to integrate the device with the automation system.

DD technology, used separately by the HART, Fieldbus Foundation, and PROFIBUS, are all essentially the same, but with small differences to accommodate their unique protocol features.

This section describes two important developments, further to DD, in integrating fieldbus devices.

17.6.1   Electronic Device Description

To make DDL technology universal, electronic device description (EDD) or EDD technology was developed. EDDL[4] similar to DDL, is also a programming language to create EDDs that is independent of the operating system. This helped decouple device developers from software developers. Device developers need not worry about configuration tools, and software developers need not worry about each individual device type. Subsequently, harmonization and enhancement of the EDDL has been undertaken by the EDDL Cooperation Team (ECT) primarily consisting of the Fieldbus Foundation, PROFIBUS Nutzerorganisation, and HART Communication Foundation through merging of their individual dialects of the DDL, leading to EDDL becoming the open international standard, IEC 61,804. Subsequently, ECT has been joined by several other organizations in this field, such as the OPC Foundation.

Fig. 17.24 illustrates the integration of EDDs with the host/controller.

The main advantage of EDDL is that it uses an electronic file written in a common language to describe an intelligent device in a machine-readable format to support a handheld communicator or software applications such as a distributed control system configuration tool, intelligent device management software, etc. Because EDDL is an open technology with international standard status, it can easily and effectively be applied to any device and any fieldbus protocol.

Apart from this, EDDL provides the following advantages for device integration:

- Full use of host features and support to device functionality
- Single universal solution for a broad spectrum of devices, protocols, tools, and tasks
- Backward compatibility for investment protection
- Consistent display of devices
- Interoperability testing of devices supporting the free choice of devices

[4]www.eddl.org.

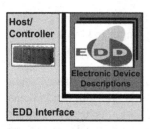

Device driver with:
- Configuration tools (Graphics based)
- Device diagnostics
- Device functionalities

Delivered by device manufacturer **(tokenized file** by EDDL), installed in host

← EDD interface **(Standardised)**

EDD (one per device):
- Is a binary executable representing device
- Contains device info such as Label, Value, Resolution, Unit description, etc.
- Configuring device is made easy
- Interoperability is addressed

EDD Interface (one per host) provides:
- Provides common executing environment for DDs
- One interface to work on many DDs of same type
- Integrated as part of host

Figure 17.24 Electronic device description (EDD)-based device integration with host.

17.6.2 FDT/DTM

Subsequently, to provide universal data exchange and to free the user from device manufacturers, field device tool/device type manager (FDT/DTM)[5] technology has been created.

FDT standardizes the communication and configuration interface among all field devices and host systems. FDT provides a common environment to access device features. Any device can be configured, operated, and maintained through the standardized user interface regardless of supplier, type, or communication protocol.

The FDT is also a piece of software that resides in the host/controller and has two components, the DTM and FDT frame, as explained subsequently.

The DTM, similar to the DD/EDD supplied by the device manufacturer, has two parts, namely the device DTM and communication DTM for fieldbus device logic and communication medium. DTM is similar to the print manager in Windows program (print drivers and their associated graphic user interface) that is installed in the computer to make the printer work. They provide a unified structure to access device parameters, configure, operate the devices, diagnose problems, etc.

DTM represents the actual device and allows access to all device functions regardless of frame in which it is hosted. Each device is configured with its DTM. DTM, residing in the host/controller, interfaces with the device on one hand over the communication protocol and provides a standardized interface (universal data interface) with the host/controller on the other hand. DTMs can be used by any FDT frame application.

The FDT frame application is also a software program that interfaces with DTMs with the following features:

- One frame can host many DTMs from many different vendors.
- Apart from the host/controller, the FDT frame can be a part of other software systems such as asset management, manufacturing execution systems, etc. or a stand-alone tool on a PC for direct communication with fieldbus devices.

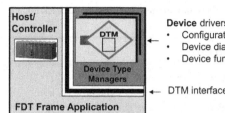

Device drivers with:
- Configuration tools (Graphics based)
- Device diagnostics
- Device functionalities

Delivered by device manufacturer (.dll file) and installed in Frame Application in host (one per device)

← DTM interface (**Standardised**)

DTM (one per device) is a driver:
- Represents actual device
- Provided by the device manufacturer
- Loaded on any Frame Application
- Has a standardized interface to Frame Application
- Has graphical user interface
- Includes complete parameters of the device
- Fits into all Frame Applications

Frame Application (one per host) provides:
- Common Environment for:
- Network Configuration
- Navigation
- User Management
- Device Management
- Database Storage
- Accepts all DTMs

Figure 17.25 Field device tool/device type manager (FDT/DTM)-based device integration with host.

- The frame provides a common environment for the management of the user, DTM, and data, network configuration, and navigation.

Fig. 17.25 illustrates the integration of DTMs with a controller/host.
Benefits of FDT technology to the user are:

- Platform independence:
  - Freedom to choose instrumentation devices and maintenance tools without constraints of communication or platform compatibility.
  - Investment protection: no need to replace existing installed devices. FDT technology incorporates existing device models.
- Fieldbus independence:
  - FDT frame application supports many protocols in both factory and process automation.
  - FDT interfaces add expanded capabilities to DDs/EDDs such as improved visualization, improved device know-how, and advanced device features.

Benefits of FDT technology to the vendor are:

- Engineering investment on existing devices is protected by building DTMs for them over their DD or EDD
- Minimum development and maintenance effort
- Reduced time, effort, and cost to introduce the device into the market and integrating it into software systems
- Ability to create tools that run on any FDT-enabled hosts
- Accepted as international standard IEC 62,453

Table 17.4 compares EDD and FDT/DTM.
The latest addition to the family of integration technologies is field device integration[6] technology. This is an integration of tried and tested EDDL and FDT technologies even though both are competing and complementary technologies in many ways, aiming to provide more value to device manufacturers, control system manufacturers, and end users.

[6]www.fdi-cooperation.com.

Table 17.4 Comparison of device description/electronic device description (DD/EDD) and field device tool/device type manager (FDT/DTM)

DD/EDD	FDT/DTM
Simple to use, text-based commands	Windows Components technology
Cannot address complex devices	Address topology of devices and projects it in FDT frame (similar to network topology)
Platform-independent	Windows platform-dependent
Consistent look and feel	Guidelines (style guides) for creating user interfaces
Upgrade is easy	Version control is major issue because of nature of development platform (com and dynamic linking)

17.7 Other Networks

All of the fieldbus networks discussed earlier used wired and slow-speed media for data exchange. To overcome the difficulty of reaching inaccessible locations for installing the instrumentation devices, wireless-based field device networks were developed. Similarly, Ethernet-based media for fieldbus networks were developed using standard components (but of industry grade) to provide a uniform network from sensor/actuator level to enterprise resource planning (ERP) level. In the following sections, two such developments, WirelessHART and Industrial Ethernet Technologies, discussed.

17.7.1 Wireless Sensor Networks

As mentioned earlier, the driving factor for the RIO concept was mainly to reduce cabling in large process plants between control systems and process points/instrumentation devices, which further led to the development of FIO. Furthermore, most often, the locations of instrumentation devices are hard to reach and are expensive for installation and cabling.

Wireless sensor technology addresses a simple, reliable, secure, and cost-effective method based on standards to interconnect control systems to process points and instrumentation devices without the need to run wires. They work with existing systems and other networks. In this section, although there are few other standards, say ISA 100, we will discuss WirelessHART[7] technology, developed by HART Communication Foundation.

WirelessHART devices are built with the following objectives:

- Built on proven and open industry standards (IEC 62,591)
- Created by industry and technology experts
- Multiple-vendor support and interoperable devices
- Use existing devices, tools, and knowledge

---

[7] http://en.hartcomm.org.

Although WirelessHART supports various network topologies such as star, mesh, and point to point, the mesh network is preferred because it is self-healing with the ability to:

- Adjust communication paths for optimal performance
- Monitor paths for degradation and repair itself
- Find alternate paths around obstructions

WirelessHART employs 2.4 GHz wireless (IEEE 802.15.4 based) for the physical layer while the data link employs secured and reliable bus arbitration with time division multiple access/carrier sense multiple access media access. The rest of the protocol structure is identical to the HART protocol. The maximum data rate is 250 kbps.

A few important limitations of WirelessHART technology are:

- Limited range owing to limited output power
- Limited local power for instruments (battery operated)
- Sharing of already used 2.4GHz frequency band
- Line of sight issues

Data communication in WirelessHART is secure because several security measures are built in to protect data integrity.

Fig. 17.26 illustrates WirelessHART devices; Fig. 17.27 shows the mesh configuration of a Wireless HART sensor network.

17.7.2 Industrial Ethernet

Ethernet technology, with its low cost of hardware extensively used in office automation, made technology an attractive option for industrial networking applications. Furthermore, the use of open protocols such as transmission control protocol/Internet protocol (TCP/IP) over Ethernet networks offers the possibility of a level of standardization and interoperability that did not remain part of the industrial networks.

Industrial Ethernet refers to the use of standard Ethernet protocols with rugged hardware that withstands the industrial environment for industrial process automation.

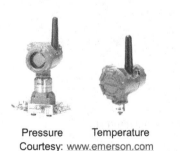

Pressure   Temperature   Temperature   Differential pressure
Courtesy: www.emerson.com   Courtesy: www.yokogawa.com

Figure 17.26 Wireless field devices.

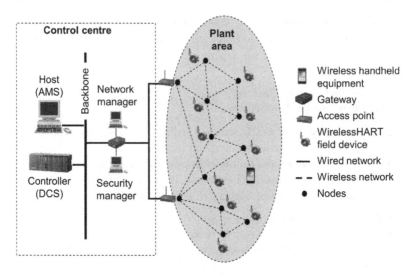

Figure 17.27 Typical WirelessHART mesh network.

Fig. 17.28 illustrates the convergence of traditional three-tier fieldbus network configuration into an Ethernet-based common network.

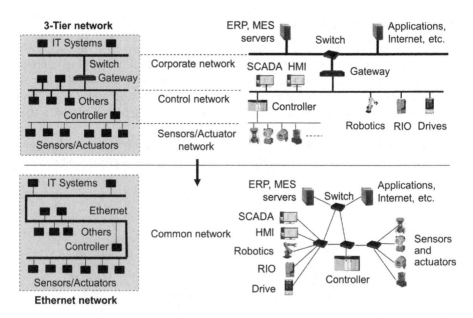

Figure 17.28 Convergence of control and information.

Some advantages of Industrial Ethernet over other networks are:

- Increased speed from the low speed of fieldbus networks to the high speed of Ethernet networks (100 Mbps and above).
- Increased distance and coverage.
- Seamless data transmission (one type of network) between the lowest level (sensors, actuators, and devices) and the top level (ERP) over uniform technology.
- Integration of office and industry world.
- Standard Ethernet access points, routers, switches, hubs, cables, and optical fiber.
- Peer-to-peer architecture over master–slave ones (simple to add and remove nodes).
- Better interoperability.

Some disadvantages are:

- Migrating existing protocols to a new protocol.
- Real-time uses may suffer for protocols using TCP owing to nondeterministic communication (difficult to estimate the response with increased traffic).
- Managing the TCP/IP stack is more complex.
- High overhead affects data transmission efficiency (minimum data size of 64 bytes).
- Prone to security intrusions from outside the plant and inadvertent or unauthorized use within the plant.

Use of full-duplex transmission and segmentation of network through switches can help reduce or eliminate collisions and make Ethernet communication more deterministic. Following are some popular examples of industrial Ethernet:

- EtherCAT[8]
- Ethernet Powerlink[9]
- EtherNet/IP[10]
- PROFINET[11]
- SERCOS III[12]

17.8 Asset Management

Asset management is a vast subject covering the management of all assets in the plant. However, in this section we are discussing the management of only the instrumentation devices. In automation systems, the instrumentation device interfaces with the process for data acquisition and control. Without reliable and consistent data, the performance of automation system is suboptimal. Hence, th instrumentation devices need to be continuously monitored and managed for reliable and

[8]www.ethercat.org.
[9]www.ethernet-powerlink.org.
[10]www.odva.org.
[11]www.profibus.com.
[12]www.sercos.org.

consistent performance. To assist this, the smart instrumentation device is built with data acquisition, control, and data exchange features, all physically and logically integrated with the device.

Before the arrival of smart devices, the host system was receiving only process data in the form of 4–20 mA and 24 VDC and nothing on either the functioning or health of the device. The diagnosis, maintenance, calibration, and configuration of instrumentation devices for their reliability and performance were reactive and off-line. Today, the smart device, with built-in diagnostics, can send maximum device data on its functioning and diagnostics online to the host for real-time analysis and predictive maintenance. The device condition's monitoring and dispatch of diagnostic data to the host is carried out online whenever the device is free from its primary function of executing automation tasks.

An asset management system at the host level sequentially gathers data sent by the devices, assigns the context, analyzes, distributes, visualizes, and finally acts. With real-time visibility into the actual health of the devices, the maintenance department achieves a quick and consistent understanding to take action.

Within the context of automation systems, process instrumentation requires considerable effort for their frequent calibration and maintenance. Device asset management helps reduce this effort and cost by proactive maintenance and provides the maximum performance and extended service life of devices. A good field device asset management system assists **proactively** in troubleshooting, predictive diagnostics, decision making, and documentation of all maintenance activities, in addition to device configuration and calibration.

For these, an asset management system must:

- Integrate diverse technologies and processes that have different data collection and data processing requirements and make them work together to improve maintenance and reduce costs.
- Monitor early warning signals or deterioration for proactive actions before a serious failure occurs.
- Support open communication standards to allow seamless work flow and tracking at all decision-making levels.
- Integrate the functionalities of devices that can also report on the conditions of related equipment to monitor their performance.

17.9 Summary

This chapter discussed the transition of I/O subsystems from CIO to RIO to FIO. The initial driving factor was cost savings related to cable, cabling, and its maintenance to move from CIO to RIO. Fieldbus I/O, physical integration of RIO, and the instrumentation device, apart from tangible cost savings benefits, also brought many intangible benefits because of their intelligent and communicability. The most important one is self-diagnostics of the devices and their report to the host for predictive maintenance. Fieldbus technology supported slow-speed three-tier networks based on multiple protocol standards. Furthermore, the technology advanced to employ Ethernet and wireless as the media for fieldbus networks. The asset management system at the host level was developed to manage fieldbus devices.

In practice, green field projects can go directly to FIO. However, in the case of expansion or retrofitting of a plant, existing or legacy instrumentation devices can be made part of FIO using RIO to support them. This approach protects the investment already made on legacy devices.

Safety Systems

18

Chapter Outline

18.1   Introduction

A **danger** or a **hazard** is always present in any process or machine, or **equipment under control**. The **risk** is the possibility of that danger happening. Risks are events or conditions that may occur any time and whose occurrence, if it does take place, has a harmful effect on people, property, and their environment. Some hard facts are:

- There is nothing like **zero** risk.
- Risks always exist but must be reduced to an acceptable level.
- **Safety** is the acceptable level of risk.

Overview of Industrial Process Automation. http://dx.doi.org/10.1016/B978-0-12-805354-6.00018-9

Hazards are basically physical objects or chemical substances that have the potential for causing harm to people, property, or their environment. Different types of hazards are:

- Physical: Objects falling or moving, and collisions, collapsing of structures, etc.
- Electrical: Flashover, shocks, burns, electrocution, etc.
- Chemical: Explosion, fire, toxic material release, result of wrong mixture of chemicals, etc.
- Mechanical: Entanglement, abrasion, grinding, cutting, thermal, pressure-releasing effects (bursting vessels, jets of gas or liquid), welding torches, etc.

A risk level for each hazard is different and needs to be quantified based on its chances (how likely), its frequency (how often), and its consequences (how bad). A tolerable risk is one that cannot be practically reduced further or the amount of risk that is accepted because any additional expense or effort does not justify its further reduction with proportionate return.

The purpose of safety is to protect people, property, and their environment from any hazard. Functional safety, a critical part of overall safety, achieves safety via safety systems (SS) to lower the probability of undesired events from happening, thereby minimizing mishaps. An SS implements the required safety functions by detecting hazardous conditions and bringing the process or machine to a safe state by ensuring that a desired action, say safe stopping, takes place. Furthermore, **functional safety** ensures that the process or machine operate correctly in response to the inputs, including safe management of its likely operator errors, hardware failures, and environmental changes. Functional safety is based on the concept of risk assessment and its reduction. For each hazard, the risk is reduced by adding layers of protection depending on the nature of risk. The SS:

- Comes into action in case the basic control system and the operator's intervention did not result in the safe state of the process or machine that is hazardous by itself or may give rise to hazardous situation.
- Independently works in parallel with the basic control system.
- Consists of one or more safety loops to monitor each hazard and takes the process to a predefined safe state as determined by the risk assessment.
- Employs dedicated instrumentation, standalone relays, modular relays, or safety programmable logic controllers (PLCs) for logic processing.

Process safety and machine safety are two separate areas with their own applications and respective standards and guidelines. This chapter discusses functional safety and SS in processes and machines, in the context of industrial automation.

18.2   Process Safety Management

The operation of many industrial processes, especially oil and gas or similar installations, involves inherent risks because of the presence of dangerous materials such as gases and chemicals. Such critical processes, while in operation, require constant monitoring and need to be moved into a **safe state** or **mitigated state** upon the occurrence of an unforeseen, unacceptable, and dangerous condition. A safe state is a process condition in which safety violations do not occur, whether the process is operating or in shutdown. After a safety breach, the safe state must be achieved within the permitted process safety time. This is to protect people, plant equipment, and the environment in and around the process from possible adverse effects that arise from such an incident.

There could be many reasons for such a hazard, either internal (say, equipment failure leading to energy buildup and subsequent explosion, accidental release of toxic gas, etc.) or external (say, fire in a charged atmosphere, etc.). Process safety management is of utmost importance to avoid an accident in the first place and mitigate loss if an accident takes place.

Process safety is a part of overall safety management and focuses on the concerns of major hazards affecting the safety of people, property, and their environment, leading to business loss. The goal of process safety management is to develop plant systems and procedures to prevent the occurrence of unwanted incidences that may lead to safety breaches. Although the operation of hazardous processes involves risks, they can be effectively managed and/or eliminated when appropriate measures are taken. Operating safely is a basic feature of plant efficiency improvement that results in higher productivity.

18.2.1 Risk

A hazard is an inherent physical or chemical characteristic that has the potential to cause harm to people, property, or their environment. To give an example, in chemical processes, it a combination of a hazardous material, an operating environment, and certain unplanned events could result in an accident. Risks are the likelihood of a specified hazardous event or an accident happening any time, and whose occurrence, if it does take place, has a harmful effect on people, property, and their environment. The risk is usually defined as the combination of the:

* Probability of the hazard occurring
* Frequency of the hazard occurring
* Severity of the hazard

18.2.2 Risk Assessment

Risk assessment is the determination of the quantitative or qualitative estimate of risk related to a well-defined situation and a recognized hazard:

* Quantitative risk assessment is based on calculations of two components of risk: the magnitude of the potential loss and the probability that the loss will occur.
* Qualitative risk assessment is based on a formal judgment of the consequence and probability using the formula: $Risk = Severity \times Likelihood$.

An **acceptable risk** is the one that judges that its cost and the efforts of implementing further countermeasure to reduce the remaining vulnerability are well understood, accepted, and tolerated. Also, assessment of the acceptability or tolerability of risks depends on how risks are perceived and the factors that influence this. Steps involved in risk analysis and assessment are:

- Identification
- History and scenarios
- Evaluation with associated hazards
- Frequency of occurrence
- Consequence and modeling
- Elimination by bringing the process to a predefined state

18.2.3 Risk Reduction

Safety is defined as freedom from unacceptable risk. This definition is important because it highlights that all industrial processes involve risk. Absolute safety, in which risk is completely eliminated, can never be achieved. Risk can only be reduced to an acceptable or tolerable level by reducing the frequency of hazardous events, the consequence, or both. Generally, the most desirable approach is first to reduce the frequency, because all events are likely to have cost implications even without serious consequences. Hence the first step toward risk reduction is to design the process such that all inherent risks are reduced to a value lower than their acceptable level by built-in features in the process itself followed by a dedicated safety management system and procedures. Risk levels can be broadly divided into:

- High, in which hazards occur and need to be **refused** altogether.
- Medium, in which hazards frequently occur and need to be **reduced** to **as low as reasonably practicable**, keeping in mind the cost-benefit of reduction.
- Low, in which hazards seldom occur and can be ignored.

Fig. 18.1 illustrates the levels of risk.

Fig. 18.2 illustrates the six steps involved in risk reduction, over and above the precautions taken while designing and engineering phases of the process.

Process safety is a blend of design, engineering, and engagement skills focused on preventing catastrophic accidents associated with the use of chemical and petroleum products, such as fire and toxic release. Safety methods employed to protect against or mitigate harm or damage to personnel, plants, and their environment to reduce risk should include:

- Adequate precautions in the design and engineering of a process
- Increased mechanical integrity of the system
- An effective basic process control system
- Safety operational procedures and training of staff
- Proactive or predictive testing of critical system components
- Safety instrumented system (discussed in the next section)
- Proper mitigating equipment and procedures

To address these measures, it is necessary to design and employ protection layers (equipment and/or administrative controls) to reduce risk in stages to an acceptable level. Protection layers start with good process and equipment design. Further layers are broadly divided into predictable (control and prevention) and unpredictable (mitigation) layers. Fig. 18.3 illustrates the protection layers.

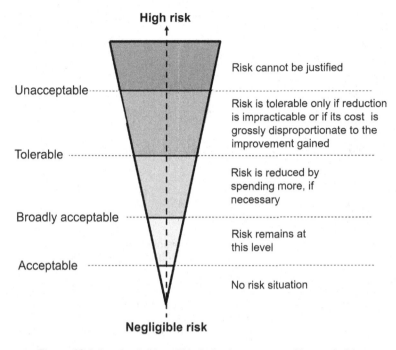

Figure 18.1 Levels of risk and level of as low as reasonably practicable.

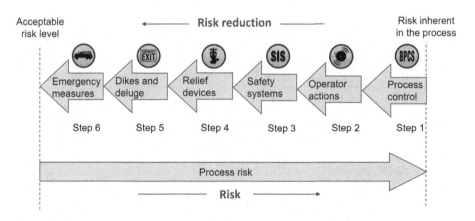

Figure 18.2 Risk reduction steps.

Fig. 18.4 illustrates actions taken at each protection layer at a broad level. Functional details of protection layers are given in Table 18.1.

These layers of protection start with safe and effective process control, extend to manual and automatic prevention, and continue with mitigation of consequences of an event:

- Protection at level 1 is provided by the basic process control system (BPCS) itself, significantly through its functionality. The maximum risk reduction that can be achieved by the BPCS is less than 10 times.

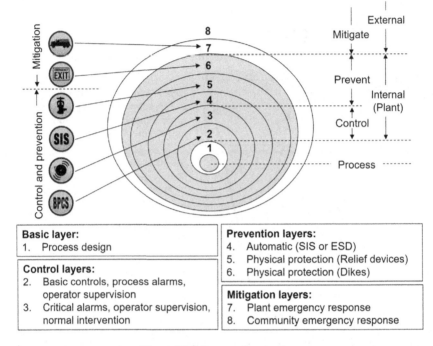

Basic layer:	Prevention layers:
1. Process design	4. Automatic (SIS or ESD)
	5. Physical protection (Relief devices)
Control layers:	6. Physical protection (Dikes)
2. Basic controls, process alarms, operator supervision	**Mitigation layers:**
3. Critical alarms, operator supervision, normal intervention	7. Plant emergency response
	8. Community emergency response

Figure 18.3 Layers of protection.

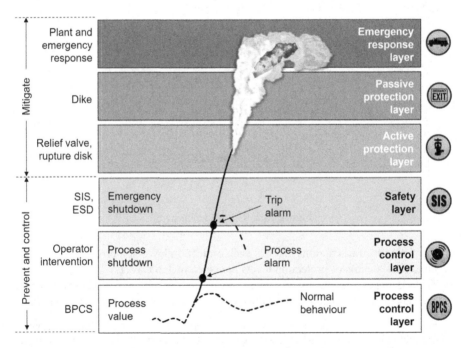

Figure 18.4 Functions at layers of protection.

Table 18.1 Layers of protection

Nature	Action	Level	Equipment	Function
Predictable	Control	Level 1	Basic process control system	Normal process automation (safe and effective process control)
		Level 2	Alarms	Operator response to critical alarms
	Prevent	Level 3	Safety instrumented system	Automated emergency shut-down actions
		Level 4	Relief	Release of excess/built-up pressure
Unpredictable	Mitigate	Level 5	Fire water and deluge	Reducing consequence after incident
		Level 6	Plant emergency response	Prevent material from reaching workers, community, and environment
		Level 7	Environment emergency response	Evacuation, firefighting, health care, etc.

- Protection at level 2 is also provided by the BPCS and associated operators. Automated shutdown routines in a BPCS are combined with operator intervention to shut the process down. Here also, the maximum risk reduction that can be achieved by the BPCS is less than 10 times.
- Protection at level 3 is provided by the safety instrumented system, typically used to bring the entire plant to a predefined safe state sequentially, in case the BPCS and operator's intervention did not result in the safe state of the process. This is discussed in the next section.

These layers are designed to **control** and **prevent** a safety-related event. If a safety-related event occurs that is not taken care of by these layers, the following additional layers are designed to **mitigate** the impact of the event:

- Protection at level 4 is an active one, with relief valves or rupture disks designed to provide a relief point that prevents a rupture, large spill, or other uncontrolled release that can cause an explosion or fire.
- Protection at level 5, a passive one, is provided by a dike or other barriers that serves to contain a fire or channels the energy of an explosion in a direction that minimizes the spread of damage.
- Protection at levels 6 and 7 are plant and environment emergency response. If a large safety event occurs, these layers respond in a way that minimizes ongoing damage, injury, or loss of life. These include evacuation plans, firefighting, etc.

The amount of risk reduction at each level depends on the nature of the risk and the amount of risk reduction that can be handled by the applicable level employed. Predictable layers are put into place to prevent hazardous occurrences whereas unpredictable layers are designed to reduce or mitigate consequences after hazardous events have occurred. Furthermore, the protection layers are further subdivided into internal (inside plant) and external (outside plant). Methods that provide layers of protection should be independent, reliable, auditable, and designed specifically for the risk involved. Overall safety is determined by how all these layers work together effectively in a coordinated manner.

18.2.4 Safety Instrumented System

Operation of any hazardous process calls for the installation of SS as a part of the process for overall risk elimination/reduction measures to deal with possible unforeseen safety breaches. The SS or safety management system that takes the process to a safe state upon detection of hazardous conditions is called the safety instrumented system or safety interlock system or safety instrumented system (SIS) or emergency shutdown system (ESD). Detected conditions may be hazardous in themselves or may give rise to a hazard eventually if no timely action is taken. SIS executes safety operations by acting to prevent the hazard or mitigating consequences in and around the process.

An SIS, a specially designed and engineered system, performs specific functions by reducing the likelihood of the impact severity of an identified emergency event and moves the process automatically either to fail-safe or maintain-safe operation mode upon detecting safety breaches. SIS is also made of the same types of automation elements (sensors, logic processor, actuators, etc.) as any process automation system. However, all elements in SIS are dedicated solely to the functioning of SIS. Furthermore, SIS operates in parallel on the same process and is independent of BPCS. This is to ensure that SIS functionality is not compromised. An SIS is considered 100% functionally safe if all failures (random, common cause, and systematic) do not lead to malfunctioning of the SS and do not result in human injury, asset loss, production loss, and environmental impact.

Fig. 18.5 illustrates the structure of SIS and its position vis-à-vis BPCS in the process plant.

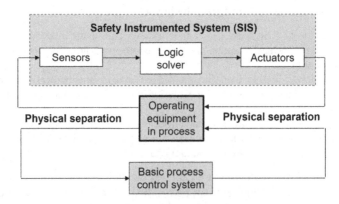

Figure 18.5 Safety instrumented system.

SIS can also have adequate redundancy to make the overall safety strategy more reliable and available. Furthermore, all components in SIS (sensors, actuators, logic processors, etc.) that are used in SIS are also specially built for highest reliability and availability. SIS is optimized for speed of operation.

Control actions are a function of the BPCS and protective actions are of the SIS. When the control function fails (or BPCS fails), the process runs out of control and there is a demand for protective action against abnormal condition. SIS responds to this demand and brings the process to a predetermined safe state.

18.2.4.1   Logic Solver

Like any control system, the SIS started with hard-wired logic solvers (relay-based or its equivalent solid state–based modules) with adequate redundancy, diagnostics, fault tolerance, and availability for performing safety-related functions.

As explained in Appendix A, relay-based and solid state–based components were employed for designing early control systems. Relay-based systems are relatively simple, less expensive, immune to most forms of electromagnetic interference, and flexible to build for different voltage ranges. However, they are bulky, prone to false trips, easily affected by environmental conditions, and unwieldy for large systems. Furthermore, any logic change called for wiring changes and drawing updates. They can handle only discrete (on–off) logic signals and not analog signals. They worked with discrete input sensors and actuators such as switches, on–off control valves, etc. On the contrary, solid-state systems, similar to relays, are also hardwired but are compact, reliable, and immune to environmental conditions.

The modules of the logic solver, whether relay-based or solid state–based, are operated in a continuous switching mode transmitting digital signals through relays or gates. The built-in diagnostic circuits on the board of each module can immediately detect if any unit stops passing the signal (malfunctioning). Normally, the detection of a failed unit leads to an alarm.

SS generally have dual or triple modular redundancy; hence failure of a single module does not cause the plant to trip but only causes an alarm. Hardware fault tolerance (HFT) provides the required redundancy. However, redundancy often reduces safety. Hence, standards must be referred to before deciding on the level of HFT.

18.2.4.2   Safety Programmable Logic Controller

Subsequently, safety PLCs replaced hardwired logic solvers. This is because of safety PLCs' ability to provide shared logic-solving facility for many safety functions within one SIS. They also offer the facilities needed by most safety functions to perform fairly simple logic combined with efficient operator interfacing and secure management of the program logic. A typical configuration of a safety PLC and its position vis-à-vis BPCS is illustrated in Fig. 18.6.

Safety PLCs are built with the highest reliability and availability, with adequate redundancy. They are specially developed for their tasks through the provision of

Operator consoles   **SIS Console**

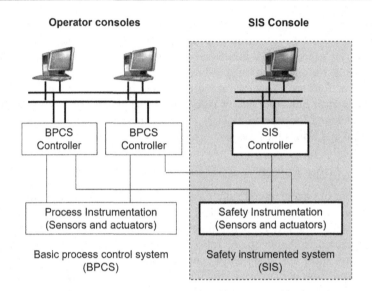

Figure 18.6 Safety instrumented system (SIS): schematic.

extensive diagnostic coverage using internal testing signals operating between scanning cycles of the application logic. Effectively, the safety PLC detects its own faults and switches itself into a safe condition before the process gets into a dangerous condition. The software of a safety PLC is specially developed to have a range of error detecting and monitoring measures to provide assurance at all times that the program modules are operating correctly. The application programs are developed with the aid of function block or ladder logic languages [International Electrotechnical Commission (IEC) 61131-3] in which each function has been extensively tested for robustness. The hardware is designed to be robust enough to withstand harsh environmental conditions.

Standard PLCs are not employed as safety PLCs because they are not designed for safety applications and have limited:

- Fail-safe characteristics
- Reliability (undetected and dangerous faults) owing to a lack of extensive diagnostics
- Stability with software versions
- Flexibility and security
- Protection in communications
- Redundancy

Fig. 18.7 illustrates typical examples in the industry of safety PLC-based SIS controllers.

- The purpose of an SIS coexisting with BPCS is to monitor potentially dangerous conditions and prevent the consequence of hazardous events. The built-in or embedded program (control strategy) that drives the SIS to prevent a safety hazard is known as the safety-instrumented function. It is designed according to a safety integrity level. Further discussion about the SIS will be continued in Section 18.2.5.

Courtesy: www.emerson.com

Courtesy: www.yokogawa.com   Courtesy: www.rockwellautomation.com

Courtesy: www.br-automation.com   Courtesy: www.schneider-electric.com

Figure 18.7 Safety instrumented system: industry examples of safety programmable logic controllers.

18.2.5 Safety Integrity Levels

Some questions such as to what extent a process can be expected to perform safely, and in the event of a failure, to what extent the process can be expected to fail safely, etc., are answered through the assignment of a target **safety integrity level** (**SIL**). SILs are measures of the safety risk of a given process, or probability of failure on demand (PFD) for the safety function. The requirements of the SIL are derived from the likely frequencies of hazardous events, the consequences of hazards, and a maximum tolerable frequency. These are determined and a safety function is engineered to bring the frequency down to a tolerable level. The risk reduction required of the safety function providing the first requirement for compliance with the safety standard is the numerical reliability measure. The numerical reliability measure is categorized by value into bands or SILs. There are four SILs based on the target reliability measure required. The higher the SIL level is, the lower the probability is of failure on demand for the SS and better the system performance is. Also, as the SIL level increases, typically the cost and complexity of the system increases as well. An SIL level applies to an entire system and does not cover individual products or components. SIL levels are used when implementing a safety instrumented function (SIF) that must reduce an existing intolerable process risk level to a tolerable risk range.

18.2.5.1 Demand and Continuous Modes

When assessing an SS in terms of failure to function, two main options exist, depending on the mode of operation. If an SS experiences a low frequency of demands, typically less than once per year, it is said to operate in **demand mode**. Hence, we can say that the **SIS and relief system** operate in demand mode, similar to an air bag system in the car that is used rarely and works in **demand** mode; while the BPCS, similar to the breaking system in a car that is used continuously, works in **continuous** mode. In view of this, for demand mode SS, it is common to calculate the average PFD, whereas the probability of a dangerous failure per hour is used for SS operating in continuous mode.

Functional safety involves identifying specific hazardous situations followed by establishing maximum tolerable frequency targets for each mode of failure. The maximum tolerable failure rate for each hazard will lead to an integrity target or SIL for each piece of equipment. The SIL defines the safety level required by each SIF. In other words, the SIL is a way to establish safety performance targets for SIS. Table 18.2 lists various SILs.

A determination of the target SIL for each hazard requires identification of the hazard involved, an assessment of the risk in the identified hazard, and an assessment of other independent protection layers that may be in place.

Also, there is no SIL for BPCS functions. BPCS functions are always SIL 0 functions and the maximum risk reduction that can be achieved is always fewer than 10 times. SIFs are implemented in the SIS to take protective actions in the event the control functions implemented in BPCS fail to take action.

18.2.6 Safety Instrumented Functions

An SIF is an identified safety function by an automated action that provides a defined level of SIL (risk reduction) for a specific hazard. SIFs are implemented as part of an overall risk reduction strategy intended to eliminate the likelihood of the occurrence of **previously identified** events that could be from a minor equipment failure or an event involving an uncontrolled catastrophic situation, say an explosion. An SIF is normally associated with the dealing of severe security breaches and not routine

Table 18.2 Safety integrity levels (SIL) and related measures

SIL	Availability	PFD avg.	Risk reduction	Quantitative consequence
4	>99.99%	10^{-5} to $<10^{-4}$	100,000–10,000	Potential for fatalities in community
3	99.99%	10^{-4} to $<10^{-3}$	10,000–1000	Potential for multiple onsite fatalities
2	99–99.9%	10^{-3} to $<10^{-2}$	1000–100	Potential for major onsite injuries or fatality
1	90–99%	10^{-2} to $<10^{-1}$	100–10	Potential for minor onsite injuries

Availability: Probability that equipment will perform the task.
PFD Avg.: average probability of failure on demand (PFD) is used to calculate safety system reliability; PFD is the probability of a system failing to respond to a demand for action arising from a potentially hazardous condition.
No standard has been defined yet for SIL 4.

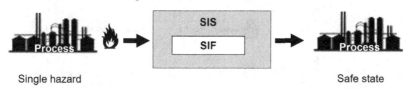

A SIF detects a specific hazard and brings the process into safe state

Figure 18.8 Safety instrumented function (SIF): single hazard.

equipment/asset protection. In other words, an SIS with an appropriate SIF is for monitoring potentially dangerous conditions and mitigating the consequence of a hazardous event.

The amount of defined risk reduction to be provided by the SIF is seen as the level of dependability of SIS. An SIF can be either a safety instrumented **protection** function that operates in continuous mode or safety instrumented **control** function that operates in demand mode. An SIF in an SIS can be compared with an application program residing in a normal PLC platform, with the difference that each SIS has its dedicated SIF with **independent** sensors, logic solver, final control elements, and SIL. Fig. 18.8 illustrates the organization of an SIF for a single hazard.

18.2.7 Safety Instrumented System (Continued)

An SIS is a combination of hardware (logic solver/safety PLC platform) and logic (SIF). There may be more than one SIF for each hazard. In other words, an SIS is a container of a set of one or more SIFs. This is illustrated in Fig. 18.9.

When an SIF has to attend to multiple potential causes for a **single hazard**, each with its own SIL, the highest SIL is generally selected for the entire SIF for execution. This situation is illustrated in Fig. 18.10.

The correct operation of an SIS requires a series of equipment in the chain to function properly:

- Sensors: They detect the hazard and provide necessary information (values of analog parameters such as temperature, flow, level, etc., and open–close status of control valve, etc.) that has the potential to create security breaches and feed to logic processor. These sensors are dedicated to SIS.
- Logic processor (PLC): They are equipped with the necessary logic; use information fed by sensors to analyze and determine whether the equipment or process is in a safe state or in emergency state by detecting abnormal operating conditions in the process (high/low value of temperature, flow, level, etc. and open–close condition of control valve, etc.); and automatically take appropriate decisions and feeds to the actuators for actions. Highly reliable logic solvers are used that provide both fail-safe and fault-tolerant operation.
- Actuators (final control elements): They implement actions decided by logic processor by driving the appropriate process equipment to change its status (open or close the control valve).

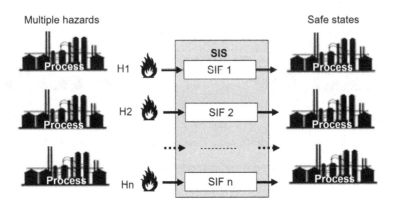

H: Hazard

Each SIF is equipped with its own **dedicated** sensors, logic solver, final elements, and **SIL**
A SIS is a combination of one ore more SIFs

Figure 18.9 Safety instrumented function (SIF): multiple hazards.

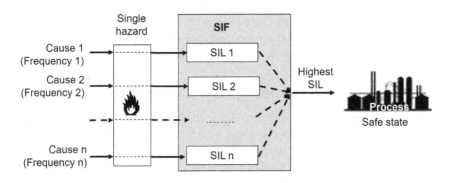

When a SIF has multiple potential causes for a single hazard, each with its own SIL, the
highest SIL generally used for the entire SIF

Figure 18.10 Safety instrumented function: multiple safety integrity levels.

They must have the ability to bring the process to a safe state or provide adequate hazard mitigation for the identified hazard.
• Support systems: They provide necessary infrastructure such as power supply, instrument air, etc., to maintain the required integrity and reliability of the entire chain of
operation.

These are required to function together in the designated order faultlessly to bring the
process into a safe state and isolate the process in the event of an emergency. It should be
kept in mind that SIS **neither improves the yield of a process nor increases efficiency,
nor reduces the cost of risk.** However, SIS does save the money by loss reduction.

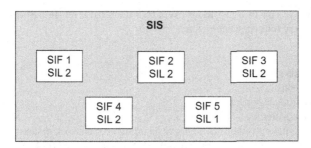

Figure 18.11 Safety instrumented system–safety instrumented function–safety integrity level relationship.

It is important to note the relationship among SIS, SIF, and SIL. Every SIS has one or more safety functions (SIFs) and each indicates a measure of risk reduction indicated by its SIL. The SIS and the equipment do not have an assigned SIL. **Basic process control is suitable for use within a given SIL environment.**

SIL defines the amount of risk reduction for a specific hazard and the level of dependability of SIF, SIF is designed based on SIL, and SIS executes the SIF whenever the particular hazard occurs. In an automation context, we can also say that an SIF is a safety loop that performs a safety function providing a defined level of protection (SIL) against a specific hazard by automatic means and brings the process to a safe state. This is illustrated in Fig. 18.11.

Before we close the discussion on risk and risk reduction, we need to revisit Chapter 3 on the use of intrinsically safe barriers (ISB) for additional standalone protection. In process plants, there is a possibility of the incidence of high-voltage surges on field signal cables (running between instrumentation devices and the control center) owing to lightning strikes, falling high voltage cable, etc. This can affect people in operation and property at the control center or at the plant. As discussed in Chapter 3, ISBs are installed to protect people and equipment in such a situation. Installation of ISBs is a part of overall process safety.

18.2.8 Safety Standards

In any process plant, there is no such thing as risk-free or 100% safe operation. Hence, one of the first tasks of the SIS designer is to perform a risk-tolerance analysis to determine what level of safety is needed to comply with safety regulations. Following are the standards and their brief descriptions.

- IEC Standard 61508 (Functional Safety of Electric, Electronic and Programmable Electronic Systems) is a general standard that covers functional safety related to all kinds of processing and manufacturing plants.
- This is an international standard relating to the functional safety of electrical, electronic, and programmable electronic safety-related systems whose failure could have an impact on the safety of persons and/or the environment. This can also be used to specify any system used for the protection of equipment or product. It is an industry best practice standard to enable

the reduction of risk of a hazardous event to a tolerable level. In particular, this standard takes care of the following equipment/systems:

- ESD
- Fire and gas systems
- Turbine control
- Gas burner management
- Dynamic positioning
- Railway signaling systems (machinery guarding and interlock systems)
- IEC Standard 61511 and International Society of Automation (ISA) S84.01 (replaced by ISA 84.00.01-2004) (Functional Safety: Safety Instrumented Systems for the Process Industry Sector), standards specific to the process industries. These standards specify precise levels of safety and quantifiable proof of compliance.

IEC standards specify four possible SILs (SIL1, SIL2, SIL3, and SIL4); however, ISA S84.01 recognizes only up to SIL3.

18.2.9   Integrated Control and Safety Systems

The BPCS is a dynamic system to manage the process in a continuously changing environment for maximum productivity whereas the SS is a static system most of the time and comes into action only when a process parameter is out of control and cannot be managed by the BPCS (or is unable to operate the process within safe limits). This means that the SS should be designed such that it reliably works when the situation demands. This means that the reliability and the availability requirements of the SS are much higher than that of the BPCS. However, integration of both of these systems has advantages considering synergy between them. The main advantage is the reduction in costs associated with maintaining two separate systems each with its own engineering, operation, maintenance, etc. This section discusses the integrated control and safety system (ICSS) within the compliance framework of standards set for SS.

18.2.9.1   Independent Operation

As illustrated in Figs. 18.5 and 18.6, BPCS and SIS are totally independent and work in parallel, except BPCS may receive some inputs from SIS at the instrumentation level for display purposes in the operator console. Here, **apart from its own/dedicated controller**, each system has independent interfaces (operator, engineering, configuration, network management, etc.). The responsibilities of control and safety personnel who manage these assets are also independent and clearly segregated. The process engineer's responsibility is to maximize plant availability and operational profit whereas that of the safety engineer is safe operation of the plant. Furthermore, according to the IEC 61511 standard, SS should be dedicated to safety-critical assets. This is because most control systems are not sufficiently robust and fail-safe to manage safety-critical functions all of the time. The standard also does not recommend implementing nonsafety functions in SS, because this may lead to greater complexity in SS and increase the difficulty of carrying out life-cycle activities such as design, validation, functional safety assessment, and maintenance.

18.2.9.2   Integrated Operation

Fig. 18.12 illustrates three approaches[1] for the integration of BPCS and SIS.

*   Interfaced: BPCS and SIS are based on dedicated hardware platforms and are connected over a communication link (gateway) for data exchange. Both BPCS and SIS are **physically and logically separate**.
*   Integrated: BPCS and SIS are based on dedicated hardware platforms and are connected over a local area network for data exchange. Both BPCS and SIS are **physically and logically separate**.
*   Common: BPCS and SIS are combined physically and functionally. They use a shared or common hardware platform [controller, input–output (I/O), etc.] with standard control and safety-related programs installed and are executed in parallel and independent of each other. **BPCS and SIS are physically not separate, but logically separate**. Here, the common cause failures could affect the reliability of the safety functions and hence must be avoided. Standards also do not recommend this configuration.

In this section, we will be discussing the integrated option, which is generally preferred. Advances in information technology have made it feasible to have ICSS to combine process control and safety functions within a common automation infrastructure while ensuring safety and security regulatory compliance. With this approach, plant personnel can view the status of the SS and its applications and combine this information with process control functions. Today, technology allows sharing of critical information not only between SS and control systems but also among other third-party

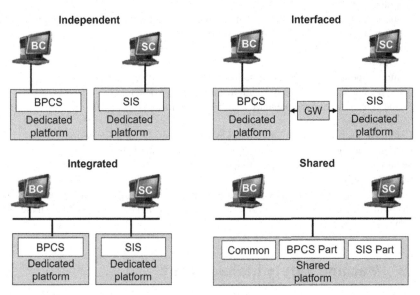

BC: BPCS Console, SC: SIS Console, GW: Gateway

Figure 18.12 Integrated control and safety systems.

[1] www.arcweb.com.

systems such as via suitable interfaces. The best and most reliable approach to integration is to maintain principles of segregation among safety and control strategies developed by different methods, groups, and vendors. The most important aspect to be considered in the process of integration is how to achieve an integrated control and safety solution with advanced control functionality, productivity, safety and security compliance. The suggested objectives are facilitating:

- Operational integration that allows a common, seamless, secure, and transparent interface to the process under consideration so that all the required information (whether of control or safety) be made available at the operational level.
- Peer-to-peer communication between safety controllers and process controllers for quick information exchange (with accurate time synchronization) from one to the other to anticipate and react to abnormal situations effectively.
- Data availability at higher levels (from the lowest level, at process I/O and at safety I/O) of operations and turning them into information that is usable for various higher-level applications.
- Common configuration tools for a single point of data entry and its replication in other databases for use at all levels in both control and SS (elimination of data mapping duplication).

Because the integration of functions in **process controllers** and **safety controllers** is not allowed, as per the IEC 61508 standard, ICSS is feasible only at the **human interface/supervisory control and data acquisition level** with a common operator console, engineering console, historian, etc., for both systems. Detailed benefits of ICSS are:

- **Engineering**:
- Common environment, tools, programming languages, and communication methods between control and SS. No additional development.
- Configurable access rights to each system ensuring safety and security.
- **Operations**:
- Common operator workplaces for both control and SS with common look and feel of graphic displays, faceplates, style of navigation, alarms, events, and trends.
- Reduced operator and maintenance training requirements.
- **Maintenance**:
- Compliance reports for automatic shutdown and reports for override actions.
- Flexible report generation and distribution with scheduling options.
- Securing historical data storage with fault tolerance and user access restrictions.
- **Security audit trail**:
- Auditing of all operations and engineering actions such as alarms, configuration changes, etc., with log reports that include date and time stamps, and the node from which the operation was performed.

18.3 Machine Safety Management

Compared with process safety, machine safety addresses safety issues in the workplace to keep people (mainly operators) near the machine safe, typically from such incidences such as electric shock, falling, crushing, temperature, radiation, noise, and chemicals associated with machines (while in operation or standstill). Conceptually, most of the discussion about process safety management applies to machine safety as well, although it is limited to machines and their operational environment. In view of this, some of the statements made in this section may look repetitive but they are made to provide continuity. Safety is now an integral part of the machine functionality rather than added as an afterthought to comply with regulations. The purpose of machine safety is to protect people from being harmed. The functional safety of the machine achieves this via SS that lower the probability of undesired events, thereby minimizing mishaps.

As per the machinery directive, the manufacturer of the machine ultimately remains responsible for the safety regulatory compliance of machines and associated safety components. It is an offense to supply machinery unless the provisions and requirements of the machinery directives (as applicable in respective regions (say, the American National Standards Institute in the United States and Conformité Européene in Europe) are met).

As mentioned under process safety, standards define safety as freedom from unacceptable risk. The most effective way to eliminate risks is to take all possible precautions during the machine design stage itself to provide functional safety. However, the reduction of all types of risks by design alone is not always possible or practical. Hence, apart from safe design, the next best options to functional safety are safeguarding, safe operating procedures, training operators about safety, and administrative measures. All of these are to ensure the safety of people and their environment. For example, stopping a machine quickly and safely not only reduces risk but increases machine life, uptime, and productivity compared with abrupt stops. Functional safety in machines usually means that SS monitor and, when necessary, override machine applications to ensure safe operation. An SS is designed to implement the required safety functions by detecting the abnormal or hazardous conditions in the machine and its environment and bringing machine operation to a safe state by ensuring that a desired action takes place. For example, an SS can monitor machine speed, direction of rotation, etc., detect deviations, if any, from the normal or expected operation, and take the machine operation to a safe state if required. Any failure in the SS itself immediately increases risks related to the machine operation.

18.3.1 Challenges

The traditional perspective has been that safety inversely affects productivity. However, the modern perspective is that safety and productivity complement each other. This is possible by considering safety in the early stages of machine design. Manufacturers are now realizing that a well-engineered SS help achieve productivity needs without compromising on operator safety.

18.3.2 Risks

Before we proceed to discussing issues related to machine safety, it should be noted that the following precautionary measures are already in place, meaning machines are equipped with:

- Appropriate markings conforming to safety standards in the region.
- Emergency stop buttons of the correct type, fit, function, and easy accessibility.

- Adequate floor area around the machine with proper lighting, free from slips, trips, debris, and foreign objects.
- Adequate guarding fitted with interlocks where required and are positively acting when required.
- Electrical enclosures that are locked and have a warning signal.

In machines, common types of hazards that cannot be eliminated and can cause serious injury to the people, if not adequately controlled are:

- Mechanical hazards such as moving parts that have sufficient force in motion.
- Nonmechanical hazards such as harmful emissions, contained fluids or gas under pressure, chemicals and chemical byproducts, electricity, noise, etc.
- Access hazards caused by poor accessibility to machines (for operation, maintenance, etc.).

The most compelling reason for machine safety is the moral obligation to avoid harming anyone dealing with the machine and to abide by the laws that require machines to be safe and sound to avoid accidents. Hence, for legal and economic reasons, safety must be taken into account over the entire life cycle of the machine (design, manufacture, installation, operation, maintenance, and eventual scrapping). Some of the costs of accidents, such as compensation to injured employees, etc., are obvious whereas many other costs are difficult to identify and quantify. The full financial impacts in case of an accident, apart from higher insurance premiums, are the loss of production, customers, and even reputation.

In view of these, investing in risk reduction in machines not only improves safety but also makes a lot of commercial sense. Some risk reduction measures, on the contrary, can actually increase productivity because machine uptime is increased.

18.3.3   Risk Assessment

Risk assessment is based on a clear understanding of the machine limits, functions, and tasks that may be required to be performed at the machine throughout its life. There are various techniques for risk assessment or estimation. However, their evaluation does not take us anywhere because they are highly subjective. For example, a risk might appear tolerable in a situation or place whereas it might be unacceptable in another situation or place. Furthermore, a history of accidents or incidents may reveal some useful indicators but it is not adequate and does not guarantee a reliable indication of accident rates that can be expected. Hence, practical approaches are necessary to arrive at risk assessment by identifying:

- Operational limits of the machines
- Possible hazards and their probability, severity, and frequency
- When and under what circumstances hazards happen, etc.
- People who might be affected

These data help priorities an assessment of risks.

18.3.4   Risk Reduction

After risk assessment, risk reduction is performed if necessary, and safety measures are selected based on information derived from the risk assessment stage. Risk reduction is defined as the measures taken to eliminate any risk throughout the foreseeable life cycle of the machine, starting from its assembly, installation, commissioning, and **operation**, until eventual disabling, dismantling, and scrapping. Among these stages, the longest is its operation stage. In general, if a risk is present and can be reduced, all efforts should be made to reduce it, subject to commercial realities (disproportionately high cost and low return, practical limitations, etc.). Risk assessment is an iterative process: identify, prioritize, quantify, redesign, and safeguard. This process is repeated to assess whether individual risks have been reduced to a tolerable level and no additional risks are introduced.

Compared with process safety, which is always in demand mode, machine safety is a high-demand application. In machine safety, the machine is brought to a safe state that can be a safe stop whenever a hazardous situation is detected. While in process safety, the entire plant is not shut down immediately. Shutdown is done in phases and in a timed manner.

18.3.4.1   Inherent Safety by Design

During the design stage, eliminate the risk by an inherent safety by design approach. Some risks can be avoided or eliminated by simple measures such as:

- Automating tasks, such as machine loading that can result in a risk.
- Using nonflammable chemicals instead of flammable, which can result in a fire.
- Avoiding sharp edges, corners, and protrusions, etc., that can cause cuts and bruises.
- Increasing gaps to avoid getting a body closer to dangerous parts (moving, heat, etc.).
- Reducing force, speed, pressure, etc., that can cause injury.

However, during this approach, care should be taken to avoid substituting one risk to eliminate another. For example, air-powered tools avoid hazards associated with electrical tools but can introduce other hazards such as injecting compressed air into the body, compressor noise, etc.

18.3.4.2   Safeguarding

Wherever **inherent safety by design** is not practicable, physical guards are placed wherever human intervention is not required. When human intervention is required for occasional or frequent access, SS are employed. Fig. 18.13 illustrates the typical structure of SS.

SS ensures that random hardware faults, systematic design errors, or human mistakes do not result in the malfunction of safety related functions with the potential consequence of the injury or death of humans, hazards to the environment, loss of equipment, or production. Contrary to the production system that focuses on **throughput**, the SS focuses on **protection** by:

- Monitoring and controlling the conditions of a machine that are hazardous within themselves or, if no action were taken, may cause hazardous situations.
- Running independently and in parallel with a machine control system.
- Working with SS (stand-alone relay, modular relay, safety controller, or safety PLC) for logic processing.
- Executing one or more **safety loops** for each hazard that monitors and controls its supply of energy, as determined by the risk assessment.

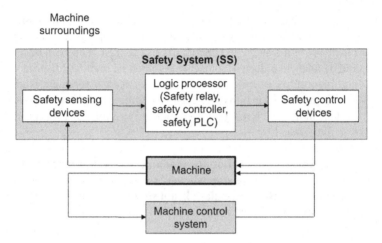

Figure 18.13 Safety system configuration.

However, an SS is as effective as its **weakest** link. Hence, there is a need to consider all aspects of the system (input, control, and output) and how they work together to meet safety requirements and standards.

The SS consists of **technical protective devices** (sensing, logic processor, and control) to accomplish the safeguarding functions ensuring safety and productivity together. Typical areas where technical protective devices required are:

- Interlocking/locking of physical guards
- Access protection
- Hazardous area protection
- Hazardous point protection
- Safe position monitoring
- Safe commands, etc.

Typical safety sensing devices are:

- Safety switches to detect position, such as moveable guards, safe machine commands, etc.
- Interlocks to prevent opening of guards
- Safety light curtain for an access or hazardous point protection
- Safety laser scanner for hazardous area protection

These sense changes in physical conditions and report them to the **safety logic processor** for further processing.

Typical **safety logic processors** are:

- Safety relays
- Safety controllers
- Safety PLC, etc.

These monitor signals from safety-sensing devices, process them with predefined built-in embedded logic, and drive **safety control devices** to control the machine or bring it to a safe state, if required.

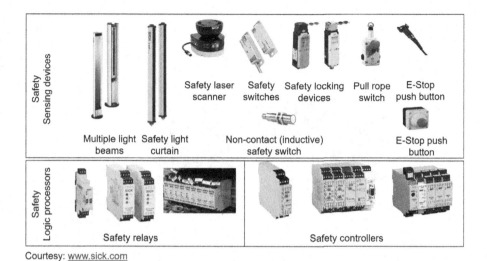

Figure 18.14 Technical protective devices: examples from the industry.

Typical **safety control devices** are contactors, starters, drives, etc. These are basically actuators that execute final commands issued by the safety logic processors to the machine. Other output devices are emergency lamps, hooters, etc., for audiovisual annunciation.

Fig. 18.14 illustrates examples of technical protective devices.

18.3.4.3   Integration of Technical Protective Devices

Wherever human interactions are part of machine jobs, technical protective devices need to be integrated to form the SS. Keeping in mind mechanical requirements such as placing the technical protective device at the right minimum distance from hazard zone for a higher response time, a protective device must be integrated into the SS. The term "safety system" describes the entire chain of monitoring signals from safety-sensing devices to processing them with its predefined, built-in, embedded logic, and to driving the safety control devices to control the machine. The entire chain of sensing devices, logic processors, and control devices is designated as functional safety and must match the intended safety level; this will be discussed in Section 18.3.5.

Fig. 18.15 illustrates an example of the integration of technical protective components in an **end-of-line packaging machine**.

18.3.4.4   Performance Level Required

Performance level required (PLr) in machine safety is similar to SIL in process safety. To determine the PLr, a risk graph is used to determine the required safety level. Parameters such as the severity, frequency, and probability of the hazard are used to determine the magnitude of risk. The result of the procedure is a PLr defined in five discrete steps. It defines the needed structure of the control for the entire safety function, the reliability of the components used, and their ability to detect faults, as well as

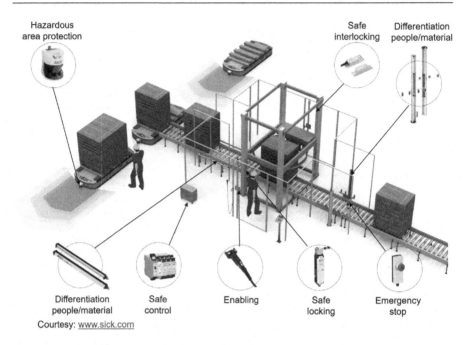

Courtesy: www.sick.com

Figure 18.15 End-of-line-packaging machine: integration of technical protective devices.

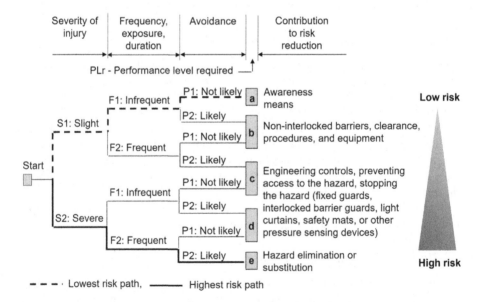

Figure 18.16 Risk graph: estimation of performance level required.

the resistance to multiple common cause faults in multiple channel control systems. This must be determined for each hazardous point.

Fig. 18.16 shows a risk graph with levels of PLr.

Table 18.3 Comparison of performance level and safety integrity level

Performance Level Required	Probability of dangerous failure per hour	Safety integrity level
a	$\leq 10^{-5}$ to $<10^{-4}$	None
b	$\leq 3 \times 10^{-6}$ to $<10^{-5}$	1
c	10^{-6} to $<3 \times 10^{-6}$	1
d	$\leq 10^{-7}$ to $<10^{-6}$	2
e	$\leq 10^{-8}$ to $<10^{-7}$	3

Note: Approximate correspondence between performance level and safety integrity level.

Various terminologies in the risk graph determine:

- Severity of injury/damage:
  - S1: Slight, usually reversible
  - S2: Severe, usually irreversible, including death
- Frequency and duration of exposure to the hazard:
  - F1: Infrequent, rare, and/or short exposure
  - F2: Frequent, continuous, and/or long exposure
- Possibility of preventing the hazard or limiting the damage caused:
  - P1: Not likely, possible under certain conditions
  - P2: Likely, possible

Table 18.3 illustrates the approximate correspondence between SIL in process safety and PLr in machine safety.

18.3.5 Complementary Measures

Some administrative measures taken are putting in place:

- Design documentation of safe operating procedures
- Education and training of operating staff
- Warning signs wherever necessary
- Machine use guidance procedures
- Personal protective equipment, etc.

In addition to these, emergency stops are provided for all machines. Operation of the emergency stop switch brings the machine to a safe position as per the defined ESD procedure. However, use of emergency switches is not considered a primary means of risk reduction. Instead, it is treated as a complementary protective measure (backup) for use in an emergency only. The switches are designed to be robust, dependable, and easily accessible. They are installed in all crucial positions where necessary. A common example is emergency stop switches installed at the two ends of an escalator.

18.3.5.1 Residual Risks

Even after all risk reduction processes, some risks remain that cannot be eliminated or further reduced. They are called residual risks. It is for the designer to judge whether the residual risk is tolerable or whether further measures need to be taken, such as providing information about those residual risks in the form of warning labels, instructions for use, etc.

18.3.5.2   Standards

Functional safety standards are intended to encourage machine designers and manufacturers to focus more on functions that are necessary to reduce each individual risk and improve the performance required for each function. These standards make it possible to achieve greater levels of safety throughout the life cycle of the machine. Many standards and technical reports provide guidance on risk assessment. Some are written for wide applicability and some are for specific applications. Following are some important standards are that are briefly discussed:

Risk assessment:

- IEC/EN 62061: Safety of machinery: Functional safety of safety-related electrical, electronic, and programmable electronic control systems

This deals with the specification and design evaluation of AC side harmonic performance and AC side filters for high-voltage direct current schemes.

- EN ISO 13849: Safety of machinery: Safety-related parts of control systems

This deals with safety-related design principles of employed control systems to establish different safety performance levels. Part 1 defines the general principles for design and part 2 describes the validation.

- ISO 12100: Risk assessments

Functional safety:

- **IEC/EN 61508:** Functional safety of safety-related electrical, electronic, and programmable electronic control systems

This standard contains requirements and provisions that are applicable to the design of complex electronic and programmable systems and subsystems. The standard is generic so it is not restricted to the machinery sector.

18.4   Summary

This chapter discussed SS in process plants and in machines. Hazards in process plants are mainly chemical whereas hazards in machines are physical, electrical, and mechanical. The purpose of safety is to protect people, property, and their environment from any hazard. SS come into action when basic control systems and operator interventions do not result in the safe state of the plant or machine. To achieve this, risks associated with the process and machines are analyzed and reduction procedures are designed and implanted through SS as per established standards. The first step in the management of safety is to take enough care while designing the process plants and machines. *Compared with process safety, which is always in demand mode, machine safety is a high-demand application. In machine safety, the machine is brought to a safe off or fail-safe situation whenever a hazardous situation is detected, whereas in process safety, everything is not shut down immediately and shutdown is done in phases and in a timed manner.*

Management of Industrial Processes

19

Chapter Outline

Overview of Industrial Process Automation. http://dx.doi.org/10.1016/B978-0-12-805354-6.00019-0

19.1   Introduction

For any industrial enterprise, the core objectives are to drive excellence in the quality of products and services, drive its own patented technology, remain in a preeminent position, and create value for all stake holders. This changed when business decisions in enterprises, as users, started taking inputs from every aspect of the supply chain. Furthermore, increases in consumerism, competition, and consumer demand for better value, etc. forced enterprises to offer flexibility, modularity, reliability, and integrated solutions. History witnessed this transformation from mechanical to pneumatic, electronic, and finally complete solutions that were information technology–based. On the other side, end users or customers wanted maximum return on the product or service purchased. In today's context, every enterprise wants to improve its market share and margins for business sustainability. Integrated automation and information technology (IT) is proving to be a great enabler in this direction.

Earlier objectives of automating the management of industrial processes were to produce goods and deliver services, conforming to quality, consistency, and cost-effectiveness, essentially to stay in the market. However, over the years, addressing new requirements and challenges, as explained earlier, have become necessary to sustain business and excel. In other words, industrial processes have to perform **much more** than just meeting quality, consistency, cost-effectiveness, and delivery standards.

Automation has had an important role in the management of all industrial processes in meeting several performance indices such as productivity, safety, and sustainability, over and above quality, consistency, cost-effectiveness, and delivery. This chapter, along with the case studies, discusses:

- Classification of industrial processes regarding their application, operation, and physical nature
- Transformation of industrial processes to meet current challenges
- Application of automation and information technology in industrial processes

19.2   Classification of Industrial Processes

Technically, all industrial processes can be broadly classified into three levels:

1. Application: Any industrial process can be either manufacturing that creates physical values to deliver goods or products as per specification or infrastructure that deliver services as per requirement.
2. Operation: Any application process, operation-wise, can be continuous, discrete, or batch.
3. Physical: Any continuous or discrete or batch process can be localized or distributed.

However, it is difficult to draw a fine line between localized and distributed processes. Furthermore, any industrial process can be of different combinations such as **batch-cum-discrete** or **batch-cum-continuous**. This essentially depends on industrial processes and their setup. For example, a batch of 100 cars of a particular model with specific colors and interiors is manufactured on an assembly line, which is a

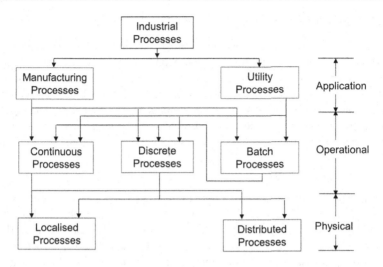

Figure 19.1 Industrial processes: classifications.

discrete process. Similarly, soap noodles required to make different brands of soap are made in a continuous process plant by one agency and given to different manufacturers who, in turn, use batch process to make different types of soap cakes using their specific recipes. Fig. 19.1 illustrates various possibilities.

19.2.1 Application

Industrial processes should be designed to meet business needs or applications. These are a customer's requirements for goods and services for their consumption while conforming to statutory and safety guidelines and local laws. As mentioned earlier, application processes have subcategories.

19.2.1.1 Manufacturing

Manufacturing is a process of transforming inputs into finished goods and products. In other words, this is a set of value-creating activities that develop, produce, and deliver goods and products to customers. These activities generally start from research and development at one end until recycling is carried out at the other end. Manufacturing processes can be further divided into either process plants or factories. These are further discussed in the following sections.

19.2.1.2 Utilities

The utility process is a costly capital investment to create assets that are used over a long time and are complex in terms of design and architecture. These processes are required for public convenience or living and for economic growth and development in a country, region, state, or area. These are further divided into civic, industrial, and backbone utilities. These are further discussed in the following sections.

19.2.2 Operational

At a high level, industrial process operations are divided into three major categories: continuous, discrete, and batch. Continuous operations are mainly concerned with processing of inputs whereas discrete operations are concerned with assembly of inputs. Batch operations can be either continuous or discrete, but of short duration. Occasionally, there are operations that use elements of both continuous and discrete type. The biggest difference between these two distinct areas is that the outcome of a discreet operation can be reversed without difficulty although that is not possible with continuous operations.

19.2.2.1 Continuous

Continuous processes are those that operate on a 24/7 basis uninterrupted with periodic and planned maintenance shutdowns. They are primarily chemical transformation (blending, and mixing, processing, etc.) or mechanical transformation (heating, pressing, forming, rolling, etc.) These are also called continuous production or continuous flow processes because the inputs that are being processed are continuously in motion, undergoing changes (chemically or mechanically) to produce end products with no interruption. The resulting product is single and unique.

In this process, there is nothing like semifinished goods, only the finished goods or end products. The production platform (industrial process) here is dedicated to the end products and cannot be used to manufacture any other product, even if it is similar. Some examples of continuous processes are power plants, cement plants, chemical/petrochemical plants, steel plants, water/sewage treatment plants, power transmission and distribution, crude transportation pipelines, etc.

19.2.2.2 Discrete

Discrete processes are for the production of distinct items (assembly of components, parts, etc.) with resulting products that are easily identifiable. This process differs greatly from the continuous process, in which the resulting product is continuous and unique. This type of process is often characterized by individual or separate unit production. Here, units can be produced in low volume of high complexity or high volumes of low complexity. Low-volume/high-complexity production calls for an extremely flexible manufacturing system that can improve quality and cut costs. High-volume/low-complexity production puts a high premium on inventory control, lead time, and reduction or limiting material cost and waste.

Here also, the production platform (industrial process) is dedicated to the end products and cannot be used to manufacture any other product, even if it is similar. Some examples of discrete processes are automotive plants, assembly of electronics modules, and mechanical modules, furniture, conveyers, packaging, etc.

19.2.2.3 Batch

Batch processes are used to manufacture a large variety of products in batches, but in relatively small quantities. This process calls for a variety of actions to be taken at different stages of batch exhibiting the characteristics of both continuous as well as discrete processes. In other words, a **batch process can be either continuous**

or discrete of relatively short durations compared with a continuous or discrete process. Here the process flow can be changed as the batch progresses and may not always follow a well-defined path; it is based on prevailing process conditions. Furthermore, the plant consumes raw material in lots and produces the output product in lots as well.

Here the production platform is not dedicated to a single product but can be used to produce similar products in batches. In other words, differing process sequences can be carried out on the same set of plant equipment to produce different products. This can be compared to a kitchen that is common for preparing many food items but is based on different recipes.

19.2.3 Physical

As discussed in Chapter 1, industrial processes that are generally either localized (spread in a relatively small physical area) or distributed (spread in a large physical area, even geographically), although it is difficult to draw a fine line between them.

19.2.3.1 Localized

The localized process is present in a relatively smaller physical area with all of its subprocesses or components **closely** interconnected. Some examples of localized processes are power plants, cement plants, chemical/petrochemical plants, steel plants, food and beverage plants, water/sewage treatment plants, factories, electrical substations, etc.

19.2.3.2 Distributed

Conversely, the distributed process is present in a relatively larger physical area (or even in a large geographical area) with its subprocesses or components distributed and **loosely** interconnected. Such a process can also be termed a **network** of many localized processes distributed over a larger physical area. Some comparative examples are power transmission/distribution networks, oil transportation pipelines, water supply/distribution pipelines, etc.

Table 19.1 provides some typical examples from industry.

As seen in the table, there are many ways in which different industrial processes can be classified. No single type of automation system can meet the requirements of all different types of processes. Hence, automation systems that manage these different types of processes are also different.

The following sections discuss the management of application processes in detail that are classified as manufacturing and infrastructure, in today's context.

19.3 Manufacturing Processes

In manufacturing processes, process plants are continuous processes whereas factories are discrete processes. This section discusses common issues associated with both process

Table 19.1 Examples of industrial processes

Application	Nature of process		Type of process		
	Localised	Distributed	Continuous	Discrete	Batch
1. Manufacturing					
1.1 Process plants					
- Chemical/Petro-chemical	X		X		
- Cement	X		X		
- Metals	X		X		
- Power generation	X		X		
1.2 Factories					
- Food and beverages	X				X
- Pharma	X				X
- Automotive	X				X
2. Utilities					
2.1 Civic					
- Power distribution		X	X		
- Gas distribution		X	X		
- Water distribution		X	X		
- Water/sewage treatment	X				
2.2 Industrial					
- Water distribution		X	X		
- Power distribution		X	X		
- Steam distribution		X	X		
- Gas distribution		X	X		
- De-mineralisation	X		X		
2.3 Backbone utilities					
- Water transportation		X	X		
- Power transmission		X	X		
- Oil/Gas transportation		X	X		

plants and factories. Specific issues associated with either process plants or factories are discussed in the corresponding sections. The manufacturing process broadly consists of:

1. Inputs: Raw material, time, people, and energy
2. Assets: Process plant or factory
3. Outputs: Product, waste, and data

These are above illustrated in Fig. 19.2.

Efficient use of resources is essential to provide a competitive edge to the enterprise to improve its sustainability and adaptability to changes, because the enterprise needs to respond quickly to continuous changes in inputs, technology, and, above all, **market dynamics**, while the end product remaining the same. Data are continuously collected by the automation system, one of the most valuable outputs to be leveraged. A lot of actionable information is hidden in this wealth of data for understanding and improving the manufacturing processes in today's context.

19.3.1 Evolution

The effort over the past four decades have been the technology adoption in industrial processes. During the 1980s, the emphasis of management was primarily on quality

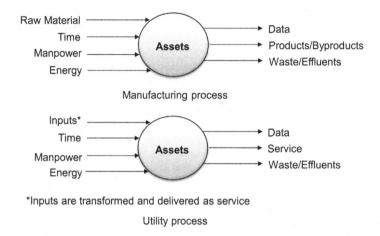

*Inputs are transformed and delivered as service

Utility process

Figure 19.2 Manufacturing and service delivery enterprises.

and results were achieved by command or through coercion. The next decade witnessed improvements in manufacturing processes:

During 1990s:

- Business process reengineering: Radical redesign of core business processes (work flows) to achieve dramatic improvements in critical performance measures such as quality, cost, service, speed, etc., within and between organizations.
- Enterprise resource planning: A business-management system that can collect, store, manage, and interpret data from all business activities, to obtain an integrated view of core business processes and track business resources in real time, using common databases. More about this is discussed in Chapter 20.
- Supply chain management: Management of the flow of goods and services, which includes the movement and storage of raw materials, work-in-process inventory, and finished goods from point of origin to point of consumption.

From 2000 onward:

- E-business initiatives: Takes the company to expand business operations on the Internet from offline to online with new ways to bring in business and market products and services.
- Collaborative manufacturing: Sharing information between business processes across internal or external partners in the value chain network so that all parties in the business relationship contribute to improve the business process as a whole.

The adoption of automation in industrial process started much earlier than 1970. In fact, the use of a centrifugal governor in 1785 in a floor mill was the first known example of industrial automation. The big impetus for the use of industrial automation came after the Second World War. Subsequently, industrial processes such as chemical plants, power plants, aviation, mass production assembly lines, etc., started using automation technology (sensors, feedback control, etc.) for their operations.

Fig. 19.3 illustrates the evolution of manufacturing processes.

Today, industrial process operations have only three layers (automation, manufacturing execution system, and enterprise resource planning) that take care of all aspects of

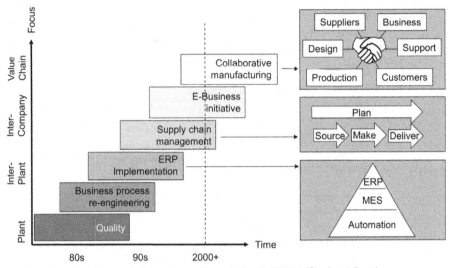

MES: Manufacturing execution system or its equivalent in Utilities (Service delivery) process,
ERP: Enterprise Resource Planning

Figure 19.3 Evolution of manufacturing strategies.

process management. The use of information in real time for better and faster decision making and the integration of information across businesses are current drivers in which visibility, mobility, etc., are further enablers. These aspects are discussed in Chapter 20.

19.3.2 Business Aspects

The following sections briefly explain various business aspects associated with the manufacturing processes that are vital for staying in business.

19.3.2.1 Drivers

In today's context of globalization, competition to stay ahead of others is the primary driver for advancements, improvements, and sustainability. This competition ensures that consumers obtain maximum benefit from products on the market. In a continuous drive to improve their offerings, manufacturing industries need to adapt to various cutting-edge technologies that help them use their key resources, **people, assets, and raw materials,** to get maximum output in an economical way, passing on the cost or performance benefits to the end user. Efficient handling of these key resources is critical to the success of any organization. A lack of synchronization among these three resources leads to a reduction in the use of assets, profits, and productivity. This calls for connecting people, assets, and raw materials to form the basic building block. An ideal solution is one that helps monitor all three key resources, defines actionable events, and assigns them to individuals responsible for follow-up and remediation.

19.3.2.2 Objectives

The business objectives of a manufacturing industry are to manufacture quality, consistent, and cost-effective products with good after-sales service that meets or exceeds customer's expectations. This is the key manufacturing objective to stay in the market and increase business. This can be achieved by:

- Offering competitive pricing while making a **profit** through cutting costs (reducing inventory, increasing productivity, automating production processes, etc.).
- Providing for a wider range of market requirements and improving a competitive advantage by establishing flexible production facilities to offer products tailored to the customer's needs.
- Innovating for achieving and sustaining market leadership or market share and establishing a brand name.

19.3.2.3 Value Creation

Value is said to be created any time an action is taken for which the benefits exceed the costs, or any time an action is prevented for which the costs exceed the benefits. The purpose of any business enterprise is value creation or value addition for all of its stake holders (customers, employees, investors, shareholders, society, environment, etc.). Because all of the stake holders are interlinked, no sustainable value can be created for one unless all are considered together. However, from a financial point of view, value is said to be created when a business earns profit, an important business objective.

19.3.2.4 Strategies

A business strategy is the one that sets objectives in an enterprise to achieve its desired goals. It can also be described as a long-term business plan. Typically a business strategy covers a period of about 3–5 years. Strategies provide overall direction to the enterprise and involve specifying the organization's objectives, developing policies, designing plans to achieve the objectives, and then allocating resources to implement the plans.

19.3.2.5 Sustainability

Sustainability or continued growth without significant deterioration is a business approach that creates a long-term stakeholder value with minimal negative impact on the environment, community, society, and economy while maintaining a healthy profit. A sustainable enterprise incorporates principles of sustainability into each of its business decisions, manufactures and supplies environmentally friendly products, and has enduring commitment to environmental principles in its business operations.

19.3.2.6 Reliability

Reliability in a business enterprise has two components: the enterprise's business health in all of its long-term operations and asset performance and availability to meet on-time delivery. This means that all assets such as machines, equipment, and automation systems in the industrial process should perform their intended or required functions consistently on demand with no degradation or failure. This is achieved through preventive and predictive maintenance.

Furthermore, reliability has a direct relationship with business sustainability. More often, organizations overlook the impact of reliability on the cost of operations, safety, and overall business performance. Reliability has a vital role in manufacturing because of its positive contribution to productivity, quality, yield, safety, and environmental performance.

19.3.2.7 Challenges

Known challenges for the enterprise are business integrity, finance, resources, competition, technology, customer loyalty, business uncertainty, risk management, right human resources, accepting and moving with changes, business sustainability, etc. However, owing to globalization, the changing business paradigm also:

- is a dynamic marketplace
- is changing product requirements
- is flexible in manufacturing
- has a shorter product life cycle with complexities
- has collaborative manufacturing through supply chain management
- has low-cost integrated solutions
- has benchmarking in the global context
- has quality assurance through consistency and predictability
- has confidence-building measures through traceability, certification, and positive assurance

19.3.3 Role of Automation

Automation technology was introduced some time earlier, primarily to monitor and control manufacturing processes on a real-time basis. However, although the plant automation system collected a lot of plant data, the following production challenges remained:

- Plant data were typically used only for to monitor, control, and visualize the process: Not leveraged to enable process visibility, analysis, and improvement through simple operator reports and dashboards to identify sources of downtime.
- Functional areas within the plant-operated silos: Limited visibility within a given functional area (i.e., process, packaging, etc.), which cannot give overall visibility across the plant with the context.

For example, the production department may know that it produced 100 units. However in the absence of quality data, the department did not know that there was a 20% scrap rate.

In view of this, it was necessary to extend automation investments to provide knowledge to all concerned roles and departments regarding:

- Operations: How are we doing against the plan?
- Production: What are the bottlenecks?
- Quality: What is the rejection rate?

- Maintenance: What are the causes of downtime?
- Engineering: What is the equipment use?
- Plant: What is the response time for a new order?
- Management: What are the key performance indices of the plant?

Thus the questions are: What are the plant data telling about the overall manufacturing process? Do the specific plant data reach every role or department? How are the plant data used to optimize production and increase profit? The answers to these are going for role-based reporting and analytics:

- Operator view: Display of real-time status and operations on equipment, etc., on operator consoles.
- Maintenance view: Display of alarms, events, history, root cause analysis, etc., on maintenance consoles and dashboards.
- Management view: Display of plant-wide performance, financials, asset use/yields, etc., on management consoles.
- These lead to the changes listed in Table 19.1.

Automation functionality expanded to meet a new manufacturing paradigm is shown in Table 19.2.

However, expanding the automation investment across the plant has to address the following challenges:

- Number of disparate systems
- Unused operation
- Islands of automation
- No communication with decision-making layer
- Laborious way to collate data
- High capital expenditure
- Poor visibility, traceability, and compliance with standards

These issues are addressed through IT-OT integration in Chapter 20.
Integrated automation systems help achieve manufacturing business objectives:

- Operational excellence
- Build superior products, achieve profits, grow the company

Table 19.2 New manufacturing paradigm

From:	To:
Manufacturing centric	Customer centric
Make to stock: forecast dependent	Make to order: demand dependent
Replenish stores	Eliminate buffers; direct to customers
Regulated yield	Optimized yield
Monolithic deployment of resources	Extended/collaborative value chain–linked deployment
Breakdown and preventive maintenance	Predictive and reliability-centered maintenance

- Combine lean manufacturing with agile manufacturing
- Cut waste and respond to market variability
- Strengthen and enhance core competencies
- Focus on operations that define the business
- Design anywhere, build anywhere
- Extend product life cycle globally
- Adopt common architectures and standards for controls, infrastructure, and equipment
- Preserve assets

These are keys to profitability and sustainability and can be achieved only through investment in integrated automation and information systems.

The next sections discuss case studies of process plants and factories with their associated automation.

19.3.4   Case Studies

In this section, we discuss case studies in the manufacturing process for both a process plant and a factory.

19.3.4.1   Process Plant: Hot Rolling Process

This example comes under the category: localized, continuous, manufacturing process (process plants). This case study is based on the automated hot strip rolling mill installed in the plant of JSW Steels,[1] India.

Hot rolling is a process in a steel plant in which the semifinished slab is reheated to a predetermined temperature and the hot slab is rolled to make the specified hot strip in a coiled form. The role of a hot strip mill (HSM) is to produce strips from semifinished steel slabs. The resulting attributes of the rolled strip are its surface quality, thickness, width, flatness, and strip profile. The sequential steps in an HSM are to reheat the semifinished steel slabs to the rolling temperatures, roll them thinner and longer through a series of rolling mill stands driven by **large electric** motors, and finally coil up the lengthened strip for easy handling and transport.

An HSM primarily consists of the following units or areas:

- Slab yard
- Reheating furnace
- High-pressure scale breaker
- Roughing mills with edger
- Crop shear
- **Coil box**
- Finishing scale breaker
- Finishing mill
- Run-out table area cooling
- Coiler
- Coil yard

[1]www.jsw.in.

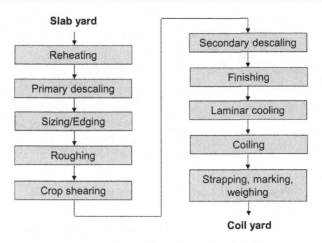

Figure 19.4 Hot rolling process: flowchart.

Rolling of steel slabs in an HSM consists of several subprocesses; at the end of them the final product, **hot rolled coil**, is produced. The process flow in hot strip rolling is illustrated in Fig. 19.4.

The main subprocesses in the hot rolling process are explained next.

Reheating of semifinished slab: The inspected and graded slabs as per the rolling schedule are placed, one at a time, on the furnace charging roller table (in the slab yard) and positioned in front of the **charge door** on the reheating furnace. Reheating of the steel slab is achieved by hot exhaust gases rushing past the slabs on the way to the recuperators. After it has been sufficiently heated, when the slab reaches the **discharge door** at the exit end of the furnace, the door opens and the slab is placed on the entry roller table that carries the slab into the roughing mill passing through a descaling unit.

Primary descaling of reheated slab: After leaving the reheating furnace, the slab passes through a descaling unit, an enclosure with highly pressurized water jets to remove the oxidized iron layer that forms at the surface of the slab in the oxygen-rich atmosphere of the reheating furnace. Descaling of heated slab is a must in HSM to attain good surface quality on the hot rolled strip.

Sizing of reheated slab: The reheated slabs are sized using a sizing press or edger that produces flatter **dogbones** leading to reduced spreading and greater sizing efficiency.

Roughing mill rolling of reheated slab: The roughing mill usually consists of one or two roughing stands in which the slab is hot rolled back and forth 5 or 7 times repeatedly to reach the minimum thickness requirement. The roughing mill also contains edger rolls that are used to roll the edge of slab and center it. A rectangular head end is critical to thread the finish mills and down coilers properly. An uneven tail can damage work-roll surfaces or cause threading problems for the next production process. Hence, the head and tail ends of every transfer bar is cropped by a pair of large steel drums, each with a shear blade extending along its length. Here the slab passes through the mills back and forth several times to

Slab yard   Roughing mill

Courtesy: www.jsw.in

Figure 19.5 Hot strip mill: slab yard and roughing mill.

obtain specified reduction as per the schedule (how much reduction, how many passes, etc.). Higher-capacity mills normally have two roughing stands and reduction is carried out in a phased manner for both. Edging on either side of the strip is also carried out for both.

Fig. 19.5 illustrates the slab yard and roughing mill.

Secondary descaling of reheated slab: Between the crop shear and the first rolling stand of the finishing mill, there is normally a second scale breaker whose task is final scale removal. Water sprays above and below the transfer bar break up the scale that has reformed (secondary scale), as well as any scale that has persisted through earlier descaling operations. A coiler located between these mills will coil the complete material coming from the last pass of the roughing mill before feeding it to finishing mills. Rolling speeds at the finishing mill will be higher than at the roughing mill. Hence rolling cannot be done with stock present in both roughing and finishing mills.

Finishing mill rolling of reheated slab: The finishing mill usually has five to seven finishing roll stands, which reduce the thickness of the transfer bar down to the gauge required. The rolling speed is set to allow the last stand to perform the final reduction at the finishing temperature, to achieve certain mechanical properties in the hot rolled strip. The finishing mills roll the rolling stock in tandem, meaning each bar is rolled through all finishing stands at once.

Laminar cooling of the hot rolled strip: After exiting the finishing mills, the strip is carried down by a large number of individually driven rolls through banks of low-pressure, high-volume water sprays that cool the red-hot strip to a specified coiling temperature and into down coilers. In addition to a highly efficient cooling model, the laminar cooling system ensures the desired coiling temperature as well as cooling to achieve the desired mechanical properties of the rolled stock.

Fig. 19.6 illustrates the finishing mill and the cooling section.

Coiling of rolled strip: The head end is deflected by a gate down to the mandrel associated with the coiler and is guided around the mandrel by pneumatically actuated wrapper rolls linked by aprons. This process includes strapping, marking, and weighing of coils.

Fig. 19.7 illustrates the coiler and coil yard.

Finishing mill   Cooling section

Courtesy: www.jsw.in

Figure 19.6 Hot strip mill: finishing mill and cooling section.

Coiler   Coil yard

Courtesy: www.jsw.in

Figure 19.7 Hot strip mill: coiler and coil yard.

19.3.4.1.1   Role of Automation

All of these functions of HSM are managed by a programmable logic controller (PLC)-based automation system. The levels of automation are:

- Level 1: Technological controls such as profile and flatness, actuators, sequencing, tracking, and communication
- Level 2: Process models such as pass schedule management, manufacturing execution system, evaluation, and, communication
- Level 3: Production planning

  Fig. 19.8 illustrates the automation system configuration.
  Fig. 19.9 illustrates some typical process displays.
  Technical benefits are:

- Reheating furnace: Better control on heating of slabs to achieve optimum rolling temperature.
- Roughing mill: Precise control of roll gaps enables better stock control in each pass and avoids excess loading on mill rolls and drive systems owing to improper roll gaps. Faster and precision gap setting increases mill output.
- Crop shear: Precise crop length reduces waste of rolling stock. This avoids **mis-rolls** in finishing mills.
- Finishing mill: Automated control of rolling speeds in stands results in better quality and dimensional accuracies of rolled product. Setting of individual stand speeds to match rolling

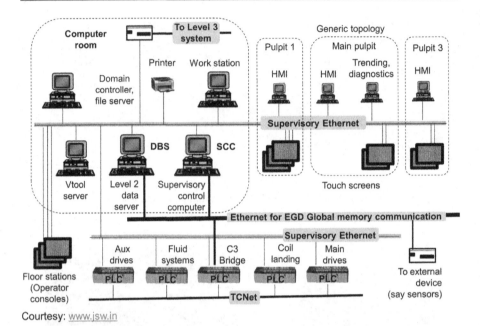

Courtesy: www.jsw.in

Figure 19.8 Hot strip mill: automation system configuration.

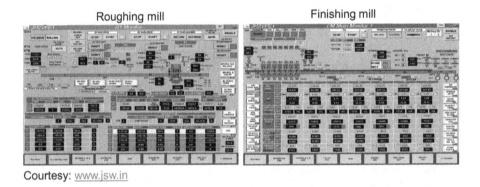

Courtesy: www.jsw.in

Figure 19.9 Hot strip mill: process displays.

program reduces setting time and mis-rolls, which occurs if speeds are set manually. This thereby increases mill output.

In addition, manual errors are avoided and higher quality of product is ensured every time, as well as good metallurgical properties, improved quality, and ease of hot strip handling right from slab yard to coil yard.

Commercial benefits are:

• Reduction of burning losses in reheating furnace while heating slabs (which would otherwise increase owing to overheating) in reduction of losses (mis-rolls) of rolling stock that will occur owing to human error (mill operator) in mill setting/mill operation (such as setting of roll gaps in roughing mill during reversing actions, mill speeds in finishing stands).

- Reduction of time in setting roll gaps for every pass in reversing action, which increases mill output.
- Precision control of **length of cut** of rolling stock by cropping shear after roughing mill results in reduction in waste and increase in mill yield.

In addition to the higher rate of production in a given time, different grades of steel can be rolled because each grade has its own mill setup coming from the automation system, increased plant availability, and higher productivity.

19.3.4.2   Factory: Cooling Module Assembly Process

This example comes under the category: localized, discrete, manufacturing process (factories). This case study is based on an automated flexible assembly system in just-in-sequence (JIS) assembly of cooling modules in the factory of MAHLE Behr GmbH & Co. KG,[2] United States.

The case study refers to the system required to supply fully assembled cooling modules that are later integrated into trucks manufactured elsewhere by the MAHLE Behr's customers. The requirement of the truck manufacturer is the supply of completed cooling modules made according to a JIS procedure. JIS means that each module is custom-built for a particular truck with line operators preassembling the correct components on each module built to order. A cooling module is made of the following components:

- Radiator
- Charge air cooler
- Condenser
- Shroud
- Oil cooler
- Surge tank
- Washer bottle
- Recirculation shield
- Small miscellaneous components

To meet the purchaser's requirements, a **flexible assembly line** capable of building any cooling module with any set of components at any time is implemented. This is with the prerequisite that all components that are needed for any module are present at the assembly line at all times.

The PLC-based automation system runs the module assembly trim line and ensures that line operators are able to build, inspect, and sequentially load the cooling modules completely and accurately into the shipping containers. A Linux-based application called X-Just-in-Sequence (XJIS) generates a bill of materials, assembly information, and build order for each shipping container of modules generating the following for the cooling module trim line:

- Bill of materials
- Assembly information
- Build order for each shipping container of modules (four modules)

[2]www.mahle.com.

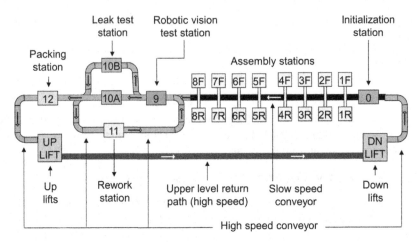

F: Front, R: Rear

Figure 19.10 Cooling module assembly setup.

With this new flexible assembly system, the information for each build order is electronically accepted directly from the purchaser (truck manufacturer) and transferred to the automation system for execution. Once a module is finished, the automation system generates and sends the applicable manufacturing and test data back to the XJIS system.

The automation system is responsible for coordinating operators' efforts at the various assembly, test, rework, and packing stations as well as for moving materials between stations by means of carriers attached to an overhead conveyor. After leaving the last assembly area, carriers move into a robotic vision inspection area where completed modules are inspected, then onto a leak test station, and finally to packing.

Fig. 19.10 illustrates the cooling module assembly setup.

Material handling is achieved by an overhead conveyor power and free system. The sequence is as follows:

1. A series of carriers ride along the power and free conveyor and are carried through a series of eight assembly areas; each assembly area has a front and rear work station.
2. After leaving the last assembly area, the carriers move into a robotic vision inspection area that also has front and rear robotic vision inspection stations. Each station has a machine vision camera that is moved into the proper inspection location by a robot.
3. After the vision inspection area the modules travel into a leak test stations. Here the conveyor is paralleled to accommodate for the increased time requirement of leak testing.
4. After the modules leave the leak test station, they are sent on to either the module packing station (modules that pass machine vision and leak tests) or a rework station.
5. At the packing station, the modules are removed and placed into the shipping containers.

After the packing station, the empty carriers make a 180° turn and move into an uplift station. The modules are lifted up to a level of 15 feet and then transported back to the assembly start area of the line where the modules enter into a down-lift station

Cooling module assembly line stations 4-8   Robotic vision inspection

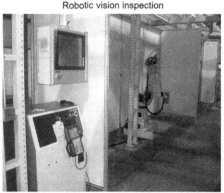

Courtesy: www.mahle.com

Figure 19.11 Cooling module assembly line stations 4–8 and robotic vision inspection.

and another 180° turn, and are thus prepared to receive a radiator at the initialization station (station 0). There are approximately 20 carriers on the line and the line produces about 100 cooling modules per shift.

Fig. 19.11 illustrates cooling module assembly line stations 4–8 and robotic vision inspection.

Role of automation system:

1. Receives via electronic data (required cooling modules, their sequence, etc.) transmitted by the customer.
2. Parses received data directly to the production automation equipment. Each assembly area is provided with proper work instructions that allow a successful build of the cooling module by the respective operators (computer-aided manufacturing).
3. Displays work instructions at each station to the operator. The operator is required to select proper components for assembly into the cooling module. As each module is selected by the operator, the automation system verifies that the correct component has been selected by barcode feedback.
4. Fastens the selected and verified component to the assembly for the electronic torque tool to verify that the fasteners are all properly tightened to a specific torque.
5. Downloads this information (steps 3–5) to the on-board radio-frequency identification database (RFID) of the module to provide a permanent record that the module is built correctly.
6. Moves the module into series of tests after the module is properly built. The set of tests depends on the module description that is read from the RFID tag.
7. Records the outcome of the tests and sends the module for either shipment or rework, with the results of the tests also being downloaded to the RFID tag.
8. Loads the good modules that pass all tests into the shipment container, along with the electronic record of the complete build and testing results.
9. Downloads the final electronic data record and conveys it electronically to the customer.

For steps 1–8, automation is a critical element to this process. Fig. 19.12 illustrates the automation system configuration.

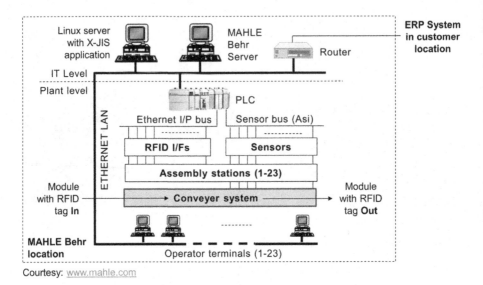

Figure 19.12 Automation system configuration.

As we see in Fig. 19.12, the role of automation in a module build is explained in the following steps:

1. RFID tags are installed at all moving modules, RFID interfaces at all stations, and sensors on the conveyor system.
2. RFID tag data are read from the **moving module** by PLC at each station during the module build and store data on the module's RFID tag after the module build.
3. At the shipping station, the RFID tag data are read from the moving module onto the **MAHLE Behr server**, where they are stored as a complete record of the product build and test results.
4. Sensors transmit conveyor status and control information continuously to the PLC.

Technical benefits are doing the following electronically:

1. Transfer of orders from customer to manufacturer
2. Flexible computer-assisted manufacturing
3. Faster part verification
4. Distribution of database (traceable for each module)
5. Assembly verification of module (machine vision)
6. Functional verification of module (pressure test)

Apart from these, no inventory is required by the customer (truck builder) because cooling modules arrive JIS, when need. Realized commercial benefits in truck manufacturing are a reduction in:

1. Work-in-progress inventory
2. Rework
3. Errors on order transmission

Furthermore, there is a close integration between truck manufacturing and cooling module assembling. Fig. 19.13 illustrates the operator panel.

Operator panel front view

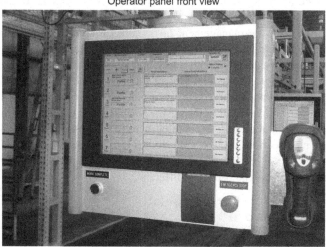

Courtesy: www.mahle.com

Figure 19.13 Operator panel.

19.4 Utility Processes

Utility processes, as mentioned, are investments made in facilities in a country, region, state, city, town, area, industrial complex, etc. for providing essential services to the public for their convenience and living, and to the industry for manufacturing process at large. They can be classified broadly as:

- **Civic utilities**: Distribution of power, water, and gas, treatment of raw water and sewage, buildings, etc., for providing essential services to the public for their convenience and living. These are generally confined to towns, cities, and metropolitan areas.
- **Industrial utilities**: Production and distribution of power, steam, compressed air, cooling water, chilled/hot water, gas, etc., and treatment of water (demineralization), factory effluents, etc., are integral to any manufacturing process. These processes are confined to industrial complexes to provide services to process plants in a manufacturing industry.
- **Backbone utilities**: These are networks for transmission of power, water, gas oil, etc., spanned over state, region, and even country for supporting/feeding civic and industrial utilities.

It is difficult to demarcate clearly among utilities because it depends on the area covered by these processes, functions performed, and end users.

19.4.1 Evolution

Traditionally, some civic utilities are treated as public services and managed, administered, and regulated as public sector undertakings (government agencies, elected bodies, cooperatives, etc.). The main consideration is to provide essential services; economic considerations are secondary (subsidized by the state). However, this practice is also moving to consider them as sustainable processes and meeting of all their key business performance indices. Currently many civic utilities are designed and operated to work as enterprises.

To be in line with the management of manufacturing processes, here also, efficient use of resources is essential to provide good service delivery to consumers while keeping the process sustainable with a low operational cost and adaptability to new changes and technology. Above all, in today's market, there are many service providers who compete to provide these services efficiently and economically as profitable enterprises. In view of these, all of the business aspects discussed earlier for manufacturing processes hold good here except that **product delivery** is replaced by **service delivery**. The next sections provide case studies for civic utility and backbone utility. Industrial utility is not specifically discussed because it is identical to civic utilities management, except the end users are different.

19.4.2   Case Studies

The following sections discuss case studies in civic utility and backbone utility processes. No case study is presented for industrial utilities because they are identical to civic utilities.

19.4.2.1   Civic Utility: Wastewater Treatment Process

This example comes under the category: localized, continuous, utility process (civic utility). This case study is based on similar projects executed by VA Tech WABAG,[3] India.

Wastewater or sewage treatment is the process of removing contaminants from waste or used water (sewage or effluent), primarily from households, offices, and industry. This process includes the removal of physical, chemical, and biological contaminants to produce environmentally safe treated wastewater (or treated effluent) for use elsewhere or for discharging into the environment. A by-product of sewage treatment is usually a semisolid waste or slurry, called sewage sludge, that has to undergo further treatment before it is suitable for disposal or landfills.

Everyday discharge sources include flushed toilets (human waste), showers and sinks, washing machines, and dishwashers. This wastewater leaves the source and travel through a network of underground pipes known as the sanitary sewer system. Wastewater treatment objectives are to manage the discharged wastewater to produce clean water that is safe for the environment and public health.

Fig. 19.14 illustrates the process flow of a typical sewage treatment plant.

Wastewater (sewage) treatment consists of subprocesses.

19.4.2.1.1   Primary Treatment
- Screening: A physical process that remove large objects that have entered the sanitary sewer system, such as bottles, rags, and pieces of wood. This protects pumps and other equipment from damage.
- Grit removal: A physical process that remove heavy inorganic solids such as sand and gravel. This protects the plant from excessive wear and tear.

[3]www.wabag.com.

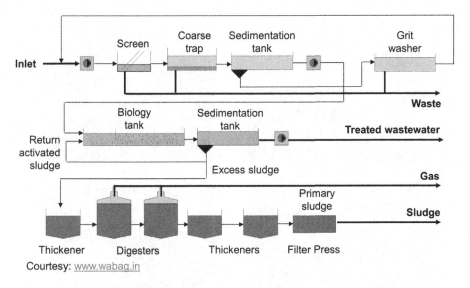

Figure 19.14 Wastewater (sewage) treatment process.

Courtesy: www.wabag.in

Figure 19.15 Wastewater (sewage) treatment plant: primary treatment.

- Sedimentation: A physical process that makes heavy particles sink, light material float, and oil and greasy material rise to the surface and be skimmed off. Sunken solids are deemed sludge. This is comparable to the treatment provided by a septic system.
- Clarification: A physical process that continuously removes solids deposited by sedimentation.

Fig. 19.15 illustrates equipment/systems involved in primary treatment.

19.4.2.1.2   Secondary Treatment
- Elimination of pollutants: A biological process in which air is pumped into wastewater to support microorganisms to consume pollutants.
- Clarification: A physical process for another round of removal of sedimentation that typically follows elimination of pollutants. This is also to conform to the minimum level of treatment required by the Clean Water Act and clarification.

Membrane bio-reactor tank

Aeration tank   Clarifier

Courtesy: www.wabag.in

Figure 19.16 Wastewater (sewage) treatment plant: secondary treatment.

Fig. 19.16 illustrates equipment nd systems involved in secondary treatment.

19.4.2.1.3 Advanced Treatment

Advanced filtration is used to improve the quality of water even further and uses additional treatment to remove suspended solids and nutrients. This is not required in all treatment plants.

• Disinfection: chlorination, ultraviolet light is added the outgoing clean wastewater.

19.4.2.1.4 Automated Functions

All of these functions of a treatment plant are managed by the controller-based automation system. The following are functions of an automated plant:

• Operation of all pumps based on inlet flow/level of tanks/duty standby/cyclic operation.
• Operation of all process equipment such as blowers/screen/filter press, etc., based on sewage flow.
• The entire aeration process, based on various parameters such bio load, air requirement, and mixed liquor suspended solids ratio, is set in the program for an efficient process.
• Advance treatment systems are all automated to produce clear effluent ready for safe use or disposal.

Fig. 19.17 illustrates the automation system configuration.
Fig. 19.18 illustrates some typical process displays.

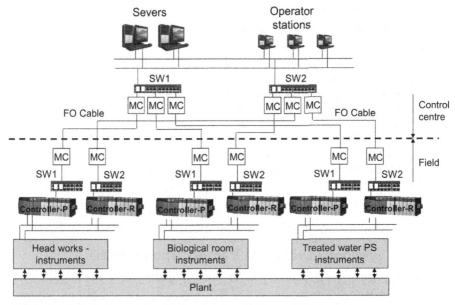

SW: Switch, MC: Media converter, PS: Pump station, P: Primary, R: Redundant
Courtesy: www.wabag.in

Figure 19.17 Sewage treatment plant: automation system.

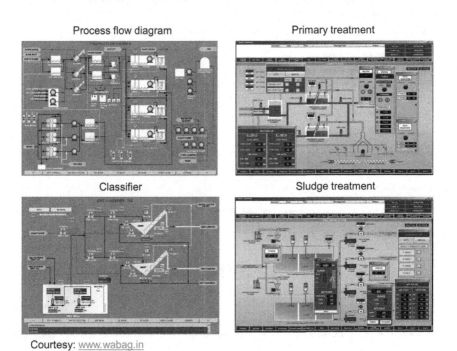

Courtesy: www.wabag.in

Figure 19.18 Wastewater (sewage) treatment plant: process displays.

19.4.2.1.5   Benefits of Automation

Following are benefits of automating the plant:

- Technical:
    - lesser requirement of operator/human intervention
    - scientific monitoring
    - entire process risks/options are predefined in the system and flawless performance of plant
    - equipment set points for limits can be clearly controlled to safeguard equipment and ensure it operates in safe zones
    - overall smart operation of plant
    - less downtimes in case of maintenance
    - systematic startup and shut down
- Commercial:
    - less staff for supervision
    - optimization of power and chemicals owing to sophisticated automation.
    - plant downtimes are minimized; hence increased productivity

19.4.2.2   Backbone Utility: Oil Transportation Pipeline

This example comes under the category: distributed, continuous, utility process (backbone utility). This case study is based on the experience of the author on executing pipeline automation projects in India.

Pipelines are employed to transport large volume of oil such as crude, petroleum products, natural gas, etc., from one place to another, spanning long distances. Compared with other modes of oil transportation, pipelines are economical, safe, fast, and environmentally friendly.

A typical oil transportation pipeline consists of:

- Pipeline for moving oil
- Originating pump station or mother station
- One or more intermediate pump stations (booster stations)
- Block valve and cathodic protection stations
- Terminal station or delivery station(s)

The crude transport pipeline is in the midstream and product transportation is in the downstream in the oil production, transportation, processing and distribution process, as illustrated in Fig. 19.19.

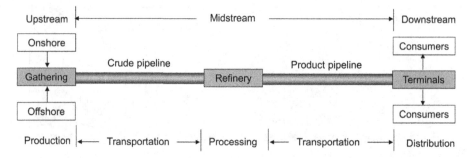

Figure 19.19 Oil production, transportation, processing, and distribution process.

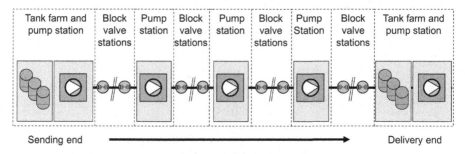

Sending end ━━━━━━━━━━━━━━━━━━━━━━━━━━━━▶ Delivery end

There may be tank farms in pump stations also for intermediate storages

Figure 19.20 Oil pipeline process.

19.4.2.2.1   Subprocesses
Following are subprocesses in a pipeline process:

- Pipeline: To move oil from source to destination(s), includes all parts of physical facilities such as line pipe, valves, pumping units, metering stations, and storage tanks.
- Tank farms: To store oil at source or at destination(s) or at intermediate pump stations as buffers, a facility of a group of tanks connected to pipeline(s).
- Pump stations: To force movement of oil from source to destinations through main and booster pumps, a facility to increase pressure of oil received through a main pipeline to pump it to the next station or terminal.
- Block valve (repeater) stations: To sectionalize pipeline for operational and maintenance purpose, a facility to block oil flow in both directions, manually and remotely operated suction and discharge valve.
- Cathodic protection station: Employed to reduce corrosion of underground pipelines, normally part of block valve station (BVS), and a facility for corrosion control of a buried pipe metal surface by making it a cathode of an electrochemical cell.

Fig. 19.20 illustrates the overall structure of the pipeline process.

In this figure, there may be tank forms at pump stations as intermediate buffers.

19.4.2.2.2   Pipeline Automation Structure
An oil pipeline is a geographically distributed process and requires a network control system (NCS) for management at an overall process level. PLCs at pump stations and remote terminal units (RTUs) at block valve stations are installed at the distributed subprocess level and are networked with the master station over a communication network. The automation level hierarchy is illustrated in Fig. 19.21.

The general automation layout is illustrated in Fig. 19.22.

Depending on the availability requirement, there may be an alternate master control center to work as a backup in case the main control center (MCC) station fails.

19.4.2.2.3   Communication
Any NCS requires a data communication network for communication between the master station and distributed station control centers (SCCs) and BVSs. The communication structure is pipeline management illustrated in Fig. 19.23.

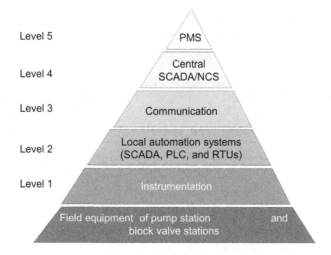

Figure 19.21 Pipeline automation hierarchy.

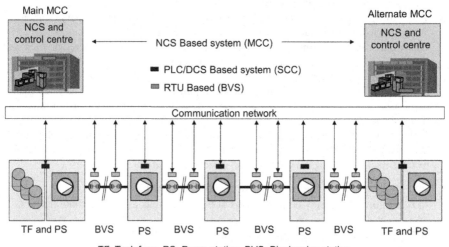

TF: Tank farm, PS: Pump station, BVS: Block valve station

Figure 19.22 Supervisory control and data acquisition/network control system configuration.

Generally, for operation reasons, configuration 1 (direct monitoring of BVS) is employed. In line with this, typical automation system configurations at the master control center, SCC, and BVS are given in the following figures.

19.4.2.2.4 Pipeline Management System

Apart from normal functions that are automated in a typical NCS, important and specific applications carried out by an NCS are:

- BVS works as a typical data acquisition and control station under direct control from MCC: acquisition of process parameter values/status such as temperature, pressure, and motorized valve and transfer data to the MCC, and control motorized valve.

Configuration 1:
BVS Are directly connected to MCC

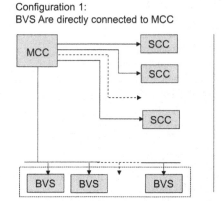

Configuration 2
BVS Are indirectly connected to MCC

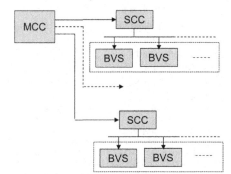

MCC: Master control centre, SCC: Station control centre, BVS: Block-valve station

Figure 19.23 Communication network configuration.

- Apart from being a data acquisition and control station under MCC, SCC also has several independent local automation functions such as monitoring all process parameters such as temperature, pressure, flow, level, and density, and executing logic functions (sequential control with interlock), proportional, integral, derivative loops, and emergency shut down operations upon abnormal conditions.
- MCC, a central supervisory monitoring and control system, does alarm and event logging, communicating with and controlling SCCs and BVSs, historical archiving, emergency control, etc.

The most important applications of pipeline management applications are given in Table 19.3.

Among these, leak detection and locating are the single most important functions carried out by the pipeline management system in real time. This calls for building a mathematical model for the entire pipeline and comparing its output with that of the actual pipeline for a **set of inputs common to both**. Any major discrepancy leads to the performance of the pipeline.

Fig. 19.24 illustrates the real-time model (RTM) of the setup.

Leak detection and locating is a family of the following application modules:

1. Pressure/flow deviation: Detects the occurrence of a leak and estimates the volume based on differences between data produced from RTM and NCS measurements. The leak can be located by generating pressure/flow profiles from possible leak locations and comparing them with actual profiles.
2. Volume imbalance: Detects the occurrence of a leak and estimates volume based on mass/volume balance. Fig. 19.25 illustrates this aspect.
3. Instrument analysis: Evaluates instrument malfunction by analyzing data received from supervisory control and data acquisition (SCADA) system. Measured data (rate, pressure, and temperature) and RTM-tuned roughness are used to calculate a best fit set of rates, pressures, and temperatures, Differences between these computed values and their corresponding measured values form instrument drifts. Instrument drift is calculated for each pipeline flow rate and pressure measurement. When computed drift exceeds user-specified thresholds

Table 19.3 Pipeline management applications

• Leak detection/location	• Predictive model
• Pressure/flow deviation	• Over/underpressure protection
• Volume imbalance	• Survival time analysis
• Instrument analysis	• Inventory analysis
• Parameter tuning	• Composition tracking
• Sensitivity study	• Temperature tracking
• Automatic look ahead model	• Pig/scrapper tracking[a]
• Pressure drop estimation	• Batch tracking[b]
• State estimation	• Pipeline performance analysis

[a]Pig/scrapper tracking: Pigs/scrappers are used to scrap the inside surface of the pipeline to remove the unwanted sticky part that is deposited while moving oil. The pig is launched at originating/any pump station, moves along with the oil, and is received at the next pump station/receiving station. The pipeline management system (PMS) predicts the location of the pig and the arrival time at the receiving points of all pigs in the pipeline network.

[b]Batch tracking: This is a case of sharing the same product pipeline for transporting different types of oil at different times depending on the requirement at the receiving end. PMS estimates the arrival of a new batch at the receiving station based on running parameters such as flow, pressure, and the profile of the pipeline. This is compared with the actual status by measuring the flow and its density of the batch. If there are any variations in estimate vis-à-vis actual and appropriate one, an alarm is generated to help the operation of the pipeline.

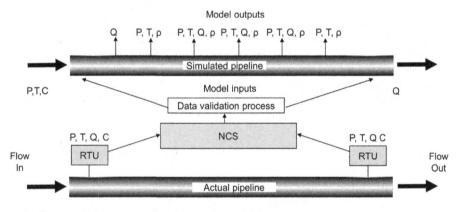

P: Pressure, T: Temperature, Q: Volume, ρ: Density, C: Viscosity

Figure 19.24 Pipeline real-time model.

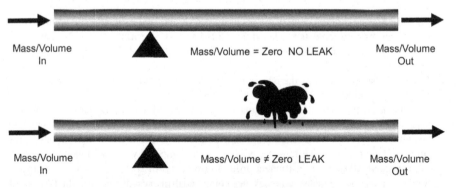

Figure 19.25 Leak detection: volume imbalance.

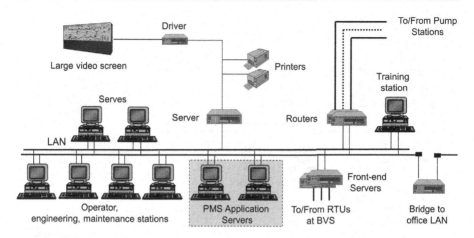

Figure 19.26 Main control center: automation system configuration.

for measurements, an alarm is generated. Poor instrumentation design provides incorrect information.
4. Parameter tuning: Updates (tunes) pipe wall roughness data and ambient temperature values for improved simulation accuracy. If calculated values of P, T, and Q from the RTM differ from data received from RTUs, adjustments are made to roughness and heat transfer coefficients. Performed for each pipe segment. Limits can be set for each tuning correction factor. Internal function of RTM.
5. Sensitivity study: This answers the following questions:
    a. How fast a leak can be detected?
    b. How accurately a leak can be located?
    c. How can leak size be detected?
    d. What is the optimal instrument spacing?
    e. What is the effect of instrument accuracy?
    f. What is the best SCADA scan rate?

Fig. 19.26 illustrates the automation system configurations at the MCC.

Fig. 19.27 illustrates the automation system configurations at the SCC and BVS.

In practice, the same pipeline can be shared to transport different varieties of oil at different times in **sequential batches**, depending on the demand at the receiving end. However, technically, the crude pipeline can only be used for different types of crude while the product pipeline is only for different types of products (kerosene, diesel, aviation fuel, petrol, etc.). In other words, a crude pipeline cannot be used as a product pipeline and vice versa. The process during batch transportation is **continuous** whereas it is **batch–cum-continuous** during transportation of many batches in sequence.

19.5 Industrial Robotics

The definition of an industrial robot is that it is a reprogrammable multifunctional manipulator designed to move material, parts, tools, or specialized devices through

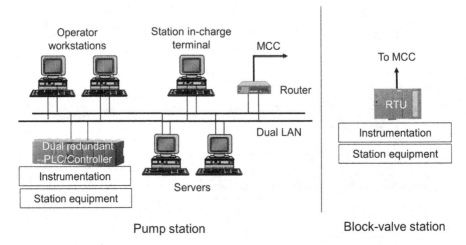

Figure 19.27 Automation system configuration in the station control center and block valve station.

variable programmed motions for the performance of a variety of tasks (Robotics Institute of America). In other words, a robot is a:

• General-purpose reprogrammable mechanical device that performs repetitive, hazardous, and strenuous human tasks.
• Specialized machine with a degree of flexibility and reprogrammability that distinguishes it from fixed-purpose automation.
• Mechanical arm programmed to perform a repetitive task in a controlled and ordered environment with an ability to modify its programmed path based on external sensory inputs.

For this, the robot interfaces with its work environment and has the ability to move its mechanical arm to perform assigned work.

Robotics is the study of robot technology that makes use of disciplines such as: dynamic system modeling and analysis, mathematics, physics, biology, mechanical engineering, electrical and electronic engineering, computer science and engineering, and **automation** (sensors, control, and actuators) technology.

Kinematics, a related subject, is the branch of classical mechanics that describes the motion of points, bodies, and systems of bodies without consideration of the masses of either objects or forces that may have caused the motion. Kinematics is an essential mathematical and modeling tool that is used to configure and program robots.

19.5.1 Structure

The control loop of the robot is to **sense**, **think**, and **act** sequentially and continuously (similar to the automation cycle explained in Chapter 1, Section 1.5) until the task is completed, as illustrated in Fig. 19.28.

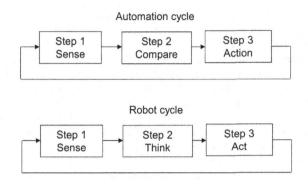

Figure 19.28 Industrial robot: control loop.

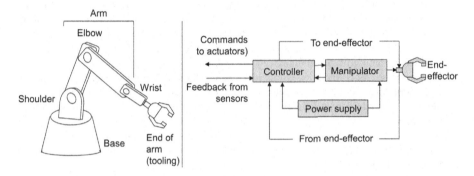

Figure 19.29 Industrial robot: overall structure and block diagram.

In the robot cycle:

· Sense: Acquisition of data from the external environment.
· Think: Analysis of data for planning the tasks to be performed.
· Act: Controlling the external environment by executing tasks.

To execute the control loop, the industrial robot is designed and integrated with several individual parts or components. Fig. 19.29 illustrates the overall structure of an industrial robot, its individual parts, and its functional block diagram.

Following are the main components of the robot:

1. **Sensors**: These are the core of robotics that provides awareness of the environment by sensing things. It is the mechanism that alerts robots. Typical parameters that are sensed by robot are: vision, position, speed and acceleration, temperature, touch, force, magnetic field, light, sound, etc.

2. **Controller**: This is the communication and information-processing device (brain) of the robot, which calculates motion trajectories (position, velocity, acceleration) of the robot joints, initiates, controls, coordinates, and terminates the motion of sequences of actuator. Also, this accepts necessary inputs to the robot and provides outputs to interface with the environment external to the robot. Processor hardware and software (operating system, application software, motion algorithms, etc.) are part of the controller.

3. Actuators: These (servomotor, stepper motor, pneumatic, hydraulic cylinder, etc.), drive the **manipulator and end-effector** independently manipulating and controlling the degrees of motion.

  a. Manipulator: the main mechanical unit of robot consisting of links and joints (prismatic and rotary) joined together in various configurations allowing the robot to perform specific classes of work. The robot axes can be divided into two categories: major axes, mainly used for positioning (waist, shoulder, and elbow), and minor axes, mainly used for orientation (wrist motions: pitch, yaw, and roll). The individual joint motions associated with these categories are referred to as degrees of freedom. Axis corresponds (mainly) to 1 degree of freedom. Typically, an industrial robot is equipped with 4–6 degrees of freedom.

  b. End-effector: This, also known as end-of-arm-tooling, is the mechanically actuated device mounted on the end of the manipulator. The end-effector is job or application specific and can be a gripper, welding gun, paint applicator, cutting torch, vacuum suction pads, deburring tool, or any other. End-effectors are actuated by electrical, pneumatic, magnetic or hydraulic sources. Apart from application-specific factors, the end-effector selection is influenced by factors such as the payload of the robot (load-carrying capacity), workspace restrictions imposed by the energy supply medium (electrical cables/pneumatic hoses), control system, sensors used on the end-effector, etc. Typical end-effectors are illustrated in Fig. 19.30.

3. Power supply: This (electric, hydraulic, or pneumatic) provides and regulates the energy to the robot that is converted to motion by the robot actuator. Electric power is the cheapest and most efficient power delivery medium and is used extensively with industrial robots. Pneumatic and hydraulic actuated robots are gradually being phased out of service because they cannot be controlled continuously and offer poor energy efficiency, However, some specific areas of application (for example, hazardous material handling) still use pneumatically/hydraulically actuated robots.

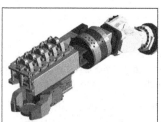
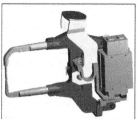

Gripper for deburring   Gripper for casting   Spot welding
plastic parts   extraction   gun

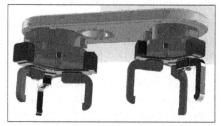

Gripper for dual material handling

Courtesy: www.difacto.com

Figure 19.30 End-effectors: industry examples.

In the following section, industrial robot is discussed as a tool in the manufacturing industry (especially in the discrete processing industry and factories) to perform routine and repetitive operations to improve productivity.

19.5.2 Specifications

All robots are designed to provide position control (controlling the end point motion (or trajectory) of the robot) and force control (controlling the end point forces and moments of the robot) and are based on kinematic and dynamic modeling:

- **Kinematic modeling** is used to establish the relation between joint rotations and end point motion. The most widely used kinematic representations are:
  - Forward kinematics: Given joint values, this defines the position and orientation (pose) of the robot's end point relative to the base coordinate frame of the robot.
  - Inverse kinematics: Given the position and orientation (pose) of the end point, this defines the robot's joint values.
- **Dynamic modeling** is used to establish the relation between external forces and moments acting on the robot, such as gravitational force, joint forces, and torque exerted by the robot actuator.

Primary specifications of robots are given:

- **Payload**: This is the maximum load a robot can carry and performing as per other specifications.
- **Reach**: This is the maximum distance a robot can reach within its work envelope.
- **Degree of freedom**: This is the number of axes a robot can support.
- **Repeatability**: This is the **precision** with which the robot can attain a pretaught position if the motion is repeated many times.
- **Drive**: This includes electric, hydraulic, or pneumatic (the most commonly used is electric).
- **Manipulator**: Serial link* or parallel link⁺.

19.5.3 Classifications

Based on **kinematic structure**, there are four main types of industrial robots based on the way the joints and axes are configured to achieve a particular kinematic structure. Each of the following types offers a different joint configuration. The joints in the arm are referred to as axes.

1. **Cartesian**: This robot is also called a rectilinear or gantry robot. A Cartesian robot has three linear joints (or a combination of them) that use the Cartesian coordinate system (X, Y, and Z). This robot may also have an attached wrist to allow for rotational movement. The prismatic joints deliver linear motions along the respective axes. This robot is normally employed in material handling, computer numeric control machine load/unload, etc.
2. **Articulated**: This robot design features rotary joints that are arranged in human arm type configuration (waist–shoulder–elbow–wrist). Here, each joint is called an axis and provides

---

*The serial link manipulator is a system that connects links serially by motor-actuated joints that extend from a base to an end-effector (shoulder–elbow–wrist structure).

⁺The parallel link manipulator is a system that uses multiple serial kinematic chains to support a single platform or end-effector.

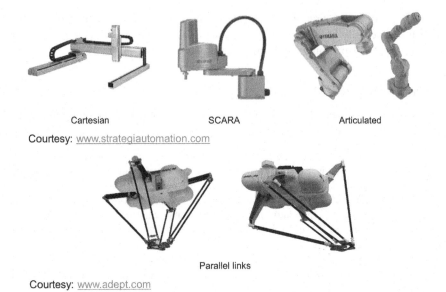

Cartesian   SCARA   Articulated

Courtesy: www.strategiautomation.com

Parallel links

Courtesy: www.adept.com

Figure 19.31 Industrial robots: industry examples.

a degree of freedom. This robot commonly has four to six axes and is normally employed for all common robot applications such as welding, painting, material handling, cutting, and deburring, and several others.

3. **Parallel links**: This robot has multiple serial arms usually connected together at the end-effector in a spider-like configuration to achieve a lightweight manipulator structure that enables very high-speed pick-and-place operations. High-speed picking of consumer goods, pharmaceutical products, confectionery, etc., are typical examples of applications of parallel links robots.

4. **SCARA**: The acronym SCARA for this robot stands for selective compliance assembly robot arm or selective compliance articulated robot arm. These robots are commonly used in assembly applications and are primarily cylindrical in design. These robots are normally employed in assembly, pick-and-place, and glue-dispensing applications.

Some industry examples of robots are illustrated in Fig. 19.31.

Following are some common applications that industrial robots can handle:

- Repetitive and routine tasks
- Hazardous and dangerous tasks
- Tasks that are difficult for humans to handle
- Handling of material
- Metal welding, cutting, and fabricating, surface finishing, etc.
- Pick-in-place of components
- Painting, finishing, etc.

Robots can operate equipment at a much higher speed and precision than humans can and are much cheaper on a long-term basis with no interaction with humans, which increases productivity.

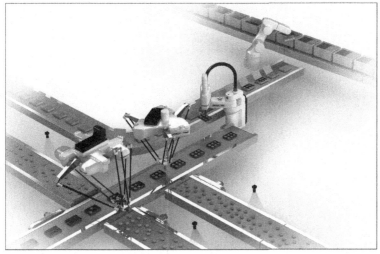

Courtesy: www.adept.com

Figure 19.32 Industrial robots: in action.

19.5.4   Advantages and Disadvantages

Advantages of employing industrial robots are:

*   Higher flexibility, reprogrammability, and kinematics dexterity
*   Quicker response time to inputs than with humans
*   Reduction in exposure to hazardous situation and accidents
*   Increased productivity, safety, efficiency, quality, and consistency of products
*   Repeatable precision at all times and can be much more accurate than can humans

The major disadvantage one can think of is that robots lack the capability to respond in emergencies unless the situation is predicted and the response is included in the system.

Fig. 19.32 illustrates robots in action.

19.6   Summary

Apart from physical classification (localized or distributed), this chapter dealt in detail with the classification of industrial processes, their evolution, and their management. General classifications are manufacturing (process plants and factories) and utilities (civic and backbone). Of these, operationally the process can be either continuous, discrete, or batch. The chapter also discussed the role of automation and case studies of each type of manufacturing and utility process. The chapter ended with a discussion on robotics and their varieties for different applications.

Information Technology–Operation Technology Convergence

Chapter Outline

20.1 Introduction

Ever since its invention, information technology (IT) has been an enabling and driving force practically in all engineering domains including automation. IT, introduced in automation technology during the 1960s[1], has completely changed the landscape of automation technology. Today, all subsystems of automation system from field level to operation level has become IT enabled and driven. The software content has overtaken the hardware content: for example, a hardwired relay logic controller moving to a soft-wired programmable logic controller. In view of this, automation technology's overall capability is increasingly being assessed from an IT perspective. Another technology that has heavily influenced and contributed to the growth of automation technology is communication and networking facilitating distributed computing. These technologies made all devices, equipment, and

---

[1]IBM 1800 Data Acquisition and Control System, 1964, is classic example.

Overview of Industrial Process Automation. http://dx.doi.org/10.1016/B978-0-12-805354-6.00020-7

systems in automation domain intelligent, communicable, and integrated from the field level to the operation level for seamless data flow in both directions. Today, the automation system represents a consistent way to manage industrial automation tasks.

This chapter refers to operation technology (OT) and information technology (IT) applicable to the following processes:

- Manufacturing processes that produce goods such as chemicals, electricity, engineering, pharma, food and beverage, etc.
- Utility processes that deliver services such as water supply, power supply, gas supply, water/sewage treatment, etc.
- Infrastructure management processes such as building management, railways, etc.

Fig. 20.1 illustrates the overall structure of the business enterprise for manufacturing processes.

Equivalent enterprise structures have been used to represent utilities and infrastructure processes. For the sake of brevity, we will restrict our discussion to manufacturing processes.

With technical and business requirements in mind, the overall management of an industrial enterprise is done at two levels: operation and business. Within the context of enterprise, the operation level is defined as OT and the business level as IT. Although, the technologies are dissimilar in nature, the enterprise needs both to achieve its overall objectives. IT and OT are also called the enterprise system and control system.

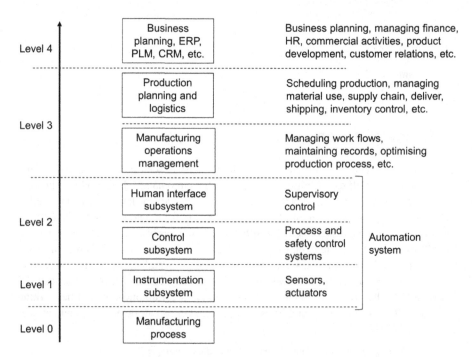

Figure 20.1 Business enterprise structure for manufacturing process.

Table 20.1 Information technology (IT) and operation technology (OT) functions

Functions	IT	OT
Purpose	• Business data processing, analysis, and applications • Human decision support	• Process data acquisition, analysis, and control • Process safety
Operating environment	• Data centers • Offices	• Process plant • Field equipment • Control centers
Data input	• Manual • Other IT systems	• Process data from plant • Operator from human interface
Data output	• Reports of summary, analysis, etc.	• Commands to plant devices • Display on human interface (alarms, events, trend, etc.) • Emergency shutdown, etc.
Owners	• Head, IT department	• Head, business unit
Connectivity	• Corporate network	• Plant and process control networks

As will be explained later in this chapter, there has been a need to bring IT-OT strategies together to reduce overall risk and increase performance of the industrial enterprise. In a way, IT-OT convergence has become logically imminent as their technology stacks are merging.

Table 20.1 illustrates the differences between IT and OT.

Over the years, IT and OT have moved toward convergence in multiple ways. The two popular directions that are discussed in the literature are:

1. Workflow and process convergence as defined by enterprise–control system convergence (IT-OT convergence).
2. Data acquisition and storage as defined by Internet of things (IoT), enabling access directly from the devices and sensors to the IT layer.

Both paths are extremely synergistic and provide highly value-added services to an enterprise. IIoT and Industry 4.0 are frameworks for converging these two developments. This chapter, apart from IT-OT convergence, provides an overview of these trends. The current context also means Internet and enterprise-wide connectivity.

20.2 Operation Technology and Information Technology

For various reasons, OT and IT systems worked independently in the enterprise even though each needed the other to achieve the overall objectives of the enterprise. Convergence of IT and OT systems is brought in to **enable** end-to end management of the enterprise in an automated and seamless way from the OT level to the IT level. This section discusses the convergence of OT and IT and the advantage that it brings to the enterprise and associated issues.

An OT system in an enterprise is composed of automation hardware and software necessary to monitor and control the processes or plants. As detailed earlier in this book, the primary activities of the automation systems are process data acquisition, process data monitoring, process data analysis/decision making, and process control. Today, traditional OT systems (hardwired, electromechanical, proprietary, dedicated, noncommunicable, nonintelligent, stand-alone, etc.) have more or less been migrated to more complex OT systems (using intelligent and communicable hardware, software, and firmware, similar to IT). Examples of OT are manufacturing execution systems (MES), supervisory control and data acquisition (SCADA), distributed control systems (DCS), network control systems (NCS), etc., which are directly related to manufacturing or service delivery process.

An IT system in an enterprise, on the contrary, is the application of computers to receive, store, retrieve, transmit, and manipulate data, often in the context of a business and commercial activities. IT includes hardware, software, and applications for transformation of data by accepting data as input, and processing and delivering a new set of data, according to some criteria, as output **with no interference** from the external world. Typical examples are enterprise resource planning (ERP), customer relationship management (CRM), and product life-cycle management (PLM).

20.3   Before Convergence

Most enterprises developed OT and IT infrastructure over a period as two different/ distinct domains and are managing and maintaining them with separate identities, technology, standards, governance methods, and organizational setup. Because of this, there is a gap between the two technologies. This forced manual and limited data exchange between IT and OT systems. The different paths OT and IT have taken during the past decades are to a large extent the result of organizational silos within an enterprise.

20.3.1   Gap

Following are some major reasons for the gap between IT and OT:

- **Personnel**: OT departments are made up of process specialists and their exchange with the IT department are limited even though OT has progressively moved to IT-like technologies at all levels from field to operation. On the other hand, IT departments are made up of computers, networks, or commercial specialists who may not be fully knowledgeable in OT.
- **Technical**: OT systems need to meet specific requirements such as real-time data acquisition, monitoring, and control of process parameters in specific environmental conditions. These aspects are not part of IT systems. OT systems call for a real-time operating system (OS) whereas IT systems for a time-sharing OS. There is no single OS architecture that meets the requirement of both OT and IT. These technical constraints led to the development of a complete set of specific OT standards covering areas such as communications, security, and process integration. Most of these have become **de facto standards** promoted by OT vendors.

- **Varying life cycles of equipment**: Typical IT infrastructure such as computers, software, and networking equipment is upgraded every 2–3 years. OT infrastructure, on the other hand, is expected to last 20 years. Use of IT framework causes a lot of heartburn in the OT world. A classic example is the Windows XP platform (heavily used in the OT world), which was regularly upgraded and finally became obsolete. The OT world is typically not able to handle such short life cycles for this. This is primarily because any piece of hardware or software is changed in the OT world only after thorough validation and testing for real time, safety, and environmental conditions. These are long-time cycle activities and cannot be carried out every 2–3 years like IT equipment.
- **Cultural and ownership**: Employee profiles in IT and OT have evolved from existing organizational separation: production-oriented in OT and business-oriented in IT. OT platforms are normally **owned** by business units responsible for profit and loss. Harmonized OT across business units in an enterprise is rare. Also, the objectives and associated performance indicators of each business unit may differ from each other. On the contrary, IT platforms and processes in all of the business units are owned and managed by a centralized organization made up of software, hardware, and infrastructure specialists.
- **Governance**: There is rarely central governance to cover the definition and execution of enterprise-wide OT strategies because responsibilities are different and scattered. On the contrary, IT decides the harmonization of IT systems throughout the company and largely imposes IT-related strategies by setting company-wide standards for software and hardware products to be used and the way infrastructure is managed.

Apart from these, OT is product and control oriented whereas IT is business and information oriented. The security and safety of OT systems and processes are big reasons for the gap.

20.3.2 Bridging the Gap

Following are some major steps adopted to bridge the gap between IT and OT:

- Establishing IT-based network infrastructure for industrial automation
- Establishing of centralized management with common organizational setup and governance practices
- Harmonizing duplicate and/or overlapping processes, data structures, strategies, policies, methods, etc.
- Reskilling employees as technology and platforms employed are becoming similar
- Defining common performance indicators at an enterprise level (key performance indices)

20.4 After Convergence

In an integrated or converged system, IT and OT systems are connected for automatic and seamless data exchange between them, with no human intervention. With this:

- OT system can monitor the process plant and send large volumes of data on the condition and status of critical assets to IT system seamlessly in real time.
- IT system can analyze or synthesize the big data received from OT, generate critical insights/visibility into the behavior of plant assets, and automate quick actions to optimize plant performance.

- In the reverse direction, OT system can execute the process-related commands generated and sent by IT.

The converged system transforms real-time data into actionable insights and uses them to drive the operational excellence of the enterprise.

20.4.1 Benefits

Tangible benefits derived from IT-OT convergence to many industries, from manufacturing to utilities, are different for different enterprises. Common benefits across enterprises provide:

- Cost reduction by deploying uniform technology, standards, and governance principles
- Risk reduction by jointly addressing cyber security issues
- Cost and time savings by allowing for smooth transition of newly developed products into existing manufacturing operations and time to market
- Better transparency with regard to costs and cost structures, leading to improving site efficiencies and to higher flexibility for shifting manufacturing between locations (produce anywhere)
- Seamless connectivity for enabling and leveraging IT technology in OT networks
- Advantages of cloud resource sharing by facilitating IT system to reach out to the cloud
- Possibilities for openings to new technology advancements in future
- Availability of cloud computing services to OT applications that can be leveraged for performing time-critical and computational-hungry applications
- Availability of cloud services for a lot of applications such as data storage, visualization, etc.

Intangible benefits accruing from IT-OT convergence are that the overall system can:

- Respond more quickly to real-time conditions
- Gain strategic visibility and insight for business process improvement
- Priorities work to reduce the risk of outages

These lead to improvement in overall organizational effectiveness, performance and productivity.

Fig. 20.2 illustrates the converged model of IT and OT.

Layer 3, or the manufacturing operations system, provides an interface between the OT and IT layers.

20.4.2 Security

This is further to the discussion in Section 16.7 on cyber security in automation systems (OT systems in the current context), when they are integrated with the external world via IT systems. Typically, OT systems demand high resiliency and availability whereas IT systems demand interconnectivity, enterprise security, and compliance. Also, both systems must coexist with the new player, the **data analysis system**, which requires connectivity for real-time data capture, sharing, and analysis for every decision in the business.

In view of this and the convergence, it is all the more necessary to secure OT systems that become more vulnerable to external cyber-attacks via the IT system.

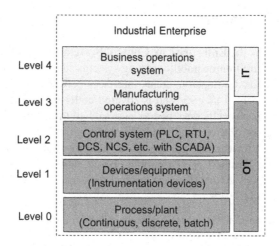

Figure 20.2 Converged enterprise structure.

Cyber security of the overall converged system becomes more complicated with complicated and highly distributed network topologies. Furthermore, effects of any cyber-attack in the IT system can trickle down to OT systems and devices, which are IT-like. The common approach to secure the OT system from the IT system is to establish firewalls and/or demilitarized zone (DMZ) layers between OT and IT systems. IT systems have their own standards and practices to cover themselves for security, which are not covered in this chapter.

However, apart from protecting individual OT and IT systems, there is a need to protect the converged/integrated system by providing end-to-end security. Following is a brief description of **security practices** that are part of the end-to-end security model of the integrated/converged system:

- Configuration management, an exclusive task for OT networks, involves configuration, base-lining, and archiving of configuration files. This includes the configuration of the programmable logic controller (PLC)/controller, human machine interface, drive, etc., and parameterization/configuration of the network and switches.
- Secure access employs network segmentation, virtual local area networks, and a DMZ to secure access between multiple layers.
- An identity service manages the user identity and associated user authorization at multiple layers.
- Application control manages the authorization of application by blocking or restricting unauthorized applications from executing in ways that put the data at risk. They typically use a white-listing practice to accomplish the task to prevent malicious application running in the environment.
- Network security provides preventative measures in hardware and software configuration of networking infrastructure against unauthorized access, misuse, malfunction, modification, destruction, or improper disclosure. This is an effective measure to create a secure platform for computers, control devices, users, and application programs to perform permitted critical functions within a secure environment.

20.5   ISA 95 Standard

This is also known as the **Enterprise or Control System Integration Standard** and is developed to support interoperability among systems supplied by vendors of IT and OT systems, standards developed to create interfaces between IT and OT for integration. This model is named ISA 95[2] and is covered under International Standard IEC 62264. This model is basically designed to automate all three types of manufacturing processes: continuous, discrete, and batch. Furthermore, this model is equally applicable to utility and infrastructure processes. Fig. 20.3 illustrates the ISA 95 model.

ISA 95, which is Purdue University model-based, defines four levels: levels 1–4. Levels 1 and 2 are basic for OT whereas level 4 is for IT. Level 3 is an interface between the OT and IT levels:

- Level 1: Instrumentation subsystem (transmitters, control valves, etc.) for sensing and manipulating the process in real time;
- Level 2: Control subsystem (PLC, remote terminal unit, DCS, NCS, etc.) and associated human interface subsystem (operator interaction, logging, maintenance/asset management, engineering, etc.) for automated control functions and SCADA. The time frame for this level is in hours, minutes, seconds, and subseconds;
- Level 3: Manufacturing execution system for control of work flow, recipe, etc., to produce desired end products while maintaining records and optimizing the process. The time frame for this level is in days, shifts, hours, minutes, and seconds;
- Level 4: Business management system for establishing basic plant schedule, production material use, delivery, and shipping, inventory management, etc. The time frame for this level is in months, weeks, and days.

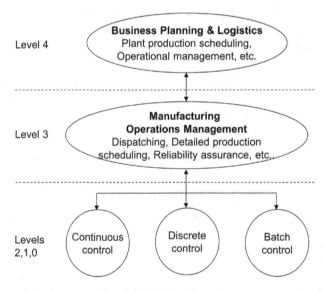

Figure 20.3 ISA 95 model.

[2]www.isa.org/isa95/.

In this, levels 1, 2, and 3 together represent OT whereas level 4 represents IT. The standard defines the terminology and the practices for:

- Delinking business processes from the manufacturing processes.
- Identifying responsibilities and functions on all levels.
- Identifying exchanged information in business to manufacturing integration.
- Defining common terminology and a consistent set of models needed for integration.
- Establishing common points for integration.
- Defining in detail an abstract model of the overall enterprise including manufacturing control functions, business process functions, and information exchange between them.
- Establishing common models and terminology for the description and understanding of the enterprise including manufacturing control functions, business process functions, and information exchange between them.
- Defining an electronic information exchange format between the manufacturing control functions and business process functions including data models and exchange.

The goal of this model is to reduce the risk, cost, and errors associated with development and implementation of interfaces every time for every new requirement and standardization of data exchange within and across business.

The standard defines structured information exchange between layers that is robust, safe, and cost-effective. The data exchange mechanism preserves the integrity of each system's information and span of control.

Fig. 20.4 illustrates interfacing details between OT and IT layers for manufacturing processes.

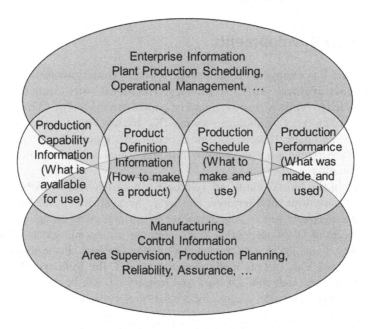

Figure 20.4 ISA 95: interfacing between manufacturing and enterprise layers.

20.5.1 Before

The situation that existed before the introduction of the standard was that:

- MES was not well defined and functions of manufacturing operation were not clearly demarcated. It was difficult to compare features of MES packages for adoption.
- It was hard to integrate business-level systems and shop floor control was industry specific.
- There were no common definitions or terminology.

20.5.2 After

- Redefined MES as manufacturing operation management with well-defined activities and made it part of OT
- Standard definitions, terminology, common requirements
- Well-understood and supported business level to shop floor control integration
- Availability of more cross-industry solutions leading to significant competition and advances in solutions
- Rapid growth in deployments

20.5.3 Benefits

Benefits of the standard are:

- End users: Drastic reduction in execution time of integration projects from years to weeks with success rate from less than 50% to over 90%.
- Vendors: Drastic reduction in integration costs because of standard format and less custom code to develop and support.
- System integrators: Standard tools and methods for integration projects.

20.6 New Developments

The Internet has changed the landscape of voice communication with voice over Internet protocol (VoIP) technology. Traditional voice communication is being replaced by VoIP. Already we have examples such as Skype, WhatsApp, etc. With the availability of technologies such as 3G, 4G, and above, higher-bandwidth communication channels have become a reality. All types of data (text, audio, and video) are treated alike and employ standard digital communication practices for transportation.

Perhaps in the time to come, all Telecoms will switch over to ISPs providing all types of communication services. The higher the bandwidth is, the better will be the service as the time taken to store the packet before forwarding becomes negligible.

The concept of IT-OT integration in an enterprise just became extended to cover integration of all intelligent and communicable **assets** distributed in the physical world over the Internet. This section briefly discusses the **IoT, the Industrial Internet of Things (IIoT),** and **Industry 4.0.** IIoT and Industry 4.0 can be considered further extensions of IoT addressing specific applications in manufacturing.

20.6.1 Internet of Things

IoT is relatively a new concept for connecting **things** (anything and everything in the physical world) that are intelligent and communicable for collecting and transporting data from one place to another on the globe over the Internet. IoT has given a new dimension to use of the Internet and has changed its landscape. Here IoT is being viewed in the context of automation technology even though IoT encompasses other domains.

To bring similarity between IT-OT convergence (discussed earlier) and IoT, we can call "things" as OT and the rest (data communication over the Internet, data processing, and decision making in cloud-based systems, etc.) as IT. The basic idea of IoT is to connect all physical devices to collect their relevant data in real time to manage the "things" better and make "things" more reliable and predictive.

The evolution of IoT is summarized thus:

- With a large number of devices with a high volume of data, manual data collection and transportation has been a tedious task.
- With cost-effective microprocessors, the devices are becoming **intelligent** and **network ready** for communication. This has made the devices function independently and with self-diagnostics to send their health status to the central system and to the right owner over wired and wireless media. This has enabled remote management, proactive/predictive maintenance, and data analysis to help decision making. These devices are also called smart devices.
- Smart devices have limited memory and processing power to handle data for analysis and decision making. Also, they need to make decisions based on multiple devices sometimes. Hence, the devices transfer data to remote locations (say, clouds) where facilities are available for processing to enable predictive maintenance, asset management, etc.

To facilitate this, all the smart devices, now named IoT-enabled devices, are designed to support Internet connectivity and are:

- Intelligent to gather information from their environment and on their own functioning/health
- Communicable over the Internet so that the collected data are transported to the concerned for further analysis and decision making
- Managed remotely over the Internet in the other direction

Fig. 20.5 illustrates the general architecture of IoT.
Summing up, IoT is:

- A network of physical objects or **"things"** embedded with intelligence in terms of electronics, software, sensors, and network connectivity, which enables them to collect and exchange data making a cyber-physical system.
- An environment in which objects, animals, people, etc., are provided with unique identifiers and the ability to transfer data over a network without requiring human interaction.
- The convergence of communication technologies (wired and wireless), micro-electromechanical systems (MEMS), and the Internet.

The concept may also be referred to as the **Internet of Everything**.

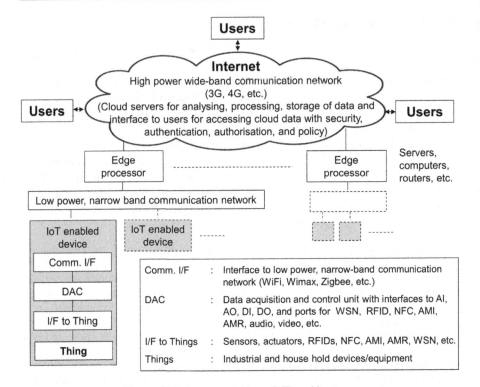

Figure 20.5 Internet of things (IoT) architecture.

20.6.1.1   Critical Issues

The next sections discuss some of associated issues:

- Security: Some critical applications need higher security, whether for data or user access, data manipulation, data integrity, or data loss. Conventional IT security aspects are being mapped to the IoT framework. However, there are not yet perfect solutions.
- Communication and connectivity: This is a big challenge because there are multiple technologies such as Wi-Fi, 3G/4G, ZigBee, Bluetooth, industrial Ethernet, etc. Furthermore, reliability is an issue especially for critical applications such as health care. This can be addressed by having redundancy built in to the communication channels.
- Data processing: A large number of devices leads to a large amount of data putting a lot of stress on the communication pipe. This can be addressed by **preprocessing** the data at the edge processor level and sending only relevant data in real time while the rest in batch is sent during off-peak hours.
- Sensors and actuators: Currently these are expensive. However, this technology is getting a huge push and is leading not only to reduce the price of sensors but also to develop new sensors. The advent of MEMS is leading to chip-sized sensors for various applications (vibration, temperature, gas sensors, etc.).
- Implementation: There are many challenges because the technology landscape is fragmented. Furthermore, far too many technologies have to be integrated right from sensors, communication, the cloud, etc.

- Standards: There are either too many standards or a lack of standards, causing a challenge in adoption and implementation. This is expected to stabilize over a period of time. The challenges here are standardization of data modeling, data models, etc.
- Vendors: There are far too many vendors and technology solution providers in the market, which makes it challenging because it is leading to **too many choices**. This can be addressed through standardization.
- Infrastructure: High-speed Internet and IoT-enabled devices are available basic prerequisites.
- Legacy equipment: Investment protection is important because factories have old machines and equipment. How to retrofit the legacy equipment and devices to make them compatible with IoT technology is also a major challenge.

20.6.1.2 Applications

The concept of IoT in simple words is using Internet power to monitor, control, and analyze the **things**. Here "things" mean the devices and equipment that can be from various segments. Table 20.2 lists the application areas.

In the context of IT-OT integration, the applications of IoT are predictive maintenance, diagnostics, prognostics, supply chain automation, asset tracking, etc.

Important points to be noted are that industrial and household devices are:

- No longer stand-alone with operational purposes alone
- Increasingly being embedded with sensors, greater processing power, and expanded memory capabilities
- Connected permanently or intermittently and together
- Able to send data to the cloud automatically with no human intervention for further analysis and management
- **Able to employ the Internet as a communication medium**.

IoT has changed the landscape of a wide array of business segments and industry sectors, bringing together a number of innovations in connectivity architecture,

Table 20.2 Typical application areas

Segment	Devices	Key benefits
Home	Appliances, lighting, heating, ventilation, and cooling	Lifestyle, security, power savings
Industry	Factory equipment, energy meters, safety equipment	Asset monitoring, predictive maintenance, increased uptime
Transportation	Tracking, speed, fuel level	Effective maintenance, monitoring, safety
Utilities	Solar inverters, energy meters, water meters	Efficient use of available resources. Renewable energy used instead of fossil fuels
Sports	Pedometer, pulse rate monitor	Improved fitness, early warning of fatigue in athletes, ability to win
Health care	Multiple body sensors	Monitoring body parameters and communicating directly for people without direct health care access helps to bring specialist support

smarter and connected devices, advanced analytics, cross-platform applications, and integration with enterprise systems. These innovations will turn **hitherto stand-alone devices** such as vehicles, white goods, consumer electronics, security systems, office equipment, etc., into **devices with intelligence and communicability** for highly integrated services. Although many IoT platforms are currently under development, all of them need to address the following:

- Device: Sensor networks and communication protocols
- Data: Models, storage techniques, real-time predictive and analytics, and deep learning
- Business: Work flow, billing, auditing and compliances, and alerts
- Services: Web interfaces, application program interface gateways, mobile applications, and support center (for alerts, notifications, and reporting)

Fig. 20.6 illustrates a typical IoT-enabled device along with its sensor modules.

This IoT-enabled device is for remotely monitoring ambient environment parameters such as temperature, humidity, air quality, fire, smoke, pressure, wind, and light This device provides connectivity to the edge processor via Ethernet, Wi-Fi, serial ports, and General Packet Radio Service. Data collected by this device can be accessed and displayed anywhere. Objectives that can be accomplished are:

- Collects environment data and sends to the remote central place
- Remotely controls auxiliary equipment to influence the ambient environment
- Maintains daily operating log

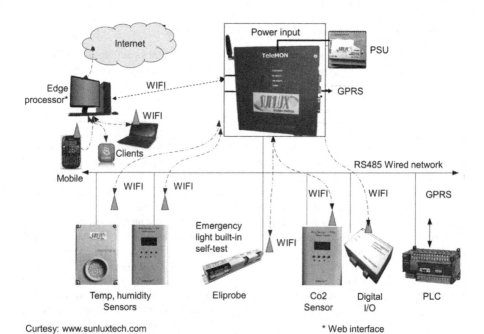

Curtesy: www.sunluxtech.com   * Web interface

Figure 20.6 Example of internet of things–enabled device and sensors.

- Analyzes data collected and displays trends
- Makes important decisions
- Gets right alarms/alerts for optimum operation
- Displays data remotely on PCs, laptops, smartphones, etc., on Web browsers

Fig. 20.6 also illustrates the nodes provided to support discrete sensors to monitor ambient environment conditions along with typical sensors.

20.6.2 Industrial Internet of Things

IoT, a general concept, makes anything and everything in the physical world communicate with each other, whereas IIoT is an application of IoT technology in the manufacturing industry. IIoT also requires all the participating devices/machines in the manufacturing setup to be intelligent and communicable.

Specific features of IIoT in the manufacturing process are:

- It captures all sensor data from the machines, facilitates machine-to-machine communication, and makes use of automation technologies that have existed in the industrial environment for years.
- **It incorporates data analytics, machine learning and big data analysis technology**.
- The driving philosophy behind IIoT is that the smart (intelligent and communicable) machines along with human intelligence are better for accurate and consistent data capturing, and subsequent dispatch.
- Analyses of the captured data enable companies to identify invisibility and inefficiency problems quickly in the manufacturing process, save time and money, and support business intelligence efforts.
- It holds great potential for quality control, sustainable practices, supply chain efficiency, operational efficiency, maintenance, and traceability.

A major concern of IIoT is interoperability between machines that use different protocols and have different architectures. The nonprofit Industrial Internet Consortium, founded in 2014, focuses on creating standards that promote interoperability and development of common architectures, including Industry 4.0, as discussed next.

20.6.3 Industry 4.0

Three earlier industrial revolutions that took place over the years are:

1. Mechanized production using water and steam power
2. Mass production using electric power
3. Electronics and IT in production

The fourth industrial revolution, **Industry 4.0**, refers to total computerization of manufacturing (**smart factory**). This was initially promoted by the German government. The whole world now uses this terminology. Industry 4.0 is a collective term for a number of contemporary technologies and concepts such as **automation, data exchange, manufacturing,** etc., put together for a **value chain organization**.

Industry 4.0 defines the following participants and their management:

1. Cyber-physical systems

This refers to integrated computation, networking, and physical processes. Each cyber-physical system can be treated as a manufacturing process managed by its own automation system. All manufacturing processes, partners in the group, are physically and widely distributed.

2. IoT

This refers to networked intelligent and communicable objects to collect (machines in this case, in the manufacturing process) and exchange data. All cyber-physical systems are linked by a communication network including the industrial Internet governed by IoT.

3. Internet of Services

This refers to internal and cross-organizational services for participants of the value chain. All services are also physically and widely distributed in distributed manufacturing processes. Required services in a manufacturing process may be available internally or externally to be borrowed from other distributed manufacturing processes.

Industry 4.0's vision and execution within the modular and structured smart factories are defined as:

- Cyber-physical systems to monitor physical processes, create a virtual copy of the physical world, and make decentralized decisions.
- IoT communicates over the Internet, cooperating with each other and with humans in real time.
- Internet of services (IoS) offers internal and cross-organizational services for use by other participants of the value chain.

Following are six design principles in Industry 4.0 that support companies in identifying and implementing Industry 4.0 scenarios:

1. Interoperability: Ability of cyber-physical systems (i.e., work-piece carriers, assembly stations, and products), humans, and smart factories to connect and communicate with each other via the IoT and IoS.
2. Virtualization: Create a virtual copy of the smart factory by linking sensor data (from monitoring physical processes) with virtual plant and simulation models.
3. Decentralization: Ability of cyber-physical systems within smart factories to make decisions on their own.
4. Real-time capability: Capability to collect and analyze data to derive insights immediately.
5. Service orientation: Offer of services (of cyber-physical systems, humans, or smart factories) via the IoS.
6. Modularity: Adapt flexibility of smart factories to changing requirements by replacing or expanding individual modules.

Industry 4.0 is about the intelligent implementation of services by the convergence of emerging technologies such as cyber-physical systems, advanced automation control, and big data analytics.

20.7 Summary

Over the past decades, most industries have developed and managed OT and IT as two different entities, maintaining separate technology, standards, governance methods, and organizational units. However, over the years, OT has been progressively migrating to IT-like technologies. In the light of this, the convergence of IT and OT is expected to bring clear and tangible advantages to companies on cost and risk reductions, enhanced performance, increased flexibility, etc. A prerequisite to achieve these benefits is an understanding of the strategic, organizational, and technological issues involved in both IT and OT and implementation of their convergence for uniform strategies, governing models, security, resource use, etc. along with reskilling of people to know about the requirements of both disciplines.

Successful IT-OT convergence helps companies exploit the potentials hidden in their supply chain by streamlining processes, increasing data transparency, and allowing for better and quicker decision making. New ideas and concepts are developing around IT-OT, providing major opportunities to leverage IT know-how to the production floor. This can make a difference when competing with peers. Considering all of this, IT-OT convergence standard (ISA 95 or IEC 62264) is developed for the interoperability of systems developed by vendors.

The chapter also briefly discussed IoT, IIoT, and Industry 4.0 as an extension of IT-OT integration.

Concluding Remarks

21

Chapter Outline

21.1 Introduction

Industrial process automation is a vast subject, even introducing it is a difficult task. Over the preceding chapters, an attempt was made to present a guided tour of the subject. This chapter briefly summarizes what we covered so far. For some time, automation systems functioned in isolation (or in silos) performing only the basic functions of industrial processes (data acquisition, data management, and process control). The primary objectives were mainly to produce goods or deliver services of quality, consistency, and cost-effectiveness. Subsequently, the objectives were extended to cover many more functions to meet technical and business challenges. In this chapter, the

Overview of Industrial Process Automation. http://dx.doi.org/10.1016/B978-0-12-805354-6.00021-9

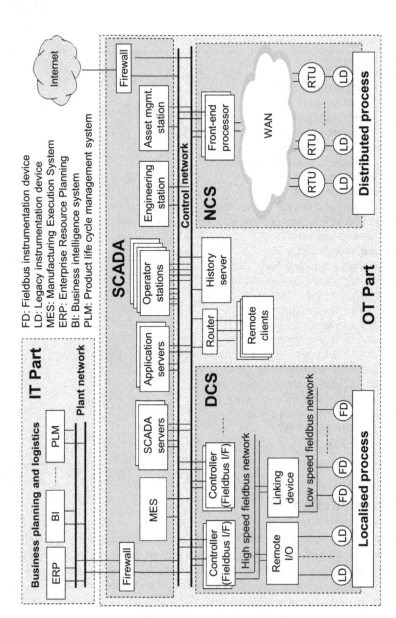

Figure 21.1 Today's automation system.

functionalities of automation systems discussed earlier are summarized in two parts: basic and extended. Fig. 21.1 illustrates today's automation system configuration.

This chapter ends with a brief discussion on some functionalities that were not covered earlier and emerging trends (what to look for in the future) in automation technology.

21.2 Basic Functionalities

Whereas Chapter 1 introduced automation in general, Chapter 2 discussed the overall structure, basic functions, and objectives of automation systems. Chapters 3–14 discussed the basic functionalities of the automation system in detail. Chapter 15 discussed customization of automation system platforms for specific applications. These are summarized below under three headings: data management, process control, and customization.

21.2.1 Data Management

Data management starts with process data acquisition and ends with decision making about whether to effect process control. Sequential steps involved are:

- **Data acquisition**: The first step in data management is to acquire process data in the control center without which no further actions are possible. Process data are received by the automation system at the control center, either locally from the controllers in centralized control systems or distributed control systems or remotely from remote terminal units (RTUs) in network control systems. Typical examples of process data are the values of analog parameters and the status of digital parameters in the process.

Input instrumentation subsystem, the interface between the process and the control subsystem, facilitates this task by collecting data from process variables (values and status) in **physical form**, converting them into its equivalent **electronic form** with **no loss of information**, and delivering them to controllers or RTUs.

- **Data conversion**: The received raw process data are converted into information for further processing.
- **Data supervision**: The next immediate task is to supervise or monitor the process parameters for any deviation from their normal/expected value or status and to generate an audiovisual alarm to alert the operator to take appropriate actions immediately.
- **Data display**: The processed information at the control center is displayed suitably on operator panels and operator stations to assist the operator to obtain better insight into the behavior of the process in real time to make appropriate decisions. This is the first part of human interface subsystem.
- **Data logging and history generation**: Data-logging stations produce on-demand, periodic, trend, and report logs. The automation system also stores process history for future use. These data can be stored on different media or systems (historian) for off-line analysis at a later date.
- **Process survey**: All of the process information available in the control center can be called up on the screens of operator stations for a survey by the operator (with no need to visit the process). The operator can see current happenings in the process whether the process is local or distributed (even geographically).

- **Process studies**: These basic online functions discussed are not enough to produce optimum results and call for running special application programs to improve the performance of the process. The collected data are used to study the behavior of the process to optimize its performance.
- **Data analysis and decision making**: To manage the process to meet its overall objectives, the automation system analyzes the received data and makes appropriate decisions to change the course of the process, if required, to meet its overall objectives, all on a real-time basis. This is the last step in data management. Actions are taken through control of the process parameters, as discussed in the next section.

21.2.2 Process Control

- This step is taken to change the course of the process based on data analysis and decision making, if necessary. Process control, or the operation to change the value or status of process parameters in the plant, is performed automatically by the automation system as per a predefined strategy. Normally the process control function is executed automatically in real time. However, operator stations are also provided with the facility for effecting manual control and actions for overriding the automatic control when necessary (say, under emergency conditions). This is the second function of the human interface subsystem.

The output instrumentation subsystem, the interface between the control subsystem and process, facilitates this task by receiving the control commands in **electronic form** from controllers or the RTU, converts them into an equivalent physical form **with no loss of information,** and forwards them to the process equipment for control of process variables (values and status).

21.2.3 Customization

- Automation systems are not designed and developed for each process and/or application. They are built on a common platform and are configured (for hardware, software, and applications) to meet the specific physical and functional requirement of the process or application.

21.3 Extended Functionalities

Over time, automation systems could no longer remain in isolation and became connected to the external environment. Furthermore, many other functionalities were also implemented to meet business and technical challenges and comply with statutory regulations. The following sections summaries some major extensions to basic automation system functionalities discussed earlier.

21.3.1 Communication and Networking

Chapter 16 discussed interconnecting automation systems with the external world (compatible systems) over local area and wide area networks for seamless data exchange between the field level and corporate level. This chapter also covered interprocess communication and issues related to cyber security.

21.3.2 Fieldbus Technology

Chapter 17 discussed the evolution of this technology, which was originally conceived to reduce field (control/signal) cabling and associated problems within the plants, by making the instrumentation devices intelligent and communicable. This extension not only fulfilled this technical and commercial requirement but also opened up several new areas such as interoperability (plug-and-play) of the instrumentation devices, remote diagnostics, asset management of devices, etc., apart from providing seamless data flow from field level to corporate level.

21.3.3 Safety Systems

Chapter 18 discussed issues related to the safe operation of process plants and machines. It is the responsibility of business enterprises to ensure the safety of people, the plants, and their environment while operating process plants and machines. This is also a statutory requirement to be complied with. Extension of the automation systems with appropriate safety systems helps prevent hazards in process plants and machines and mitigate them if they take place.

21.3.4 Management of Industrial Process

After going through the evolution of industrial process management, Chapter 19 discussed the broad classifications of industrial processes as manufacturing (process plants and factories) and utility (civic and backbone) along with their management and case studies. The chapter also briefly discussed robotics and its application in factory automation.

21.3.5 Information Technology–Operation Technology Convergence

Chapter 20 discussed the convergence of operation technology (the plant and its automation system) and information technology (IT) (business systems). Integrated management became necessary for seamless data movement from the field level to the enterprise level to meet business and technical challenges, especially for the conversion of business decisions into operational decisions and their implementation on a real-time basis. This chapter also discussed new topics such as Internet of things (IoT), industrial Internet of things (IIoT), and Industry 4.0.

21.3.6 Additional Functionalities

Some additional functionalities that are relevant in automation systems but not discussed in preceding chapters are:

- **Abnormal situation management**: This is a situation in which an industrial process is disturbed and the automation system cannot manage. There could be reasons including human error. The goal of abnormal situation management is to make the operator more effective in handling the situation by employing advanced alarm management, ergonomic operator displays, early event detection, operations information solutions, and procedural automation tools.

- **High-performance graphics**: This displays information making use of the data in context not only by showing the process value but also where it is relative to **what is good**. Abnormal conditions are brought out clearly, effectively, and sparingly by using color consistently and graphics with a proper hierarchy.
- **Time synchronization**: Industrial automation requires high availability for time synchronization with zero master takeover time or zero sync path switchover time in case of master or link failure. High availability is primarily achieved by redundancy: redundant master or two sync messages over two disjoint sync paths simultaneously. This lets each end station always receive multiple sync messages so that in case of a single point of failure, it is guaranteed that at least one sync will be received.
- **Collaborative screens**: This presents the analysis of the data collected and insights across multiple screens or on large screen displays to help businesses enterprise get a competitive edge.
- **Fire and gas (F&G) control systems**: In addition to process safety systems discussed in Chapter 18, F&G detection systems are deployed to monitor process plant activity continuously and, in case of hazardous conditions, initiate appropriate actions. These systems are designed to work independently, from the detection of hazardous gases up to a proper plant shutdown. This is an additional layer of protection over and above emergency shutdown. F&G systems are extensively employed in gas turbine installations.
- **Closed-circuit TV for process plants**: This technology is also increasingly used in the industrial sector to reduce accidents at **blind spots** or to monitor processes remotely for additional safety.
- **Fiscal metering and allocation metering**: Flow metering solutions for custody transfer and allocation metering systems are critical for both business transactions and meeting regulatory compliance especially for transferring liquid and gaseous fuels across facilities. Liquid and gas metering systems along with meter proving systems and online density measurement systems are important components in this category.
- **Remote monitoring and maintenance**: This allows operating conditions to be determined and faults eliminated, in some cases resulting in big savings for industrial production (minimizing labor costs, filling the knowledge gap resulting from nonavailability of experienced operators, etc.).

21.4 Emerging Trends

Automation technology is either technology driven or application driven. Automation technology is built out of the following **enabling** technologies:

- Engineering technology (sensors, control elements, control engineering, electronics, embedded, digital signal processing, electrical and drives, micro-electro-mechanical systems, and nanoelectro-mechanical systems, etc.)
- IT (computer, communication, networking, etc.)

Automation technology is continuously adopting advances in these technologies. Similarly, the automation technology is being deployed in newer and challenging application areas to make entire manufacturing and service delivery processes increasingly efficient, productive, safe, secure, environmentally friendly, compliant with statutory regulations, and so on. To make all of this happen, the future automation system will be an information-enabled platform with integrated safety and security supported by industrial Ethernet as the dominant network from the field level to the corporate level and the standard for common data exchange and integration.

In this section, a few trends are discussed broadly in three parts: information technology driven, engineering technology driven, and application driven.

21.4.1 Information Technology Driven

Automation systems are an integral part of today's business enterprises and are interconnected with the world. As we have seen in the preceding sections, information, communication, and networking technologies have had enabling roles in the evolution of automation technology. In line with this trend, future automation systems will adopt more and more new functionalities from IT. Some emerging trends in this direction are given next.

21.4.1.1 Cloud Computing

This refers to the outsourcing of IT resources available **elsewhere (cloud)** for data storing/accessing, program sharing, etc., over the **Internet** instead of using their **own** local resources: in other words, moving IT facilities out of the industrial enterprise to an external service provider. This has become necessary because IT facilities in the enterprise cannot manage the continuous demand for in-house processing of large data in real time for effecting improvements. This has become even more relevant with the arrival of IoT, IIoT, and Industry 4.0. Process data historians are early adapters of cloud computing for storage and easy retrieval.

21.4.1.2 Big Data

This is defined as data sets that are too large and complex, and are growing, changing quickly, and cannot be captured and processed locally by commonly used software tools within an acceptable response time. The continuous increase in data production and gathering technologies, such as IoT-based systems, calls for high information storage and processing capacity. In the light of this, big data have now extended their reach to several domains including industrial process automation, even though the full potential remains mostly untapped. Remote monitoring and diagnostics discussed earlier heavily leverage the big data capability.

21.4.1.3 Data Analytics

This refers to analyzing data and deriving insights into operations including hidden patterns. Several statistical and data analytics software tools are available to perform such analysis. In recent times, with the availability of computational resources, large sets of data are being collected and analyzed through big data analytics. The challenge for the industry is to create the data required to derive meaningful insights from the analysis. Once this is done and validated with actual test data, it can be implemented in the field to perform activities such as predictive maintenance, improving asset efficiency, and enhancing the quality of manufactured products.

21.4.1.4   Data Mining

Too much data are coming to the organization too quickly for users of any type to effectively parse them and get to the most relevant insights. In other words, the volume of data:

- Is large for comprehensive analysis
- Has a range of potential correlations and relationships among a multiplicity of data sources
- Massive for human data analysts to test all hypotheses
- Complicated to derive all of the value buried

A classic example connecting these four trends is the development of monitoring and diagnosis algorithms using large chunks of data (big data), stored and executed remotely (cloud computing), converting data into information (data analytics) using statistical tools (data mining).

21.4.1.5   Artificial Intelligence

The idea behind artificial intelligence (AI) is to figure out how people do what they do, program computers to do it, and thereby replace human intelligence with computers. Machine learning, a subset of AI, uses analysis of past experience and behavior to deal with unfamiliar situations and predict future events, and to do so on a massive scale using big data techniques so that machine learning becomes the core to every such application. A self-driving car is an example; robot operation and maneuvering are other classic examples.

21.4.1.6   Diagnostics, Prognostics, and Remote Maintenance

IoT, IIoT, and Industry 4.0 ensure that there is no dearth of data on the operation of a plant and associated equipment. These data need to be converted into information to monitor the (productivity, reliability, availability, and maintainability) key performance indicators on a 24/7 basis. A model, empirical, AI-based framework to diagnose and prognosticate the health of process and equipment is a growing trend. Remote monitoring and diagnostic centers are common occurrences for many large enterprises with geographically distributed operational facilities. These are prevalent in power plants, continuously monitoring rotating machinery to capture data precursors to failures.

21.4.2   Engineering Technology Driven

These technologies are deployed in automation system to build various hardware subsystems:

- Sensors, control elements, and control engineering
- Electronics, embedded systems, and digital signal processing
- Electrical systems and drives
- Micro-electro-mechanical systems and nano-electro-mechanical systems, etc.

Some trends in this direction are discussed next.

21.4.2.1 Networked Control Loops

Control loop performance is significantly affected by the availability and reliability of the signal transmission in the loop. In traditional control loops, availability of data is generally taken to be high and the primary focus has been on the other key control loops elements: sensors, actuators, and controllers. With industrial communication networks slowly gaining prominence, the performance of control loops need to be analyzed considering data drops, latency, and other communication-related issues. Enhanced proportional, integral, derivative controllers operating over wireless networks accounting for data loss and latency are slowly becoming available on the market.

21.4.2.2 Internet of Things

The introduction to IoT and related extensions (IIoT and Industry 4.0) have been discussed in Chapter 20. This technology, already in full swing, is going to dominate in the future in both automation and consumer areas. The major challenge is in establishing the stability, reliability, security, and troubleshooting of IoT networks.

21.4.2.3 Smart Sensors

Future sensors will be IoT enabled, low-power, low-cost, micro/nano technology based, intelligent, communicable, integrated with wireless communication, suitable for peer-to-peer communication, and for measurement of all types of physical parameters. In addition to the sensing element, smart sensors integrate signal conditioning, embedded algorithms, and digital interface. This simplifies the addition of new sensors into a system and is popular in both consumer and industrial applications.

21.4.3 Application Driven

As discussed in Chapter 19, most traditional applications of automation discuss either manufacturing process or utility process. In the future, apart from this, automation systems will be deployed for many challenging applications such as smart grids, smart buildings, smart cities, and smart factories that are close to people in their daily lives. This is expected to add a huge amount of convenience to the everyday life of people.

21.5 Summary

After a long guided tour on industrial process automation, this concluding chapter of the book summarized the overall functions of the automation system, both basic and extended. The chapter also briefly discussed some additional functionalities that were not covered earlier and concluded with emerging trends (what to look for in the future) in automation technology.

Appendix A: Hardwired Control Subsystem

A.1 Introduction

In the early days, automation systems were purely based on pneumatic, mechanical, and hydraulic components for both measurement and control. All of these were slow, bulky, sluggish, and less reliable, and they required more maintenance, space, and power. Over time, automation technology moved to electrical, electronic, and finally processor-based/information technology. Advances in electronics, information, communication, and networking technologies had a vital role in making the entire control subsystem more compact, power-efficient, reliable, flexible, communicable, and self-supervised.

Control subsystems started out with hardwired systems in line with the technology available at the time. This appendix details the beginning of modern control subsystem implementation, which was mainly hardware-based or hardwired for both discrete and continuous processes.

Throughout our discussion of the implementation of various technologies of control subsystems, we will employ the example of a water-heating process with its instrumentation and human interface subsystems.

A.2 Discrete Control

As discussed in Chapter 6, discrete process automation systems (open and sequential control with interlocks) handle only the discrete inputs and outputs from the instrumentation and human interface subsystems.

A.2.1 Relay Technology

Modern programmable control systems started with relay technology for automating assembly lines in automobile industries. In this technology, the simple electromechanical relay, as illustrated in Fig. A.1, is the central control element for implementation of the automation strategy.

A simple electromechanical relay consists of a coil of wire surrounding a soft iron core (electromagnet), an iron yoke to provide a low reluctance path for magnetic flux, a movable iron armature, and a pair of contacts. The armature hinged to the yoke is mechanically linked to a moving contact, and it is held in place by a spring action.

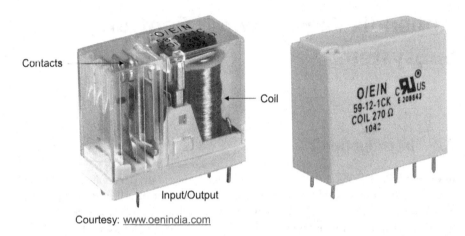

Courtesy: www.oenindia.com

Figure A.1 Typical electromechanical relay.

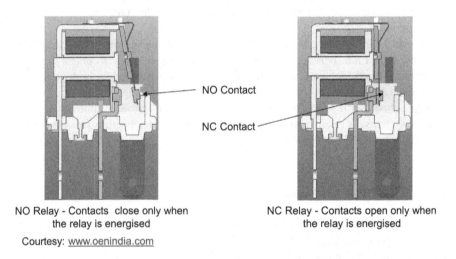

NO Relay - Contacts close only when the relay is energised

NC Relay - Contacts open only when the relay is energised

Courtesy: www.oenindia.com

Figure A.2 Construction of normally open and normally closed relays.

In the de-energized state, there is an air gap in the relay magnetic circuit leaving the armature in its normal state. On the contrary, in the energized state, the air gap in the relay magnetic circuit gets closed, moving the armature to its forced state. The armature stays in the forced state as long as the relay remains energized.

Depending on the mechanical arrangements of the contacts, the de-energized state of the relay is called the normal state, which can either have the contacts open [normally open (NO)] or the contacts closed [normally closed (NC)].

Fig. A.2 illustrates the mechanical arrangement of two contacts in NO and NC relays.

Normally under de-energized conditions, the contacts remain open in NO relays and remain closed in NC relays. When the relay is energized, the mechanical action closes the contacts in NO relays allowing the electrical signal to pass through, and it

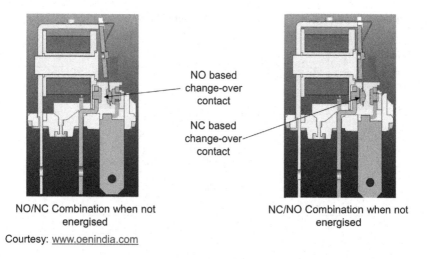

Figure A.3 Construction of CO relay.

opens the contacts in NC relays not allowing the electrical signal to pass through. *NO relays function as discrete signal buffers or signal repeaters whereas NC relays function as signal inverters*. These relays basically **repeat** the discrete/digital input signal with or without inversion, depending on whether the relay is NO or NC.

A general-purpose relay with three contacts, called a changeover (CO) relay, is illustrated in Fig. A.3.

CO relays are flexible because they can be wired either as NO or NC relays.

There are also relays with additional or parallel NO, NC, CO, or combinations of these to facilitate multiple (isolated) outputs. However all these relays have single inputs to energize or de-energize the relay. In our discussions, only NO and NC relays are employed to describe the relay-based strategies.

Figs. A.4 and A.5 illustrate the physical, logical (truth table), and timing relationships between the inputs and outputs for both NO and NC relays.

A relay with NO contact retains the open contact under no input. When it receives a voltage input, the coil gets energized and operates the output contacts to close, to transmit the true signal to the next stage, working as a **buffer/repeater**.

A relay with NC contact retains the closed contact under no input. When it receives a voltage input, the coil gets energized and operates the output contacts to open or has no signal transmission to the next stage, working as an **inverter**.

Figs. A.6 and A.7 illustrate the realization of two input AND and two input OR circuits using a pair of NO relays. NOT circuits are realized by NC relays.

Using the same approach, it is possible to realize the gates for a variety of logical functions, such as NAND, NOR, XOR, and XNOR. With these, we can derive any combinatorial logic for the implementation of a discrete automation strategy.

One more relay component that must be considered for the implementation of relay-based automation strategy is the **timer relay**. This relay, when energized, changes its output state after a programmed delay, as illustrated in Fig. A.8.

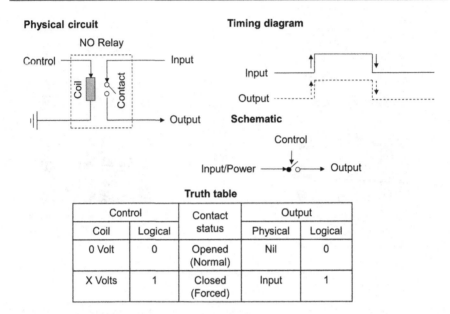

Figure A.4 Input-output relation in normally open relay (buffer/repeater).

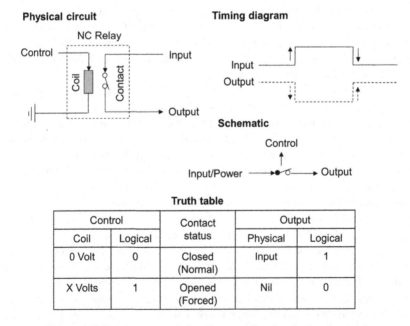

Figure A.5 Input–output relations in normally closed relay (inverter).

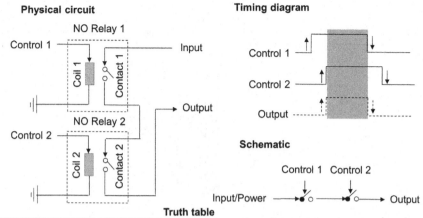

Physical circuit

NO Relay 1

Control 1 — Coil 1 — Contact 1 — Input

NO Relay 2

Control 2 — Coil 2 — Contact 2 — → Output

Timing diagram

Control 1
Control 2
Output

Schematic

Control 1   Control 2

Input/Power → Output

Truth table

Control 1		Control 2		Contact status		Output	
Coil 1	Logical	Coil 2	Logical	Contact 1	Contact 2	Physical	Logical
0 Volt	0	0 Volt	0	Opened (Normal)	Opened (Normal)	Nil	0
0 Volt	0	X Volts	1	Opened (Normal)	Closed (Forced)	Nil	0
X Volts	1	0 Volt	0	Closed (Forced)	Opened (Normal)	Nil	0
X Volts	1	X Volts	1	Closed (Forced)	Closed (Forced)	Input	1

Figure A.6 Realization of two-input AND gate.

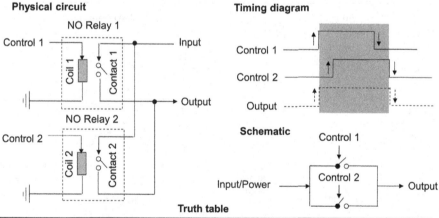

Physical circuit

NO Relay 1

Control 1 — Coil 1 — Contact 1 — Input

NO Relay 2

Control 2 — Coil 2 — Contact 2 — → Output

Timing diagram

Control 1
Control 2
Output

Schematic

Control 1

Input/Power → Control 2 → Output

Truth table

Control 1		Control 2		Contact status		Output	
Coil 1	Logical	Coil 2	Logical	Contact 1	Contact 2	Physical	Logical
0 Volt	0	0 Volt	0	Opened (Normal)	Opened (Normal)	Nil	0
0 Volt	0	X Volts	1	Opened (Normal)	Closed (Forced)	Input	1
X Volts	1	0 Volt	0	Closed (Forced)	Opened (Normal)	Input	1
X Volts	1	X Volts	1	Closed (Forced)	Closed (Forced)	Input	1

Figure A.7 Realization of two-input OR gate.

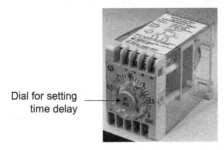

Dial for setting
time delay

Courtesy: www.gicindia.com

Figure A.8 Timer relay with delay time setting.

This is like any other relay but with a facility to program the delay time. The delay can be either on-delay or off-delay. Timer relays can be of either NO, NC, or CO configurations.

A typical wiring schematic for implementing the relay-based automation strategy using relay components is illustrated in Fig. A.9.

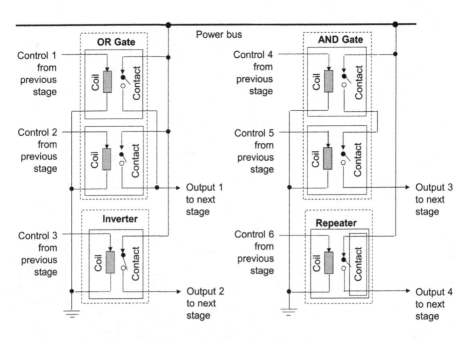

Figure A.9 Electrical wiring scheme for relay strategy.

Here a control signal moves from one stage to the other while satisfying the logic. The output from the previous stage becomes the control input for the next stage(s). In other words, implementation of relay logic means interconnecting various relays in a predefined and programmed manner.

A.2.1.1 Control Strategy Implementation

As discussed, the relay-based automation strategy is applicable only for discrete process automation. Now that we have introduced the relay components for all logic functions (repeater, inverter, NOT, AND, OR, timers, etc.), we will look into the implementation of relay-based automation strategies for the discrete automation cases discussed in Chapter 6.

Fig. A.10 illustrates the water-heating process as a discrete process with its instrumentation and human interface subsystems.

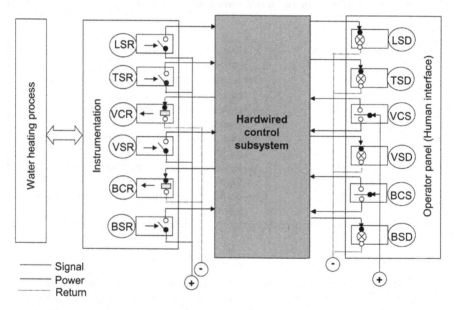

Figure A.10 Water heating process with instrumentation and human interface.

The internal structure of the control subsystem will be discussed in subsequent sections.

The functions and interconnections of the instrumentation devices and the human interface components with the control subsystem are explained in Table A.1.

Because the control subsystem is built of hardwired relay components, it is a hardware strategy. Before we start discussing various relay-based automation strategies, we will look into a few basic implementation schemes that are required. One such requirement is related to the issue of commands to the process.

Table A.1 Instrumentation devices and their interconnections

Between control subsystem and instrumentation (process)		
Discrete instrumentation devices		
1	LSR	Level status relay (preset value reached/not reached)
2	TSR	Temperature status relay (preset value reached/not reached)
3	VCR	Valve control relay (command to open/close valve)
4	BCR	Breaker control relay (command to close/open breaker)
5	VSR	Valve supervision relay (opened/closed status of valve)
6	BSR	Breaker supervision relay (closed/opened status of breaker)
Between control subsystem and human interface (operator panel)		
Discrete panel components		
1	LSD	Level status indication lamp (on/off for reached/not reached)
2	TSD	Temperature status indication lamp (on/off for reached/not reached)
3	VCS	Valve control switch (momentary commands to open/close)
4	BCS	Breaker control switch (momentary commands to close/open)
5	VSD	Valve status indication lamp (on/off for opened/closed)
6	BSD	Breaker status indication lamp (on/off for opened/closed)

In most cases, the commands are generally of momentary type, meaning that after the issue of the commands, the switch returns to its neutral position. This is required to meet the following conditions:

- The switch position should not interfere with the functioning of the logic to issue commands automatically to the process. If the bi-stable command switch is employed, it remains in one of its two positions and blocks the issue of automatic commands by the logic.
- For safety reasons, if the bi-stable command switch remains in the closed position, it can issue the command to the process upon powering the control subsystem.

For further processing of commands, the issued commands need to be remembered. Hence there is a need to latch (or remember) the momentary commands as maintained commands, as illustrated in Fig. A.11.

The scheme's timing diagram and its applications to open or close the valve and close or open the breaker are illustrated in Figs. A.12 and A.13.

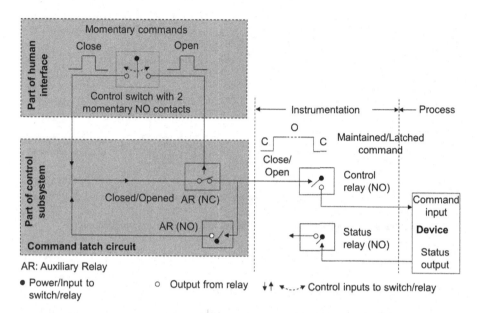

Figure A.11 Schematic of command latching.

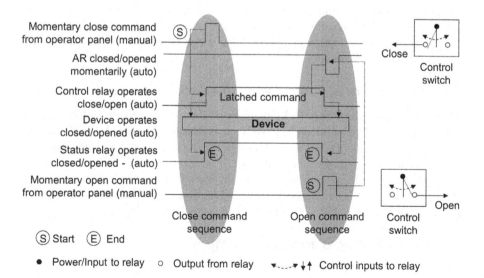

Figure A.12 Timing diagram of command latching.

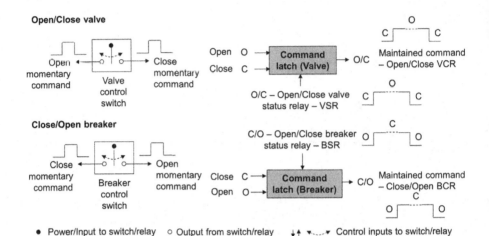

Figure A.13 Application of command latching to valve and breaker control.

A.2.1.2 Open Loop Control: Discrete

Fig. A.14 illustrates the implementation of simple discrete open loop control of level and temperature in the water-heating process, as discussed in Chapter 6.

Fig. A.15 illustrates the associated timing in this strategy.

Relay buffers and drivers are provided as a part of the control subsystem to isolate the control subsystem from the process and to repeat (reshape and/or strengthen) the incoming signals in both directions. As seen here, there is absolutely no check on how long the valve is open or how long the breaker is closed, because this depends on the operator's manual interaction.

A.2.1.3 Sequential Control With Interlocks: Discrete

In sequential control with interlocks, there are no changes in either the instrumentation subsystem or the human interface subsystem, as illustrated in Fig. A.10. The change is only within the hardwired control subsystem (relay strategy implementation) to achieve sequential control with the interlock.

As illustrated in Fig. A.16, the control subsystem built for sequential control with interlocks starts the heating process provided the interlock condition is satisfied, as discussed in Chapter 6. Fig. A.17 illustrates the associated timing sequences.

A.2.1.4 Advantages and Disadvantages

Advantages of the relay-based strategy are that it:

- is simple
- is easy to implement
- is easy to troubleshoot
- does not require highly skilled persons for design, operation, and maintenance

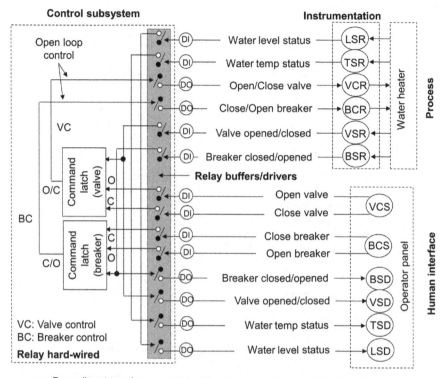

Figure A.14 Open loop control.

Disadvantages are that it:

- occupies more space (bulky)
- consumes more power
- is low and has sluggish response
- is unreliable (sensitive to environmental conditions such as temperature, humidity, dust, and vibration)
- is not maintenance-free (needs periodic preventive maintenance)
- is inflexible for modifications and extensions because the scheme is hardwired

A.2.2 Solid-State Technology

Some of the disadvantages of the relay-based strategy can be overcome by solid-state technology. In this case, the electromechanical relay-based logic elements are replaced by their equivalent solid-state elements on a one-to-one basis while maintaining all of the technical features. Hence the cases discussed previously for discrete process automation (open loop control and sequential control with interlocks) remain the same except for the change in control elements from a relay-based to solid state, as shown in Fig. A.18.

Using the basic logic functions, one can realize additional solid-state logic functions, such as NAND, NOR, and XOR.

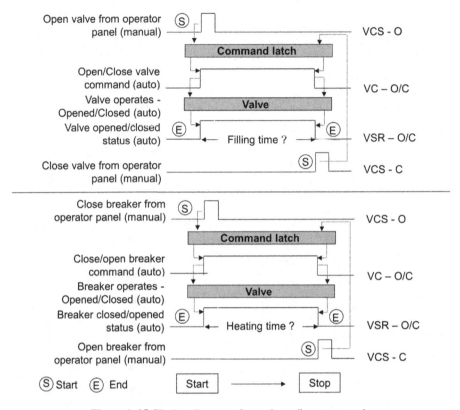

Figure A.15 Timing diagram of open loop discrete control.

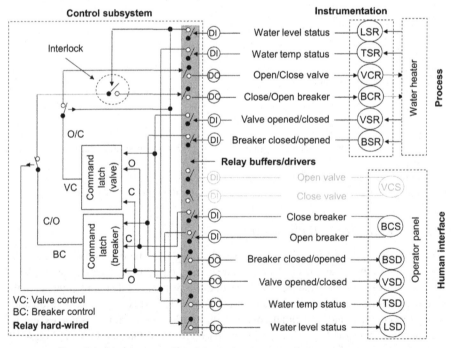

Figure A.16 Sequential control with interlock.

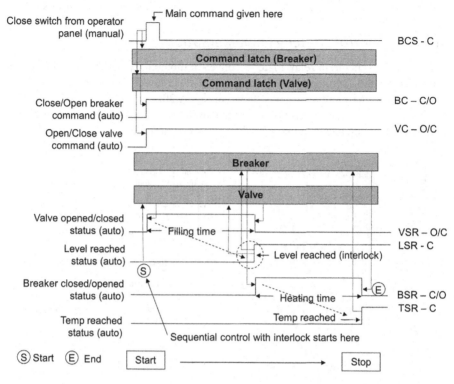

Close switch from operator panel (manual)
Main command given here

BCS - C

Command latch (Breaker)

Command latch (Valve)

Close/Open breaker command (auto)
BC – C/O

Open/Close valve command (auto)
VC – O/C

Breaker

Valve

Valve opened/closed status (auto)
Filling time
VSR – O/C

Level reached status (auto)
Level reached (interlock)
LSR - C

S

Breaker closed/opened status (auto)
Heating time
E
BSR – C/O

Temp reached
TSR – C

Temp reached status (auto)
Sequential control with interlock starts here

S Start   E End   Start ————————————→ Stop

Figure A.17 Timing diagram of sequential control with interlock.

Function	Basic relay scheme	Basic solid state scheme
Buffer/Driver/ Repeater	Control ↓ Input —▶●/○—▶ Output	Input ▶ Output
Inverter/NOT gate	Control ↑ Input —▶●—○—▶ Output	Input ▶● Output
2 Input AND gate	Control 1 Control 2 ↓ ↓ Input ▶●/○ ▶●/○▶ Output	Input 1 Input 2 ⟩ Output
2 Input OR gate	Control 1 ↓ Input ▶ [Control 2 ↓] ▶ Output	Input 1 Input 2 ⟩ Output

● Power/Input to relay   ○ Output from relay   ↓↑ Control inputs to relay

Figure A.18 Solid-state equivalents of relay logic elements.

A.2.2.1   Control Strategy Implementation

We devised a scheme in relay logic to convert a momentary command into a maintained command (command latch). Fig. A.19 illustrates a similar scheme for solid-state logic. Its timing diagram is shown in Fig. A.20.

The following sections discuss implementation of various control strategies with solid-state logic elements in line with their relay counterparts.

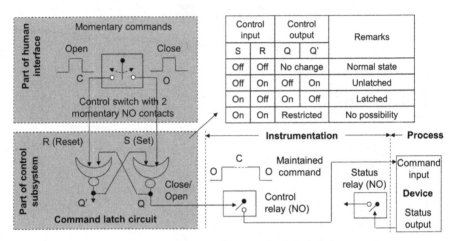

Figure A.19 Schematic for command latching.

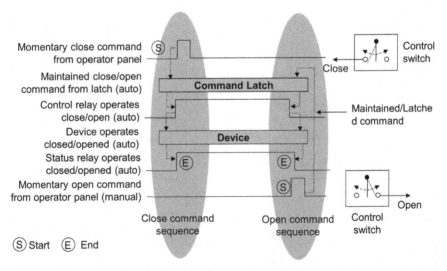

Figure A.20 Timing diagram for command latching.

A.2.2.2 Open Loop Control: Discrete

Fig. A.21 illustrates the implementation of a solid-state strategy for discrete open loop control of the water-heater process. All of the aspects previously discussed for the relay-based strategy for open loop control are applicable here, too.

A.2.2.3 Sequential Control With interlocks: Discrete

Fig. A.22 illustrates the implementation of a solid-state strategy for sequential control with interlocks of the water heater process. All of the aspects previously discussed for a relay-based strategy for open loop control are applicable here, as well.

In the previous illustrations, the change is only in the hardwired control subsystem (relay-based strategy); there is no change in either the instrumentation or the human interface subsystem.

A.2.2.4 Advantages and Disadvantages

Solid-state technology is advantageous because it:

• has all the advantages of the relay-based strategy
• is compact (occupies less space)

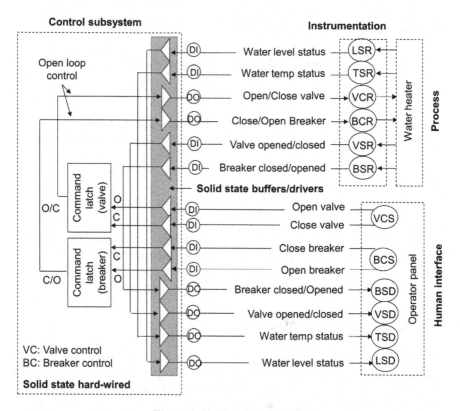

Figure A.21 Open loop control.

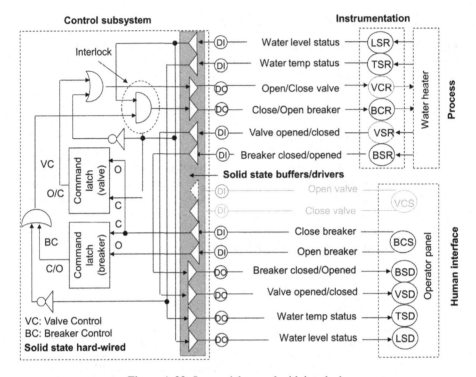

Figure A.22 Sequential control with interlock.

- consumes less power
- has faster response
- is maintenance-free (no need for periodic preventive maintenance)
- is reliable (not sensitive to environmental conditions such as temperature, humidity, dust, and vibrations)

A disadvantage is that:

- it is inflexible for modifications and extensions because the scheme is hardwired (as in the relay-based strategy)

Early solid-state logic elements were made of discrete components (circuits of resistors, capacitors, diodes, transistors, etc., mounted and wired on a board or a printed circuit board). Later, discrete component-based elements moved to integrated circuits mounted on a printed circuit board.

A.3   Continuous Control

As we know, continuous process automation systems deal with continuous processes, which have only continuous inputs and outputs. The example used for the implementation of these strategies is again the water-heating process with analog instrumentation devices and analog panel components instead of discrete components used for discrete process automation.

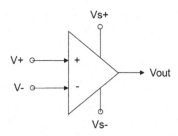

Figure A.23 Typical operational amplifier (operational amplifier).

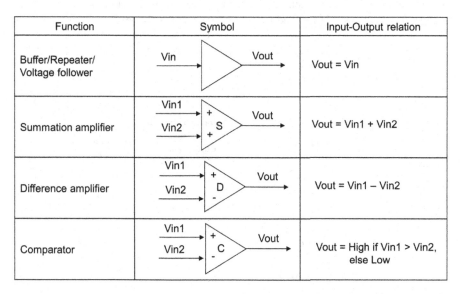

Function	Symbol	Input-Output relation
Buffer/Repeater/ Voltage follower	Vin → ▷ → Vout	Vout = Vin
Summation amplifier	Vin1 → + Vin2 → + S → Vout	Vout = Vin1 + Vin2
Difference amplifier	Vin1 → + Vin2 → − D → Vout	Vout = Vin1 − Vin2
Comparator	Vin1 → + Vin2 → − C → Vout	Vout = High if Vin1 > Vin2, else Low

Figure A.24 Control elements of continuous process automation.

A.3.1   Solid-State Technology

In this technology, the most basic control element is a solid-state **operational amplifier**, (op amp), shown in Fig. A.23.

An op amp is basically a DC-coupled high-gain electronic voltage amplifier with a differential input and usually a single-ended output. An op amp produces an output voltage that is typically much higher than the difference in input voltages, on the order of 106. With some additional circuitry, the op amp can produce various control functions, as shown in Fig. A.24.

A.3.1.1   Control Strategy Implementation

Solid-state components are applicable only to implementing automation strategies for a continuous process. We have discussed solid-state components for all continuous functions (repeater/driver/buffer, summation, integration, differentiation, etc.). Now we will look into implementing solid-state automation strategies for the continuous process automation cases discussed in Chapter 6. Here the control subsystem is built of wired solid-state components. Hence this is also a hardware strategy.

A.3.1.2 Open Loop Control: Continuous

Fig. A.25 illustrates the water-heating system as a continuous open loop control with instrumentation and human interface subsystems.

Details of the instrumentation devices and human interface panel components employed in the example are given in Table A.2.

Fig. A.26 illustrates the implementation of the open loop continuous control of a water-heating process. As seen here, desired values for level and temperature are set independently of each other. The water heater produces the level and temperature as

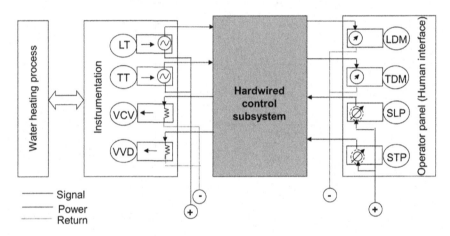

Figure A.25 Water heater automation: continuous.

Table A.2 Instrumentation devices and their interconnections

Between control subsystem and instrumentation (process)		
Analog instrumentation devices		
1	LT	Water level transmitter
2	TT	Water temperature transmitter
3	VCV	Variable control valve
4	VVD	Variable voltage drive
Between control subsystem and human interface (operator panel)		
Analog panel components		
1	LDM	Water level display meter
2	TDM	Water temperature display meter
3	SLP	Desired water level setting potentiometer
4	STP	Desired water temperature setting potentiometer

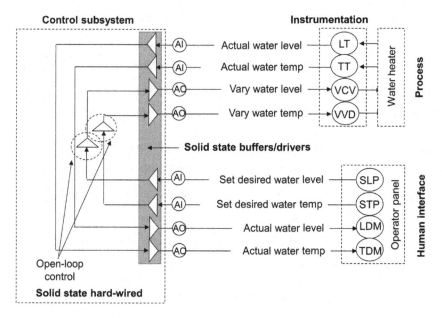

Figure A.26 Open loop control: continuous.

per reference values. The outputs (actual level and actual temperature) stay at the set-point values until another change is made.

Fig. A.26 illustrates the implementation of the open loop continuous control of a water-heating process.

As seen here, desired values for level and temperature are set independently of each other. The water heater produces the level and temperature as per reference values. The outputs (the actual level and actual temperature) stay at the set-point values until another change is made.

A.3.1.3   Closed Loop Control: Continuous

The implementation of a continuous closed loop control, as illustrated in Fig. A.27, is a true continuous control, because the output continuously follows the reference value. As seen in this scheme, the instrumentation and human interface requirements are similar to an open loop control, whereas the automation strategy is different.

A.4   Hybrid Control

Hybrid control deals with a combination of discrete and continuous processes that has both continuous and discrete input–output. The example used to implement this strategy is once again the water-heating process with input analog instrumentation devices, input and output analog panel components, and discrete output instrumentation devices.

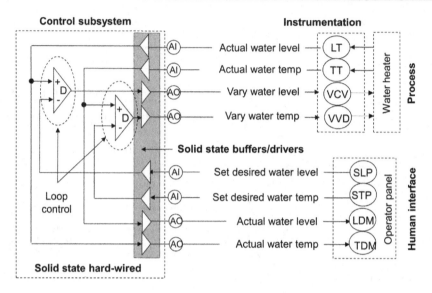

Figure A.27 Closed loop control: continuous.

A.4.1 Solid-State Technology

In a hybrid control, solid-state technology must be used because the process has both discrete and continuous signals. This is discussed in the following sections.

A.4.1.1 Control Strategy Implementation

We will now revisit two hybrid strategies, two-step control and two-step control with dead-band, as discussed in Chapter 6. In these strategies, the control subsystems are built of wired solid-state components. Hence these are also hardware strategies.

A.4.1.2 Two-Step Control

The two-step control is a combination of both discrete and continuous control and is a cost-effective approximation of continuous control. In fact, with two-step control we can theoretically achieve true continuous control, but not practically because of hardware limitations of the final control elements.

We discussed two-step control strategy (with its two variants, with and without dead-band) and its advantages and disadvantages. Here, we will discuss its implementation using solid-state control components for both variants.

Fig. A.28 illustrates the water-heating process with its instrumentation and human interface subsystems.

Contrary to open loop control, this closed loop control scheme employs both discrete and continuous instrumentation devices and continuous panel components in the human interface subsystem.

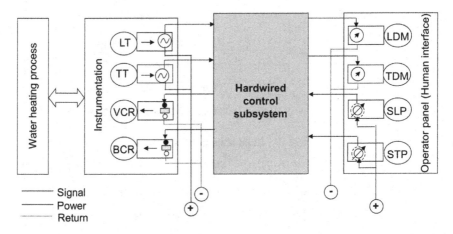

Figure A.28 Water heater automation: hybrid.

Table A.3 Instrumentation devices and their interconnections

Between control subsystem and instrumentation (process)		
Analog instrumentation devices		
1	LT	Water level transmitter
2	TT	Water temperature transmitter
Discrete instrumentation devices		
1	VCR	Valve control relay
2	BCR	Breaker control relay
Between control subsystem and human interface (operator panel)		
Analog panel components		
1	LDM	Water level display meter
2	TDM	Water temperature display meter
3	SLP	Desired water level setting potentiometer
4	STP	Desired water temperature setting potentiometer

A list of analog and digital instrumentation devices and analog human interface panel components is given in Table A.3.

Fig. A.29 illustrates the implementation of a simple two-step control strategy using solid-state control components.

The intention here is to maintain the water level and water temperature as per the set desired values. Here, the outputs (or results) follow the reference values in small discrete steps based on the accuracy or resolution of the instrumentation devices. The system is basically unstable because the control subsystem responds to even minor changes in the actual level and temperature.

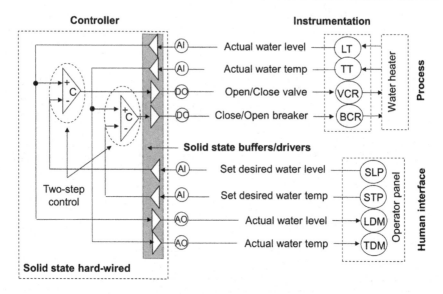

Figure A.29 Two-step control.

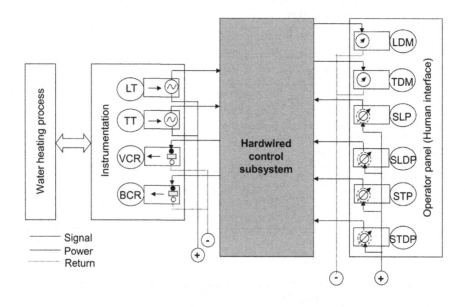

Figure A.30 Water heater automation: hybrid.

A.4.1.3 Two-Step Control With Dead-Band

As shown in Fig. A.30, implementation of the two-step control with dead-band requires some additions to the human interface subsystem (operator panel) to set the desired dead-bands for both level and temperature control; the rest is identical to simple two-step on–off control.

Table A.4 Instrumentation devices and their interconnections

Between control subsystem and instrumentation (process)		
Analog instrumentation devices		
1	LT	Water level transmitter
2	TT	Water temperature transmitter
3	VCR	Valve control relay
4	BCR	Breaker control relay
Between control subsystem interface (operator panel)		
Analog panel components		
1	LDM	Water level display meter
2	TDM	Water temperature display meter
3	SLP	Desired water level setting potentiometer
4	SLDP	Desired water level dead-band setting potentiometer
5	STP	Desired water temperature setting potentiometer
6	STDP	Desired water temperature dead-band setting potentiometer

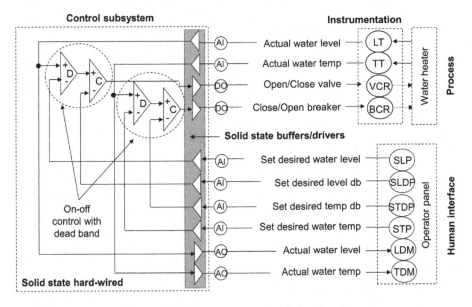

Figure A.31 Two-step control with dead band.

A revised list of analog instrumentation and analog human interface panel components is given in Table A.4.

Fig. A.31 illustrates the implementation of the two-step control strategy with deadband using solid-state components.

As seen here, the higher the dead-band is, the coarser the control is, whereas the lower the dead-band is, the finer the control is.

Appendix B: Processor

B.1   Introduction

The control subsystem/data acquisition control unit (DACU) has four primary subsystems: power supply, processor, input–output (I/O), and communication. Of these, the processor subsystem is the heart of the controller because it controls and coordinates all other subsystems as well as the operations of the DACU. In this appendix, the construction and functions of the processor module are discussed in detail. We will discuss a hypothetical processor to explain the basic concepts involved.

B.2   Hardware Structure

In this section, our example is an 8-bit hypothetical processor, as shown in Fig. B.1.

The processor discussed here is based on early architecture to explain the concepts for beginners in a more simple way.

B.2.1   Bus

The power system module placed on the bus provides power to all other functional modules. Furthermore, the bus provides the path for communication between the processor module and the functional modules that are placed on the bus, as illustrated in Fig. B.2.

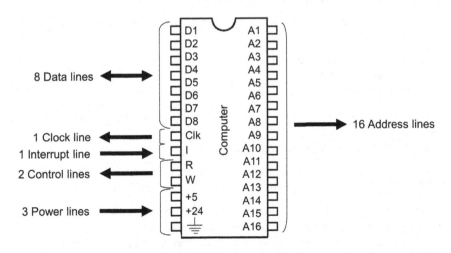

Figure B.1 Hypothetical microprocessor chip.

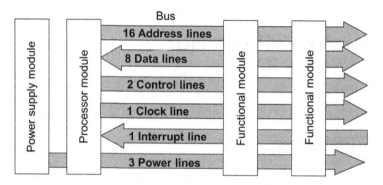

Figure B.2 Processor bus.

In this setup, the following lines are provided on the bus:

- Power lines to supply power to the processor and other functional modules
- Address lines to address the memory locations and registers in the functional modules
- Data lines to carry data from memory/functional modules to the processor and vice versa
- Control lines to facilitate the read data from the memory/functional modules into the processor and the write data from the processor into the memory/functional modules
- Interrupt line to allow the functional modules to interrupt the processor to draw its attention for some higher-priority work
- Clock lines to provide processor-generated clock signals for the functional modules to synchronize their operations with the processor

In this example, only the important lines on the bus essential to understanding the basics of processor function are considered. In practice there are many more lines on the bus for additional functions.

B.2.2 Address Space and Distribution

In our processor, 16 address lines are available that can address a maximum of 2^{16} or 65,536 locations with ranges as follows:

- Binary, 0000 0000 0000 0000 to 1111 1111 1111 1111
- Octal, 000000 to 177777
- Decimal, 0 to 65,535

Fig. B.3 illustrates the total address space available and its allocation in the processor.

Address space allocations are made as follows:

- The first 61,440 addresses are for locations in the memory modules.
- The last 4,096 addresses are reserved for registers in the functional modules.

Fig. B.4 illustrates the allocation of the address space for addressing the memory locations and the registers in various functional modules.

With this arrangement, the processor can have a maximum of 15 memory blocks, each with up to 4,096 locations and 256 functional modules, and each with up to 16 registers.

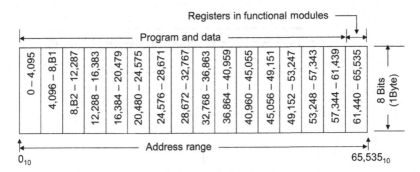

Figure B.3 Address apace.

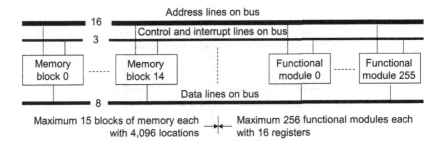

Figure B.4 Address space allocation.

B.2.3 Interfacing of Modules With Bus

The following subsections explain the interfacing of the functional modules of the controller with its bus.

B.2.3.1 Power Supply Module

Fig. B.5 illustrates the interfacing of the power supply module with the bus.

The power supply module receives the external power supply (230V AC or 48/24V DC), generates regulated +5V DC and +24V DC, and feeds this to the bus for processor and functional modules connected to the bus.

B.2.3.2 Processor Module

Fig. B.6 illustrates the interfacing of the processor module with the bus.

The processor module manages the address lines, data lines, and control lines for read and write operations from and into the memory and registers in the functional modules. In addition, the processor receives the interrupt signal (generated by other functional modules) over the bus and provides the clock signal on the bus for use by other functional modules.

B.2.3.3 Memory Module

Fig. B.7 illustrates the interfacing of the memory module with the bus.

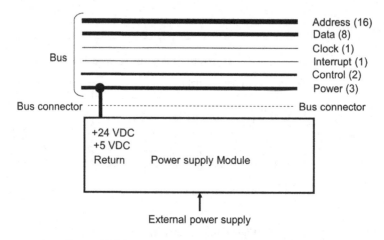

Figure B.5 Power supply module interfacing with bus.

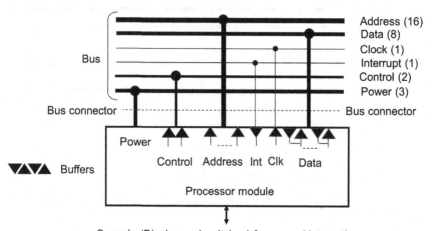

Figure B.6 Processor module interfacing with bus.

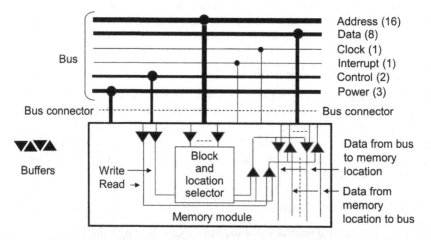

Figure B.7 Memory module interfacing with bus.

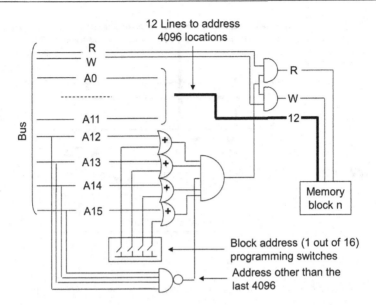

Figure B.8 Scheme for selection of memory block/location.

The memory module uses the address lines, data lines, and control lines to place the data on the bus from its location and write the data from the bus into its location.

Fig. B.8 illustrates the scheme for the application of the last four address lines (A12, A13, A14, and A15) for the selection of memory blocks.

The other 12 address lines (A0–11) use the facility built into the memory block to select the location within the block.

B.2.3.4 Functional Modules

Fig. B.9 shows the interfacing with the bus of the functional modules, such as the *I/O module, communication module, and watchdog modules.*

Here also, the module register address selection is similar to that of a memory block. The selected functional module, under read operation, responds to the processor request over control lines to place the data on the bus to be received by the processor. Similarly, under write operation the module receives the data from the bus placed by the processor. Furthermore, the functional modules can send the interrupt signal provided they are allowed by the processor to interrupt.

Here the selected functional module performs two operations on the bus: it sends the data to the processor from one of its registers and receives the data from the processor for storage in one of its registers, as illustrated in Fig. B.10.

B.2.3.5 Bus Extension Modules

As illustrated in Fig. B.11, the bus extension (parallel) module has no logical functions to perform and simply amplifies and drives the bus signals in both directions over the media. It is a physical link between two racks, extending the bus. In the case of bus

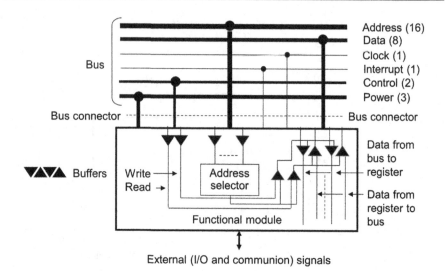

Figure B.9 Functional module interfacing with bus. *I/O*, input–output.

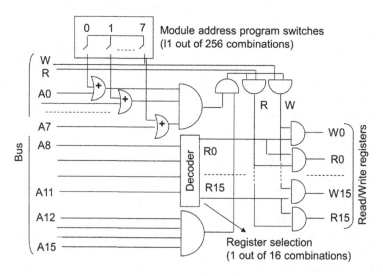

Figure B.10 Sequence of operations for writing process outputs. *R*, read; *W*, write.

extension (serial) module, the only difference is that serial/parallel and parallel/serial converters are present at the input and output stages.

B.2.4 Operations on the Bus

The processor performs two operations on the bus: it receives the data from the memory/register in a functional module and sends the data to the memory/register in a functional module.

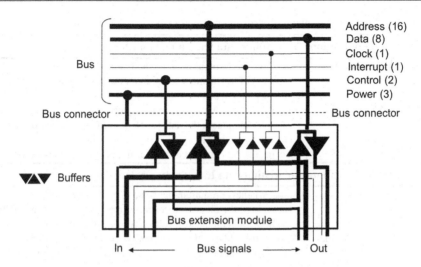

Figure B.11 Interfacing of bus extension module with bus.

The processor is always the bus master; all the other modules (memory, watchdog, I/O, communication, etc.) are like the slaves. Only the processor can read data from and write data into the registers from the other modules over the bus. The following sections explain the read/write operation on the bus for memory locations as well as the registers in the functional modules.

B.2.4.1   Memory Module

Table B.1 illustrates the sequence of operations in reading from and writing into a memory location in a memory module.

B.2.4.2   Input–Output Module

Table B.2 illustrates the sequence of operations in reading from and writing into a register in the I/O module.

B.2.4.3   Communication Module

Table B.3 illustrates the sequence of operations in reading from and writing into a register in the communication module.

B.2.4.4   Watchdog Module

The watchdog works the same way as the I/O modules.

B.2.4.5   Bus Extension Modules

As discussed earlier, the bus extension (parallel and serial) modules have no logical functions; they simply amplify and drive bus signals in both directions over the media. They are a physical link between two racks, extending the bus. Hence, they have no role in operating the bus.

Table B.1 Memory module read/write operation on bus

Sequence of operations for reading data from a memory location		
1	Processor	Starts, whenever required, the read operation by placing the address of the memory location on the bus followed by a read request on the bus
2	Memory block	Accepts the address and the read request sent by the processor on the bus and places the data (content of the selected/addressed location) on the bus
3	Processor	Accepts the data on the bus and terminates the read operation
Sequence of operations for writing data Into a memory location		
1	Processor	Starts, whenever required, the write operation by placing the address of the memory location on the bus. Then it places data to be written on the bus, followed by a write request
2	Memory block	Accepts the address, the write request, and the data sent by the processor on the bus, and then writes the data into their selected/addressed location

Table B.2 Input–output module read/write operations on bus

Sequence of operations for reading process inputs		
1	Process	Electronically maintains the latest values/states of its parameters at the input channels of the input–output (I/O) module
2	I/O module	Converts[a] the electronic inputs into data and transfers them to its data-in register. The data-in register always has the latest input data. This is done irrespective of whether the latest data is read by the processor.
3	Processor	Starts, whenever required, the read operation by placing the address of the data-in register of the I/O module on the bus followed by a read request
4	I/O module	Accepts the address and the read request sent by the processor on the bus, and places the data (content of the selected data-in register) on the bus
5	Processor	Accepts the data on the bus and terminates the read operation
Sequence of operations for writing process outputs		
1	Processor	Starts, whenever required, the write operation by placing the address of the data-out register of the I/O module on the bus. It places data to be sent on the bus, followed by a write request
2	I/O module	Accepts the address, write request, and the data sent by the processor on the bus, transfers them to its addressed data-out register, converts[b] the data into electronic signals, and drives the output channels
3	Process	Accepts the electronic outputs received from the output channels of the functional module for further operations

[a]Level in digital input, analog to digital in analog input, serial to parallel in pulse input.
[b]Level in digital output, digital to analog in analog output, parallel to serial in pulse output.

Table B.3 Communication module read/write operation on bus

Sequence of operations for reading the message from the medium		
1	Medium	Feeds the electronic pulses serially into the communication interface (I/F)
2	Communication I/F	Collects the electronic pulses serially, converts them into bits, assembles them (serial to parallel conversion) into a message of 8 bits, and transfers them to its data-in register. The data-in register always has the latest message until it is overwritten by the next message. This is done irrespective of whether the latest message is read by the processor.
3	Processor	Starts, whenever required, the read operation by placing the address of the data-in register of the communication I/F on the bus followed by a read request on the bus
4	Communication I/F	Accepts the address and the read request sent by the processor on the bus and places the data (content of the data-in register) on the bus
5	Processor	Accepts the data on the bus and terminates the read operation
Sequence of operations for writing the message to the media		
1	Processor	Starts, whenever required, the write operation by placing the address of the data-out register of the communication I/F on the bus. It places data to be sent on the bus, followed by a write request.
2	Communication I/F	Accepts the address, the write request, and the data sent by the processor on the bus, transfers them to its data-out register, converts (parallel to serial) the data into electronic pulses, and sends them on the medium serially
3	Medium	Accepts the electronic outputs received from the communication I/F for further transmission

B.2.5 Internal and External Buses

For technical reasons, the processor generally supports two buses, internal and external. The internal bus is an integrated part of the processor subsystem. Modules such as memory, watchdog, and communication, which require higher attention, are placed on the internal bus, whereas the I/O modules are on the external bus. These buses have no difference logically as far as communication between the processor and the functional modules is concerned. They differ only in their physical placement. Operations over the internal and external bus are the same. The internal bus increases the speed of communication with memory, watchdog, communication, module, etc.

Appendix C: Hardware–Software Interfacing

C.1 Introduction

In this appendix, the hardware and software interfacing of various functional modules is discussed. Without this interface, software in the processor cannot access the functional modules to perform process data acquisition and process control. In other words, without the integration of hardware and software, no data transfer takes place between the processor and the functional modules.

C.2 Architectural Aspects

This section explains the architectural aspects of the processor, which are relevant for hardware interfacing with the software.

C.2.1 Address Distribution

As seen in Fig. C.1, the processor provides for 4096 addresses (address range) to address the registers of the functional modules with the following distribution:

- Maximum of 256 functional modules
- Maximum of 16 registers per functional module

Generally these addresses can be freely allotted to any functional module through the address selection facility within the functional module.

C.2.2 Processor Registers

We discussed the following about processor registers:

- Program instructions reside in the program memory area.
- Data reside either in the data memory area or in the registers of the functional modules, because the latter are identical to the memory locations.
- Execution of stored instructions in the program memory operates on the data in the data memory and/or the registers in the functional modules to produce results.

The processor is provided with the following registers:

- Program counter: A register to indicate the memory location address of the instruction under execution by the processor. Upon completing executing every instruction, the program counter increments by either 1 or 3 automatically (see Section C.2.4) to indicate the address of the memory location of the next instruction in the program.

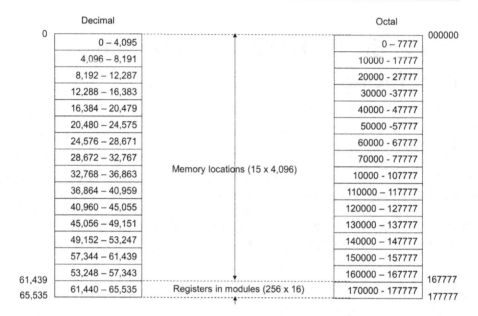

Figure C.1 Address distribution.

- Accumulator: A register to perform the following functions:
  - Hold the data read from memory/registers temporarily during its manipulation
  - Hold the result before transferring it to memory/registers
  - Act as one of the two operands in arithmetic/logical operations

C.2.3   Data Range in Memory/Registers

As illustrated in Fig. C.2, the processor, an 8-bit machine, can accommodate a maximum of 8 bits (1 byte) of information content or supporting data that range as follows:

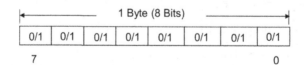

Figure C.2 Data range in memory/registers.

- Binary, 00 000 000 to 11 111 111
- Octal, 000 to 377
- Decimal, 0 to 255

The content of a byte in the memory can be an instruction, an address, or data (arithmetic or logical) whereas the content in a register is always the data.

C.2.4 Instruction Formats and Sets

Here basic instruction examples have been selected to address the concepts in the working of processor for automation strategy programming. In practice, the processors have an extensive repertoire of instructions.

For the program (instructions stored in the memory) to operate on the data in the memory/registers of the functional modules, we need to define the instructions for various operations for data movement, logical, arithmetic, and control operations. The following sections illustrate the various instruction formats.

The general format of the instruction has two fields, operation code and address. Furthermore, some instructions do not need the address field whereas others need one (2 bytes), as illustrated in Fig. C.3.

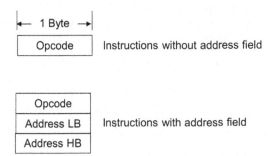

Figure C.3 Instruction formats.

Instructions without address fields occupy one location in the program memory and make the program counter increment by 1; upon completion they go to the next instruction in the program. Similarly, instructions with address fields increase the program counter by 3; upon completion of their execution they go to the next instruction in the program.

C.2.4.1 Data Movement Operations

Table C.1 lists data movement instructions and their operations.

C.2.4.2 Logical Operations

Table C.2 lists logical instructions and their operations.

C.2.4.3 Arithmetic Operations

Table C.3 lists arithmetic instructions and their operations.

C.2.4.4 Control Operations

Table C.4 lists control instructions and their operations.

Table C.1 Instructions for data movement operations

Operation code	Mnemonic	Address	Remarks
0008 (00 000 000)	CLA	–	Clear accumulator
0018 (00 000 011)	LDA	X	Load content of location X in accumulator
0028 (00 000 100)	STA	X	Store content of accumulator in location X
0038–0078	Reserved		

Table C.2 Instructions for logical operations

Operation code	Mnemonic	Address	Remarks
0108 (00 001 000)	ANA	X	Logically AND location X content with accumulator content
0118 (00 001 001)	ORA	X	Logically OR location X content with accumulator content
0128 (00 001 010)	CMA	–	Logically complement content of accumulator
0138 (00 001 111)	SRA	–	Logically shift right by 1 bit of content of accumulator
0148 (00 001 100)	SLA	–	Logically shift left by 1 bit of content of accumulator
0158–0178	Reserved		

Table C.3 Instructions for arithmetic operations

Operation code	Mnemonic	Address	Remarks
0208 (00 010 000)	ADA	X	Add content of location X to accumulator content
0218 (00 010 001)	SBA	X	Subtract content of location X from accumulator content
0228 (00 010 010)	INA	–	Increment content of accumulator by 1
0238 (00 010 011)	DCA	–	Decrement content of accumulator by 1
0248–0278	Reserved		

Table C.4 Instructions for control operations

Operation code	Mnemonic	Address	Remarks
0308 (00 011 000)	JMP	X	Jump to location X
0318 (00 011 001)	JNZ	X	If content of accumulator is non-zero, jump to location X
0328 (00 011 010)	JMZ	X	If content of accumulator is zero, jump to location X
0338 (00 011 011)	CSR	X	Call subroutine after saving the calling program status in stack
0348 (00 011 011)	RSR	X	Return from subroutine after restoring the called program status from stack
0358 (00 011 100)	HLT	–	Halt program execution
0368, 0378	Reserved		

C.2.5 Program Interfacing With Functional Modules

Each functional module is provided with a few addressable registers to interface with the program:

- Status registers to hold information related to the functioning and control of the module and its operation
- Data registers to hold data received from the process and data to be sent to the process

The general formats of both status and data registers are illustrated in Fig. C.4.

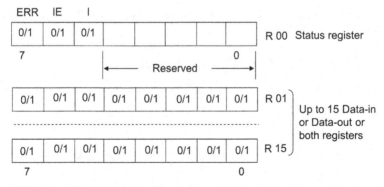

ERR: Error bit (0: module healthy and 1: module faulty) - set by module
IE: Interrupt enable bit (0: Disable and 1: Enable) - set by processor
I: Interrupt bit (0: No interrupt and 1: Interrupt) - set by module

Figure C.4 General format for status and data registers in functional module.

C.2.5.1   Status Register

Explanations of bits 7, 6, and 5 in the status register are as follows:

- Bit 7, or error (ERR) bit, indicates the health of the functional module. If 0, the module is healthy and if 1, the module is not healthy (or faulty). This bit is set internally by the module and is read only by the processor. The module clears this bit (or sets it to 0) internally once the fault is cleared or the module is healthy. The processor has no control over this bit except that it can read it.
- Bit 6, or interrupt enable (IE) bit, indicates whether the functional module is enabled to interrupt the processor. If 0, the module is not enabled and if 1, the module is enabled. The processor controls this bit (setting it to 0 or to 1) and the functional module has no control over this bit.
- Bit 5, or interrupt (I) bit, indicates whether an interrupt in the module is pending. If 0, no interrupt is pending and if 1, an interrupt is pending. The interrupt is generated by the functional module (setting it to 1), and the module clears the bit internally once the processor reads it (resetting it to 0). This is also a read-only bit by the processor. The bit is controlled by the module and the processor has no control over it.

C.2.5.2   Data Registers

The data registers are of two types:

- Data-in registers to hold the data received from the process. The functional module controls the contents in this register and the processor has no control over it.
- Data-out registers to hold the data sent to the process. The processor controls the contents in this register and the functional module has no control over it.

The following sections illustrate the specific formats of status and data registers for different types of functional modules.

C.2.6   Interfacing of functional modules with software

The following sections explain the status and the data registers in input–output modules for interfacing them with the software.

C.2.6.1   Digital Input and Digital Output

Fig. C.5 illustrates the formats of status and data registers in a digital input module.

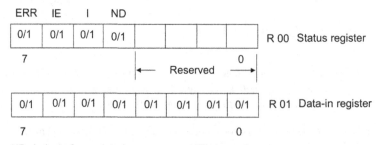

ND: Arrival of new data in one or more I/P channel

Figure C.5 Status and data registers in digital input module.

This module supports eight inputs and hence requires two registers: one for status (R00) and one for data-in (R01). Although the explanations for bits ERR and IE are the same as explained earlier, bit I gets set provided bit IE is set by the processor and one of the following conditions take place:

- Module is or becomes faulty (bit ERR gets set)
- Arrival of new data in one or more channels (ND) (bit ND gets set)

Fig. C.6 illustrates the formats of status and data registers in a digital output module.

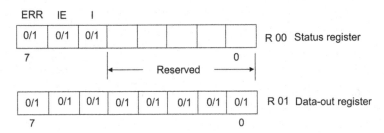

Figure C.6 Status and data registers in digital output module.

This module supports eight outputs and hence requires two registers: one for status (R00) and one for data-out (R01). Although the explanations for bits ERR and IE remain the same, bit I gets set provided bit IE is set by the processor and the module is or becomes faulty (bit ERR gets set).

C.2.6.2 Analog Input and Analog Output

Fig. C.7 illustrates the formats of status and data registers in an analog input module. This module supports four inputs and hence requires five registers: one for status (R00) and four for data-input (R01–04). Although the explanations for bits ERR and IE remain the same as explained previously, bit I gets set provided bit IE is set by the processor and the module is or becomes faulty (bit ERR gets set).

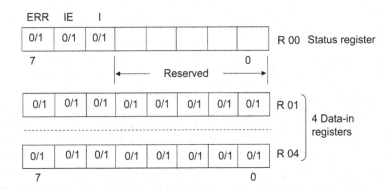

Figure C.7 Status and data registers in analog input module.

428 Appendix C: Hardware–Software Interfacing

Fig. C.8 illustrates the formats of status and data registers in an analog output module.

This module supports two outputs and hence requires three registers: one for status (R00) and two for data-output (R01 and R02). Although the explanations for bits ERR and IE remain the same, bit I gets set provided bit IE is set by the processor and the module is or becomes faulty (bit ERR gets set).

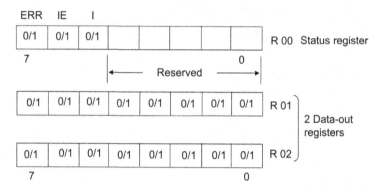

Figure C.8 Status and data registers in analog output module.

C.2.6.3 Pulse Input and Pulse Output

Fig. C.9 illustrates the formats of status and data registers in a pulse input module.

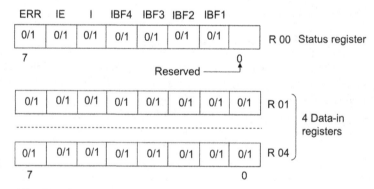

IBFn : Data-in register n status (Counter n full/Not full)

Figure C.9 Status and data registers in pulse input module.

This module supports four inputs and hence requires five registers: one for status (R00) and four for data-input (R01–04). Although the explanations for bits ERR and IE remain the same, bit I gets set provided bit IE is set by the processor and the module is or becomes faulty (bit ERR gets set) or any counter becomes full [data-in register n (IBEn status) gets set]).

Fig. C.10 illustrates the formats of status and data registers in a pulse-output module.

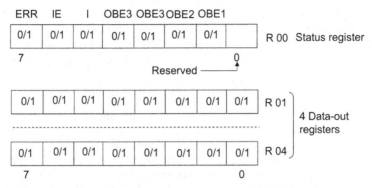

OBEn : Data-out register n status (Counter n empty or not empty)

Figure C.10 Status and data registers in pulse output module.

This module supports four inputs and hence requires five registers: one for status (R00) and four for data-input (R01–04). Although the explanations for bits ERR and IE remain the same as explained, bit I gets set provided bit IE is set by the processor and the module is or becomes faulty (bit ERR gets set) or the counter becomes empty (IBEn gets set).

C.2.6.4 Communication

Fig. C.11 illustrates the formats of status and data registers in a communication module.

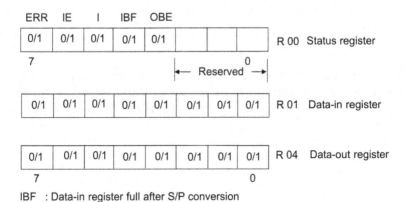

IBF : Data-in register full after S/P conversion

OBE : Data-out register empty after P/S conversion

Figure C.11 Status and data registers in communication module.

This module supports one bidirectional communication channel and hence requires three registers: one for status (R00), one for data-in (R01), and one for data-out (R02). Although the explanations for bits ERR and IE remain the same, bit I gets set provided bit IE is set by the processor and one or more of the following conditions are met:

- The module is or becomes faulty (bit ERR gets set).
- Send (data-out) register becomes empty with the completion of parallel (P) to serial (S) conversion or after pushing out the 8-bit parallel data serially to the communication medium [bit OBE (data-out register empty after P/S conversion) gets set].
- Receive (data-in) register becomes full with the completion of serial to parallel conversion or after assembling the serial data received from the communication medium into 8-bit parallel data in the data-in register [bit IBF (data-in register full after S/P conversion) gets set].

C.2.6.5 Watchdog

Fig. C.12 illustrates the formats of status and data registers in a watchdog module.

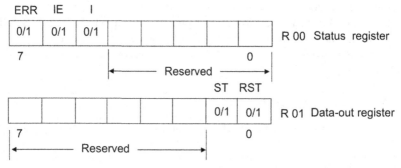

RST: Reset re-trigger mono-stable multi-vibrator reset (Write only by processor)
ST : Announce non-fatal fault (Write only by processor)

Figure C.12 Status and data registers in watchdog module.

This module also requires two registers: one for status (R00) and the other for data-out (R01). Although the explanations for bits ERR and IE remain the same, bit I gets set provided bit IE is set by the processor and the module is or becomes faulty (bit ERR gets set).

Appendix D: Basics of Programming

D.1 Introduction

This appendix gives some insight into the basics of programming and its methods in lower levels (machine and assembly-level) to access various registers in input–output (I/O) modules to perform data acquisition and control.

D.2 Lower-Level Programming

Lower-level programming methods use machine language and assembly language, as discussed in subsequent sections.

D.2.1 Machine Level

The starting point of programming is at the machine level. This calls for full knowledge of the processor architecture and its instruction set. Primarily, coding at the machine level is done with combinations of 0s and 1s. Loading of the program and the data is done manually. This method, which is not suitable for large programs, was used to input small programs to boot the machine and hand over control to an interactive terminal. To facilitate this, computers were provided an interactive console (hardwired) with switches to input combinations of 0 and 1s into the addressed memory locations and console lamps to display the contents of the addressed locations. This approach has been subsequently replaced by a booting program (to start the computer on powering on) that is resident in the nonvolatile memory of the computer, which gets activated upon **power-on** of the machine.

D.2.2 Assembly Level

Because it is inconvenient and causes errors to work with 0s and 1s for coding, assembly-level programming replaced machine-level programming. Assembly-level programming also calls for full knowledge of the processor architecture and its instruction set. The only difference is that mnemonics are used to abbreviate various instructions, addresses, labels, and data while coding the program. The **assembler**, a program supplied by the data acquisition control unit (DACU) vendor, converts the assembly-level program into its machine equivalent executable code. Here, each line in the assembly-level program gets converted into one machine level instruction (one-to-one). Assembler programs are effective in developing system programs, because they can exploit the architecture of the processor.

D.3   Programming Examples

In the following examples, it is assumed that the first 256 locations (addresses from 000000_8 to 000377_8) in the memory are allocated for the user program (instructions) and the next 256 locations (addresses from 001000_8 to 002777_8) are reserved for user data. Furthermore, the program starts from location 0 (000000_8) automatically on **powering-on** of the processor.

When the processor wants to read from or write into memory locations or registers in the functional modules to execute an instruction, the address becomes part of the instruction. As the address is 16 bits, it is stored along with the instructions in 2 bytes (lower bytes followed by higher bytes) consecutively, as seen in the following programming examples. The programs illustrate both the machine level and its assembly language equivalent.

The following procedures are adopted in the programming examples discussed in this appendix:

- An image of the input data read from the registers in the input modules is created in input image buffer in the memory (data acquisition).
- The program operates on the latest available data in the input image buffer and stores the result in output image buffer (OPIB) in the memory.
- The contents of OPIB are written into the registers of the output modules to effect process control.

With these arrangements, the registers in the I/O modules are delinked from the program. **In other words, the program always interfaces with the input image and output image buffers.** Following are programming examples that employ this approach to handle I/O for process data acquisition, data analysis and decision, process control, and communication.

Furthermore, in all of the programs, before the input and output scans, the health of the I/O modules is checked and the watchdog is activated if the modules are found to be faulty. The schematic and address allocation for the watchdog module are given in Fig. D.1.

This is common to all the subsequent programming examples.

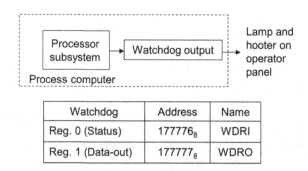

Watchdog	Address	Name
Reg. 0 (Status)	177776_8	WDRI
Reg. 1 (Data-out)	177777_8	WDRO

Figure D.1 Watchdog module: schematic and address allocation.

D.3.1   Programming With Digital Input–Output

Problem: to acquire discrete inputs from the process and display them on indication lamps on the operator panel

This is an example of reading the status of discrete inputs from the process using a digital input (DI) module and driving the indication lamps on the operator panel using a digital output (DO) module.

Fig. D.2 illustrates the schematic and allocation of the address to DI and DO modules.

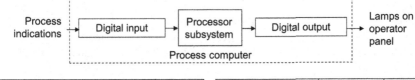

Digital input	Address	Name	Digital output	Address	Name
Reg. 0 (Status)	170000₈	DIR0	Reg. 0 (Status)	170010₈	DOR0
Reg. 1 (Data-in)	170001₈	DIR1	Reg. 1 (Data-out)	170011₁	DOR1

Figure D.2 Digital input/digital output (DI/DO) modules: schematic and address allocations.

Fig. D.3 is a program flowchart.

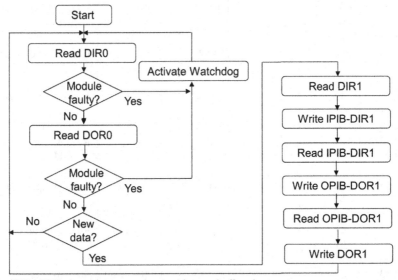

IPIB, OPIB: Input image and output image buffers in memory

Figure D.3 Flowchart: digital input/digital output (DI/DO) module programming.

Table D.1 shows the program in assembly level with its machine equivalent machine code.

Table D.1 Machine/assembly-level program for digital input–output

Memory location address[a]		Assembly-level instruction			
	Content	Label	Instruction		Remarks
Program area					
000000	001	STR	LDA	DIR0	Read status register
000001	000				DIR0
000002	170				
000003	010		ANA	XXX	Extract bit 7 (diagnos-
000004	400				tic bit) from DIR0
000005	000				
000006	031		JNZ	WDG	If module is faulty,
000007	060				jump to activate
000010	000				watchdog
000011	001		LDA	DOR0	Read status register
000012	010				DOR0
000013	360				
000014	010		ANA	XXX	Extract bit 7 (diagnos-
000015	000				tic bit) from DOR0
000016	001				
000017	031		JNZ	WDG	If module is faulty,
000020	060				jump to activate
000021	000				watchdog
000022	001		LDA	DIR0	Read status register
000023	000				DIR0
000024	360				
000025	010		ANA	YYY	Check for new data
000026	201				
000027	001				
000030	032		JMZ	STR	Go to start if no new
000031	000				data
000032	000				
000033	001		LDA	DIR1	Read data-in register
000034	000				DIR1
000035	361				
000036	002		STA	IPIB-DIR1	Store in input image
000037	100				buffer
000040	001				
000041	001		LDA	IPIB-DIR1	Read from input
000042	101				image buffer
000043	001				
000044	002		STA	OPIB-DOR1	Store in output image
000045	100				buffer
000046	001				

Table D.1 Machine/assembly-level program for digital input–output—cont'd

Memory location address[a]	Assembly-level instruction			
	Content	Label	Instruction	Remarks
000047	001		LDA OPIB-DOR1	Read from output
000050	100			image buffer
000051	001			
000052	002		STA DOR1	Store in data-out
000053	011			register DOR1
000054	360			
000055	030		JMP STR	Repeat program
000056	000			
000057	000			
000060	001	WDG	LDA ZZZ	Load mask for
000061	002			extraction of watch-
000062	000			dog bit
000063	002		STA WDR1	Drive watchdog
000064	111			
000065	111			
000066	030		JMP STR	Repeat program
000067	000			
000070	000			
Data area				
000400	200	XXX		For extraction of diagnostic bit
000401	010	YYY		For extraction of new data bit
000402	002	ZZZ		For driving watchdog bit
000500		IPIB-DIR1		
000501		OPIB-DOR1		

[a]Address represents program counter in program area.

D.3.2   Programming With Analog Input–Output

Problem: to acquire one continuous input from the process and display it on a meter on the operator panel

This is an example of reading the value of a continuous parameter from the process using an analog input module (channel 0) and driving it onto the display meter on the operator panel using an analog output module (channel 0). Fig. D.4 illustrates this schematic, Fig. D.5 displays the flowchart, and Table D.2 shows its program in assembly-level languages.

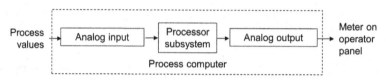

Analog input	Address	Name
Reg. 0 (Status)	170020_8	AIR0
Reg. 1 (Data-in 0)	170021_8	AIR1
Reg. 2 (Data-in 1)	$1700E_8$	AIR2
Reg. 3 (Data-in 2)	170023_8	AIR3
Reg. 4 (Data-in 4)	170024_8	AIR4

Analog output	Address	Name
Reg. 0 (Status)	170030_8	AOR0
Reg. 1 (Data-out 0)	170031_8	AOR1
Reg. 2 (Data-out 1)	170032_8	AOR2

Figure D.4 Analog input/analog output (AI/AO) programming: schematic and address allocations.

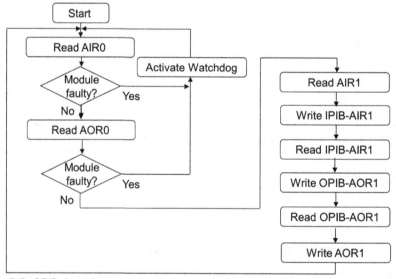

IPIB, OPIB: Input image and output image buffers in memory

Figure D.5 Flowchart: analog input/analog output (AI/AO) module programming.

Table D.2 Assembly-level program for analog input–output

Assembly-level instruction			
Label	**Instruction**		**Remarks**
Program area			
STR	LDA	AIR0	Load status register AIR0
	ANA	XXX	Extract bit 7 (diagnostic bit)
	JNZ	WDG	If module is faulty, jump to activate watchdog
	LDA	AOR0	Load status register AOR0
	ANA	XXX	Extract bit 7 (diagnostic bit)
	JNZ	WDG	If module is faulty, jump to activate watchdog
	LDA	AIR1	Load data-in register AIR1
	STA	IPIB-AIR1	Store in input image buffer IPIB-AIR1
	LDA	IPIB-AIR1	Read input image buffer IPIB-AIR1
	STA	OPIB-AOR1	Store in output image buffer OPIB-AOR1
	LDA	OPIB-AOR1	Load output image buffer OPIB-AOR1
	STA	AOR1	Store in data-out register AOR1
	JMP	STR	Repeat program
WDG	LDA	YYY	Load mask for extraction of watchdog bit
	STA	WDR1	Drive watchdog
	JMP	STR	Repeat program
Data area			
XXX	'200'		Mask for extraction of diagnostic bit
YYY	'002'		Mask for driving watchdog bit
IPIB-AIR1			Input image
OPIB-AOR1			Output image

D.3.3 Programming With Pulse Input–Output

Problem: to acquire one pulse input from the process and display it on a counter on the operator panel

This is an example of reading the pulses from the process using a pulse input module (channel 0) and driving them onto the display counter on the operator panel using a pulse output module (channel 0). Fig. D.6 illustrates the schematic, Fig. D.7 demonstrates the flowchart, and Table D.3 shows its program in assembly-level language.

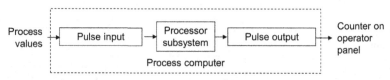

Pulse input	Address	Name		Pulse output	Address	Name
Reg. 0 (Status)	170040_8	PIR0		Reg. 0 (Status)	170050_8	POR0
Reg. 1 (Data-in 0)	170041_8	PIR1		Reg. 1 (Data-out 0)	170051_8	POR1
Reg. 2 (Data-in 1)	170042_8	PIR2		Reg. 2 (Data-out 1)	170052_8	POR2
Reg. 3 (Data-in 2)	170043_8	PIR3		Reg. 3 (Data-out 2)	170052_8	POR3
Reg. 4 (Data-in 4)	170044_8	PIR4		Reg. 4 (Data-out 3)	170052_8	POR4

Figure D.6 Pulse input/pulse output (PI/PO) programming: schematic and address allocations.

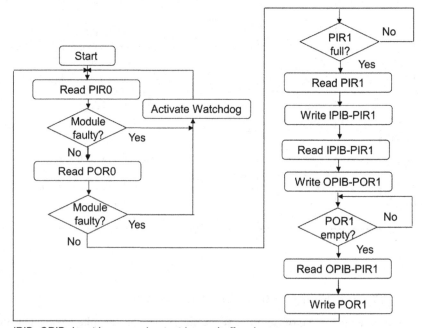

IPIB, OPIB: Input image and output image buffers in memory

Figure D.7 Flowchart: pulse input/pulse output (PI/PO) module programming.

Table D.3 Assembly-level program for pulse input–output

Label	Instruction		Comments
Program area			
STR	LDA	PIR0	Load status register PIR0
	ANA	XXX	Extract bit 7 (diagnostic bit)
	JNZ	WDG	If module faulty, jump to activate watchdog
	LDA	POR0	Load status register PIR0
	ANA	XXX	Extract bit 7 (diagnostic bit)
	JNZ	WDG	If module faulty, jump to activate watchdog
LOOP1	LDA	PIR0	Load status register PIR0
	AND	YYY1	Check for counter full
	JZ	LOOP1	If counter not full, wait
	LDA	PIR1	Load data-in register PIR1
	STA	IPIB-PIR1	Store in input image buffer IPIB-PIR1
	LDA	IPIB-PIR1	Load input image buffer IPIB-PIR1
	STA	OPIB-POR1	Store in output image buffer OPIB-POR1
LOOP2	LDA	POR0	Load status register POR0
	ANA	YYY2	Check for counter empty
	JZ	LOOP2	If counter not empty, wait
	LDA	OPIB-POR1	Load output image buffer OPIB-POR1
	STA	POR1	Store in data-out register POR1
	JMP	STR	Repeat program
WDG	LDA	ZZZ	Load mask for extraction of watchdog bit
	STA	WDR1	Drive watchdog
	JMP	STR	Repeat program
Data area			
XXX	'200'		Mask for extraction of diagnostic bit
YYY1	'002'		Mask for counter full
YYY2	'002'		Mask for counter empty
ZZZ	'002'		Mask for driving watchdog bit
IPIB-PIR1			Input image
OPIB-POR1			Output image

D.3.4   Programming With Communication

Problem: to receive the message from medium 1 and send it to medium 2

This is an example of reading the incoming message from medium 1 using communication module 1 and send it to medium 2 over communication module 2.

Fig. D.8 illustrates the schematic, Fig. D.9 shows the flowchart, and Table D.4 shows its program in machine and assembly-level languages.

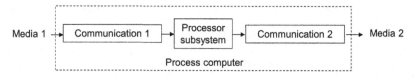

Communication 1	Address	Name	Communication	Address	Name
Reg. 0 (Status)	170060_8	CMR0-1	Reg. 0 (Status)	170070_8	CMR0-2
Reg. 1 (Data-in)	170061_8	CMR1-1	Reg. 1 (Data-in)	170071_8	CMR1-2
Reg. 2 (Data-out)	170062_8	CMR2-1	Reg. 2 (Data-out)	170072_8	CMR2-2

Figure D.8 Communication module - Schematic and address allocations.

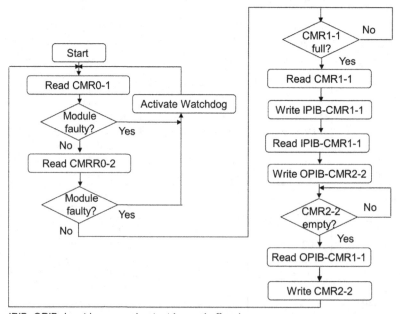

IPIB, OPIB: Input image and output image buffers in memory

Figure D.9 Flowchart: communication module (CM) programming.

Table D.4 Assembly-level program for communication

Label	Instruction		Remarks
Program area			
STR	LDA	CMR0-1	Load status register CMR0-1
	ANA	XXX	Extract bit 7 (diagnostic bit)
	JNZ	WDG	If module faulty, jump to activate watchdog
	LDA	CMR0-2	Load status register CMR0-2
	ANA	XXX	Extract bit 7 (diagnostic bit)
	JNZ	WDG	If module faulty, jump to activate watchdog
LOOP1	LDA	CMR0-1	Load status register CMR0-1
	AND	YYY1	Check for buffer full
	JZ	LOOP1	If message not full, wait
	LDA	CMR1-1	Load data-in register CMR1-1
	STA	IPIB-CMR1-1	Store in input image buffer IPIB-CMR1-1
	LDA	IPIB-CMR1-1	Load input image buffer IPIB-CMR1-1
	STA	OPIB-CMR2-2	Store in output image buffer OPIBCMR2-2
LOOP2	LDA	CMR0-2	Load status register CMR0-2
	ANA	YYY2	Check for buffer empty
	JZ	LOOP2	If counter not empty, wait
	LDA	OPIB-CMR2-2	Load output image buffer OPIB-CMR2-2
	STA	CMR2-2	Store in data-out register CMR2-2
JMP	STR		Repeat program
WDG	LDA	ZZZ	Load mask for extraction of watchdog bit
	STA	WDR1	Drive watchdog
	JMP	STR	Repeat program
Data area			
XXX	'200'		Mask for extraction of diagnostic bit
YYY1	'010'		Mask for buffer full
YYY2	'020'		Mask for buffer empty
ZZZ	'002'		Mask for driving watchdog bit
IPIB-CMR1-1			Input image
OPIB-CMR2-2			Output image

D.3.5 Programming With Interrupt

Problem: to service an alarm from the process immediately upon its occurrence and drive the watchdog (audiovisual alarm)

This is similar to the first example in which process indications are acquired using a DI module and sent to lamps on a mimic panel using a DO module; this occurs through continuous scanning for new data. The only difference is that the hardware interrupt facility is employed to perform the job immediately provided **the DI module is enabled to interrupt the processor.** In programming with interrupt, the following occurs:

- Processor enables the DI module to interrupt.
- Processor keeps executing the current program.
- DI module, upon change of state in any of its inputs, sets a new data bit in its status register and interrupts the processor.
- Processor, after completion of the execution of the current instruction, checks for pending interrupts before moving on to the execution of the next instruction in the current program.
- Processor, on recognition of an interrupt, suspends further execution of the current program.
- Processor automatically saves the latest execution status of the current (interrupted) program (contents of program counter, accumulator) in the first-in/last-out stack in the memory.
- Processor branches to general interrupt service routine to identify the source of the interrupt (DI module in the current case).
- Processor branches to DI service routine.
- Processor reads the data-in register of the DI module and stores it in the input image buffer in the memory.
- Processor returns to the interrupted program by restoring the status of the interrupted program from the first-in/last-out stack in the memory (resumption of the interrupted program).

Fig. D.10 is a flowchart and Table D.5 shows the program.

It is also possible to use the feature of module interrupt when there are faults (by setting its diagnostic bit) to prevent module faults from being checked in the previous program.

D.4 Assembling of Program

Automation programs written (coded) in assembly-level language need be converted into their machine-executable equivalents for execution by the processor. This process is called **assembling** and is carried out by a special program called an **assembler**, which is supplied by the DACU vendor. Technically, this conversion can be done using the DACU itself in offline mode, as illustrated in Fig. D.11.

However, unlike a general-purpose computer, the DACU is optimized for the real-time execution of automation functions. It has limited resources for programming environment, so it is not used for the assembling process. This process, done offline

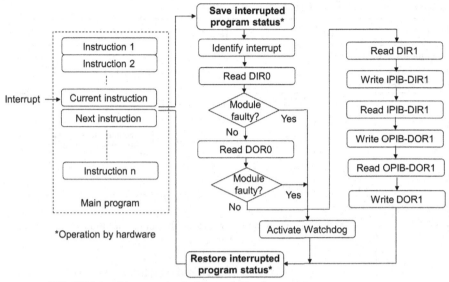

Figure D.10 Flowchart: interrupt programming. *DI*, digital input; *DO*, digital output.

in a **host machine,** is called **cross-assembling**, and it uses **cross-assembler** software supplied by the DACU vendor.

The host machine is also called the programming device or programming terminal (normally a personal computer or a handheld programmer). Vendor-supplied cross-assembler programs specific to the DACU run on the host machine to produce the machine code specific to the DACU. Here the DACU is called the **target machine.** The cross-assembling process is illustrated in Fig. D.12.

Apart from helping the programmer code the program, the host or the programming terminal converts the assembly-level program into a machine-executable program and downloads the resultant machine code into the target machine. The host also provides the following additional facilities to the programmer:

- Editing
- Debugging
- Simulation and testing
- Troubleshooting
- Documentation and reporting
- Storing

The handheld programmer, which is computer-based, is compact and more useful as field equipment for troubleshooting at the plant or on the shop floor.

Table D.5 Assembly-level program for interrupt

Label	Instruction		Remarks
Program area			
Main program			
	AAA	XXX	Instruction n-2
	BBB	YYY	Instruction n-1
Interrupt→	**CCC**	**ZZZ**	**Instruction n**→ Saves interrupted program status in stack and jumps to ISR →
	DDD	PPP	Instruction n+1
	EEE	QQQ	Instruction n+2
Interrupt service routine			
ISR	**Programming instructions for finding the source of interrupt**		
	....		
	....		
	JMP	DISR	Jump to digital input service routine
Digital input interrupt service routine			
DISR	LDA	DIR0	
	ANA	XXX	Extract bit 7 (diagnostic bit)
	JNZ	WDG	If module is faulty, jump to activate watchdog
	LDA	DOR0	Load status register DOR0
	ANA	XXX	Extract bit 7 (diagnostic bit)
	JNZ	WDG	If module is faulty, jump to activate watchdog
	LDA	DIR1	Load data-in register DIR1
	STA	IPIB-DIR1	Store in input image buffer IPIB-DIR1
	LDA	IPIB-DIR1	Read input image buffer IPIB-DIR1
	STA	OPIB-DOR1	Store in output image buffer OPIB-DOR1
	LDA	OPIB-DOR1	Load output image buffer OPIB-DOR1
	STA	DOR1	Store in data-out register DOR1
	RTS	**STR**	**Return to main program**
WDG	LDA	YYY	Load mask for extraction of watchdog bit
	STA	WDR1	Drive watchdog
	RTS	**Return**	→ Restores interrupted program status from stack and returns to interrupted program →
Data area			
XXX	'200'		Mask for extraction of diagnostic bit
YYY	'002'		Mask for driving watchdog bit
IPIB-DIR1			Input image
OPIB-DOR1			Output image

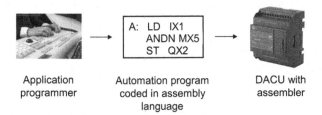

Assembler program, residing in DACU, converts the application program in assembly level language **off-line** into its machine executable equivalent and stores the same in DACU for its execution **on-line**

Figure D.11 Assembling process. *DACU*, data acquisition and control unit.

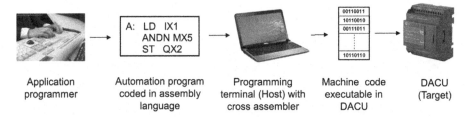

Figure D.12 Cross-assembling and downloading process. *DACU*, data acquisition and control unit.

D.5 Higher-Level Programming

Coding the automation program at the assembly level makes the coded automation program executable only on the specific DACU platform for which the program has been coded and not on any other platforms. Hence programming of the machine is done using higher-level languages that are easily understood by the programmers, which makes the program **portable** for use on any other machines and platforms with a minimum of adaptation.

Programming of the DACU for automation functions is done by using higher-level languages covered by the International Electrotechnical Commission 61,131-3 standard, as discussed in Chapter 9. Like the assembler program, the compiler program is developed by the platform vendor. It converts the higher-level language program into its machine-executable equivalent and downloads it into the DACU (target machine). Unlike an assembler, which coverts each line of the program into one machine instruction, the compiler converts each line of the program into a sequence of many instructions. Compiling process is not efficient as an assembling process because the former does not exploit the hardware features of the platform. The processes of compiling, cross-compiling, and downloading are identical to an assembling process except for the input program (assembly level or higher level).

Higher-level programming is dealt with in Chapter 9.

Index

Printed in the United States
By Bookmasters